Chemical Kinetics of Small Organic Radicals

Volume II
Correlation and Calculation Methods

Editor

Zeev B. Alfassi, Ph.D.
Professor
Department of Nuclear Engineering
Ben Gurion University of the Negev
Beer Sheva, Israel

CRC Press, Inc.
Boca Raton, Florida

Library of Congress Cataloging-in-Publication Data

Chemical kinetics of small organic radicals.

Includes bibliographies and indexes.
1. Free radical reactions. 2. Free radicals (Chemistry) 3. Chemical reaction, Rate of
I. Alfassi, Zeev B.
QD471.C49 1988 547.1'394 87-21812

Direct all inquiries to CRC Press, Inc., 2000 Corporate Blvd., N.W., Boca Raton, Florida, 33431.

International Standard Book Number 0-8493-4362-3 (v. 1)
International Standard Book Number 0-8493-4363-1 (v. 2)
International Standard Book Number 0-8493-4364-x (v. 3)
International Standard Book Number 0-8493-4365-8 (v. 4)

Library of Congress Card Number 87-21812
Printed in the United States

INTRODUCTION

The reactions of organic radicals is involved in many processes both in the liquid and the gaseous phase. These reactions are responsible for processes such as coal liquefaction and gasifications, air and water pollution, stratosphere chemistry, combustion, gaseous products formation in several synthetic processes, and consequently detonation hazards, CCl_4 toxication, and so on. The knowledge of the rate constants of these reactions and their temperature dependence is important both from theoretical points of view to check various theories on chemical reactivities and from practical points of view in order to predict the behavior of unknown new systems.

These volumes will discuss the subject of reactions of free radicals both from its general point of view as general methods of formation and detection, thermochemisry, and structure and from the point of view of specific radicals giving detailed measurement of rate constants of reactions of the more common radicals. The important reactions of each radical will be given correlation between rate constants, and other chemical data will be drawn in order to enable estimation of unmeasured rate constants for other reactions. The possibility to predict the behavior of complex chemical systems by numerical simulation of all the possible reactions using known reaction constants will be given in general. A special treatment will be given to the modeling of air pollution and combustion.

THE EDITOR

Prof. Zeev B. Alfassi, Ph.D. was born in Tel Aviv, Israel. He received his undergraduate and M.Sc. degrees from the Hebrew University in 1964 and 1965, repectively. He moved to the Soreq Nuclear Research Center in collaboration with the Weizman Institute and received his Ph.D. degree in 1970. Prof. Alfassi stayed on with the Soreq Nuclear Research Center as a research associate and later as a senior scientist, while lecturing for the Department of Nuclear Sciences at Ben Gurion University in Beer Sheva, Israel. He became a senior lecturer in 1974 and was named an Associate Professor in the Department of Nuclear Engineering in 1980. Later, Prof. Alfassi was named Professor and Chairman of the department,

Prof. Alfassi has been a visiting professor and scientist at many universities in the U.S. and Germany and has written many journal articles and technical reports.

CONTRIBUTORS

Zeev B. Alfassi, Ph.D.
Professor
Chairman of Department of Nuclear Engineering
Ben Gurion University of the Negev
Beer Sheva, Israel

Tibor Bérces, D.Sc., Ph.D.
Professor
Department of Homogeneous Reaction Kinetics
Central Research Institute for Chemistry
Hungarian Academy of Sciences
Budapest, Hungary

Kristie A. Bull, B.A.
Deparment of Chemistry
Stanford University
Stanford, California

Jeffrey S. Gaffney, Ph.D.
Staff member
Isotope and Nuclear Chemistry Division
Los Alamos National Laboratory
Los Alamos, New Mexico

Jenn-Tai Hwang, Ph.D.
Professor
Department of Chemistry
National Tsing Hua University
Hsinchu, Taiwan, Republic of China

Joel E. Keizer, Ph.D.
Professor and Director
Institute of Theoreticl Dynamics and
Department of Chemistry
University of California
Davis, California

Ferenc Marta
General Director
Department of Physical Chemistry
Central Research Institute for Chemistry
Hungarian Academy of Sciences
Budapest, Hungary

TABLE OF CONTENTS

Volume I

Volume II

Volume III

Volume IV

Chapter 7

THE EFFECT OF DIFFUSION ON CHEMICAL REACTION RATES

Joel Keizer

TABLE OF CONTENTS

I. INTRODUCTION

Since the pioneering work of Harcourt and Essen[1] in the 1860s, the rates of bimolecular chemical reactions have been characterized by their rate constants. The rate constant, of course, is the proportionality factor between the product of the concentrations of reactants and the rate of reaction, expressed as the change in concentration of reactants per unit time. Thus, for the generic bimolecular reaction:

$$A + B \xrightarrow{k^{obs}} \text{products} \quad (1)$$

the bimolecular rate constant is defined by the equality:[2]

$$d\bar{\rho}_A/dt = -k^{obs}\bar{\rho}_A\bar{\rho}_B \quad (2)$$

where the $\bar{\rho}$ represents bulk densities or concentrations. For rapid reactions, the bimolecular rate constant is large and, in the gas phase, k^{obs} is related to the binary collision rate.[3] Just how large k^{obs} is depends on two factors: one chemical and the other physical. The chemical factor involves the molecular details of the reaction process and may depend on such things as the overlap of electron densities. The physical factor is related to the accessibility of the reactants to each other. In the gas phase, the physical factor is determined by the number of collisions between A and B per unit time, which can be calculated from the kinetic theory of gases.[2,3] Of course, even the most rapid chemical reaction cannot exceed the rate at which reactants collide.

In condensed phases, the bimolecular collision rate is not always what one expects based on simple gas-phase kinetic theory. If it were, one would anticipate that the recombination rate constant for ethyl radicals would be the same in the gas phase and solution. Instead, one finds in solution a recombination rate constant[4] of about $1 \times 10^9\ M^{-1}\ sec^{-1}$, which is only 5% of the same rate constant in the gas phase.[5] Since radical recombination is an intrinsically rapid process, the reducton of the rate constant in solution must be understood in terms of a modification of the collision rate due to the ongoing reaction. The lower collision rate for a rapid reaction in solution is not, however, attributable to a change in the velocity distribution of reactants, which is still given by the equilibrium Maxwell distribution. Rather, rapid reactions in solution induce a nonequilibrium change in the spatial distribution of reactants. The possibility of such a reaction-induced nonequilibrium effect was first discussed by Smoluchowski.[6] As was clear from this early theoretical work, these reductions in the collision rate are appreciable for only the most rapid chemical reactions.

According to Smoluchowski, rate constants in solution have an upper bound determined by the rate at which reactants can diffuse together. The upper bound obtained by Smoluchowski, expressed in units of cubic centimeters per molecule per second is

$$k^{obs} = 4\pi D'R \quad (3)$$

where D′ is the relative diffusion constant of the reactants and R is the average relative separation at which reaction occurs. For typical small molecules in the usual solvents, D′ is about $10^{-5}\ cm^2\ sec^{-1}$ and R is of the order of 5×10^{-8} cm. Using these numbers and converting the expression in Equation 3 to units of $M^{-1}\ sec^{-1}$ using the factor 6.02×10^{20} gives a value of $k^{obs} = 4 \times 10^9\ M^{-1}\ sec^{-1}$. This provides a typical upper limit for bimolecular

Table 1
SOME RAPID REACTIONS AFFECTED BY DIFFUSION

Reaction	$k^{obs}/M^{-1}sec^{-1}$	Solvent	Ref.
Radical Recombination			
$I^{\cdot} + I^{\cdot} \rightarrow I_2$	8×10^9	CCl_4	8
$CH_3^{\cdot} + CH_3^{\cdot} \rightarrow C_2H_6$	1.2×10^9	H_2O	9
Radical + ion			
$CH_3\dot{C}HOH + OH^- \rightarrow CH_3\dot{C}HO^- + H_2O$	7×10^9	H_2O	9
$Co^{2+} + e^-(aq) \rightarrow Co^+$	1.2×10^{10}	H_2O	9
Neutralization			
$OH^- + NH_4^+ \rightarrow NH_3 + H_2O$	3.4×10^{10}	H_2O	11
$H^+ + C_6H_5COO^- \rightarrow C_6H_5COOH$	3.5×10^{10}	H_2O	11
Proton transfer			
$NH_4^+ + NH_3 \rightarrow NH_3 + NH_4^+$	1.3×10^9	H_2O	12
$OH^- + H_2O \rightarrow H_2O + OH^-$	5×10^9	H_2O	12
Molecule + ion			
$I_2 + I^- \rightarrow I_3^-$	4×10^{10}	H_2O	12
$HgBr_2 + Br^- \rightarrow HgBr_3$	5×10^9	H_2O	13
Photochemical			
$(CH_3)_2CO^* + NC(CH_2)_2CN \rightarrow$ oxetane	3.5×10^9	Acetone	14
Naphthalene* + $CH_3COCOCH_3 \rightarrow$	1×10^{10}	Cyclohexane	14
Naphthalene* + $CH_3COCOCH_3^{\cdot}$	5×10^8	Paraffin (ℓ)	14
Complex Formation			
$Cd^{2+} + EDTA^{3-} \rightarrow CdEDTA^{2-} + H^+$	4×10^9	H_2O	13
Pyridine + 3-hydroxypyrene $\rightarrow$ H-bonded complex	8.6×10^9	Benzene	12
Quenching			
$CH_3COCOCH_3^{\cdot}$ + naphthalene $\rightarrow$ $CH_3COCOCH_3$ + naphthalene	1×10^{10}	Benzene	14
$(C_6H_5)_2CO^* + (C_2H_5)_3N \rightarrow (C_6H_5)_2CO + (C_2H_5)_3N$	2×10^9	Benzene	14
Excited-state electron transfer			
$Ru(dipy)_3^{2+*} + Fe_{aq}^{3+} \rightarrow Ru(dipy)_3^{3+} + Fe_{aq}^{2+}$	1.9×10^9	H_2O	14
Excited-state electron transfer			
$Ru(dipy)_3^{2+*} + Co(ox)_3^{3-} \rightarrow Ru(dipy)_3^{3+} + Co(ox)^{4-}$	7×10^9	H_2O	15

rate constants in solution. Reactions with rate constants of this order of magnitude are usually called diffusion controlled.

Although many reactions of interest to chemists are not diffusion controlled, there is a great variety of reactants which are rapid enough to be at least partially effected by diffusion. Table 1 presents a list of some of the most important types of rapid reactions for which diffusion effects have been implicated, along with examples of each type. The particular examples listed in Table 1 are characterized by small energy barriers for reaction, typically less than a few kilojoules per mole. Consequently, at room temperature, almost all collisions have sufficient energy for reaction. Measured activation energies[7] for these reactions, on the other hand, are of the order of 15 kJ/mol. This is typical of the activation energy for diffusion in many solvents, as would be predicted by the Smoluchowski expression in Equation 3.

All the bimolecular rate constants for the reactions listed in Table 1 fall in the range of values 2×10^9 to 4×10^{10} M^{-1} sec^{-1}. These values are close to the value of 4×10^9 M^{-1} sec^{-1} estimated using the Smoluchowski theory, and are typical of the largest rate constants measured in solution. The explicit effect of diffusion on the reactions in Table 1 is illustrated dramatically by the photochemical energy transfer between naphthalene and biacetyl. Using liquid cyclohexane as the solvent for this reaction, the observed rate constant is measured to be $1 \times 10^{10}\ M^{-1}\ sec^{-1}$. Switching the solvent to liquid paraffin, however, reduces the rate constant to $5 \times 10^{8}\ M^{-1}\ sec^{-1}$. This reduction can be attributed to the higher viscosity of paraffin, which is responsible for decreasing the diffusion constants of the reactants.

Although the reactions in Table 1 represent many important reaction types, such as proton transfers and complex formation, not all reactions of these general types are affected by diffusion. For example, transfer of a proton from water to the barbiturate anion has an observed rate constant[16] of $1 \times 10^{5}\ M^{-1}\ sec^{-1}$, which is well below the range where diffusion might be implicated. On the other hand, the examples in Table 1 suggest that participation of ions, radicals, or excited molecular states is at least necessary for diffusion effects to be important. Another example shows that these characteristics are not sufficient for diffusion to be important. Even though the decomposition reaction of bicarbonate into CO_2, i.e., $H^+ + HCO_3^- \rightarrow H_2O + CO_2$, involves oppositely charged ions, the reaction rate is controlled by the linear to trigonal rearrangement of bicarbonate.[17] Indeed, the rate constant for this reaction is slower by five orders of magnitude than that for formation of carbonic acid,[18] $H^+ + HCO_3^- \rightarrow H_2CO_3$, which is, in fact, diffusion controlled.

In the remaining sections of this chapter, we focus on some of the modern theoretical developments dealing with diffusion effects on rapid reactions. Section II outlines the Smoluchowski approach and its generalizations. These modifications of the original theory allow the effects of rotation, molecular interactions, and increased concentrations of reactants to be treated. In Section III, recent developments based on the nonequilibrium thermodynamic theory of fluctuations are described. This theory shows that the validity of the Smoluchowski approach is restricted to dilute solution and long lifetimes. General expressions for the effect of increased concentrations, unimolecular lifetimes, and molecular interaction are given. This theory is not plagued by the divergence that hampers application of the Smoluchowski theory to bimolecular reactions in membranes, and its application to reactions in two dimensions is explained. Other theoretical treatments which have been applied to rapid reactions are reviewed elsewhere.[19] These include the use of mean passage time, Fokker-Planck equations, and numerical simulations. In the final section, we review the current status of comparison of theory and experiment and discuss the prospect for future comparisons.

II. SMOLUCHOWSKI'S APPROACH AND ITS GENERALIZATIONS

A. Basic Theory

1. Noninteracting Molecules

Smoluchowski's theory of diffusion effects on rapid reactions appears as an afterthought in a paper on aggregation of colloid particles.[6] As a consequence, it appears to be a rather crude model for describing bimolecular chemical reactions between molecules. Nonetheless, the physical picture which it provides is very clear and the simple expressions it gives for noninteracting molecules are easy to understand. Smoluchowski considered the symbolic bimolecular reaction in Equation 1, with the rate constant defined as in Equation 2. He assumed further that every bimolecular encounter of A and B which achieved a relative separation R would lead to reaction. This separation R is called the encounter radius, which should approximately equal the sum of the average molecular radii of the two molecules. To simplify the calculation, the molecules were also assumed to be spherical.

To calculate the reaction rate, Smoluchowski placed an observer on one of the molecules, say, the A molecule. The observer then kept track of the flux of B molecules towards A. Since all of the B molecules which touched the encounter radius R would react, the number of reactions occurring per second is simply the flux, j_R, of B molecules crossing a sphere of radius R around an A molecule multiplied times the number of A molecules. In terms of the number density, the reaction rate would then be

$$d\bar{\rho}_A/dt = -j_R\bar{\rho}_A \tag{4}$$

where j_R is the number of B molecules crossing the sphere of radius R around an A molecule per unit time. To calculate the relative flux of B molecules seen by an observer on A, Smoluchowski noted that, in an inhomogeneous solution, this flux is caused by diffusion. Since the observer sits on the A molecule, it is necessary to describe the relative diffusion of A and B. According to Fick's law,[20] the relative diffusion flux involves the sum of the diffusion constants, $D' = D_A + D_B$, and is

$$\mathbf{j} = -D'\nabla\rho_B \tag{5}$$

where ∇ is the gradient operator. This implies that the time-dependent density of B molecules around a central A molecule satisfies the diffusion equation:

$$\partial\rho_B/\partial t = D'\nabla^2\rho_B \tag{6}$$

According to Equation 5, the number of B crossing the encounter radius per second is

$$j_R = 4\pi R^2 D'(\partial\rho_B/\partial r)_R \tag{7}$$

since A and B are spherically symmetric. Combining Equations 4 and 7 then gives:

$$d\bar{\rho}_A/dt = -4\pi R^2 D'(\partial\rho_B/\partial r)_R\, \bar{\rho}_A \tag{8}$$

From Equation 8, one deduces that the bimolecular rate constant is given by the expression:

$$k^{obs} = 4\pi R^2 D'(\partial\rho_B/\partial r)_R/\bar{\rho}_B \tag{9}$$

To complete Smoluchowski's calculation, it is necessary to obtain $\rho_B(r,t)$ and evaluate the derivative $(\delta\rho_B/\delta r)_R$. To do so requires solving the diffusion Equation 6 which, since it is a partial differential equation, means that both boundary and initial conditions must be introduced. The boundary conditions adopted by Smoluchowski are appropriate for reaction with random replacement of reactants; i.e., for every product molecule formed, one each of the reactant molecules A and B is added at random locations to the solution. Thus, at large distances from the central A molecules, one has $\lim_{r\to\infty} \rho_B(r) = \bar{\rho}_B$, which is the constant bulk density of B. At the encounter radius, on the other hand, $\rho_B(R) = 0$, since each encounter results in reaction. If at time zero the B molecules are distributed with the constant density $\bar{\rho}_B$ around the central A, the diffusion Equation 6 can be solved with the above boundary conditions to obtain:[6]

where erfc is the complementary error function. To evaluate the rate constant for reaction, it is necessary to calculate the derivative $(\delta\rho_B/\delta r)_R$ as required by Equation 9. This can be done using Equation 10, which allows Equation 8 to be written:

$$d\bar{\rho}_A/dt = -4\pi D'R[1 + R/(\pi D't)^{1/2}]\, \bar{\rho}_A\bar{\rho}_B \tag{11}$$

The observed rate constant is seen to depend on time and to be given by the formula:

$$k^{obs} = 4\pi D'R[1 + R/(\pi D't)^{1/2}] \tag{12}$$

For small molecules, D' is of the order of 10^{-5} cm^2 sec^{-1} and R is of the order 10 Å. Thus, after a microsecond or so, the second term in Equation 11 can be neglected, and the rate constant takes on its steady-state value:

$$k^{obs} = 4\pi D'R \tag{13}$$

This is Smoluchowski's expression for the rate constant of a rapid bimolecular reaction.

Collins and Kimball[21] developed an extension of the Smoluchowski theory which applies to reactions with rates that are not completely diffusion controlled. This was done by relaxing the boundary condition that the density of B at the encounter radius must vanish. This boundary condition is clearly correct for only infinitely fast reactions. The more realistic boundary condition of Collins and Kimball involves the total flux at the encounter radius and is

$$4\pi R^2 D'(\partial \rho_B/\partial r)_R = k^o \rho_B(R, t) \tag{14}$$

which k^o is the bimolecular reaction rate constant when a B is at the encounter radius. This is sometimes called the radiation boundary condition by analogy to a related problem in heat conduction. Combining Equations 14 and 9 then gives:

$$4\pi R^2 D'(\partial \rho_B/\partial r)_R = k^o \rho_B(R, t) = k^{obs}\bar{\rho}_B \tag{15}$$

which shows that k^o will generally be larger than k^{obs} since the density of B at the encounter radius will be smaller than its bulk value.

It is easy to solve the diffusion equation with the Collins-Kimball boundary condition. Since the transient terms vanish on a rapid time scale, it suffices to consider only the final steady state. At steady state, the flux of B through the surface of any sphere around A must equal the flux through the surface at the encounter radius. Thus, for the steady state, Equation 15 implies that

$$\partial \rho_B/\partial r = k^{obs}\bar{\rho}_B/4\pi r^2 D' \tag{16}$$

Integrating this equation then gives:

$$\rho_B(r) = \bar{\rho}_B(1 - k^{obs}/4\pi D'r) \tag{17}$$

where the boundary condition $\rho_B(\infty) = \bar{\rho}_B$ was used. The boundary condition in Equation 14 is required to calculate an expression for k^{obs}. Using Equation 17 to calculate the derivative in Equation 14 and rearranging the resulting equation leads to

$$k^{obs} = 4\pi D'Rk^o/(k^o + 4\pi D'R) \tag{18}$$

This is the Collins-Kimball generalization of the Smoluchowski expression for k^{obs}.

When the reaction rate constant at the encounter distance is large, the Collins-Kimball expression reduces to that of Smoluchowski. Specifically, when $k^o \gg 4\pi D'R$, k^o dominates in the denominator of Equation 18 and the equation reduces to $k^{obs} = 4\pi D'R$. This is the limit of diffusion control. The opposite limit, i.e., $k^o \ll 4\pi D'R$, describes an intrinsically

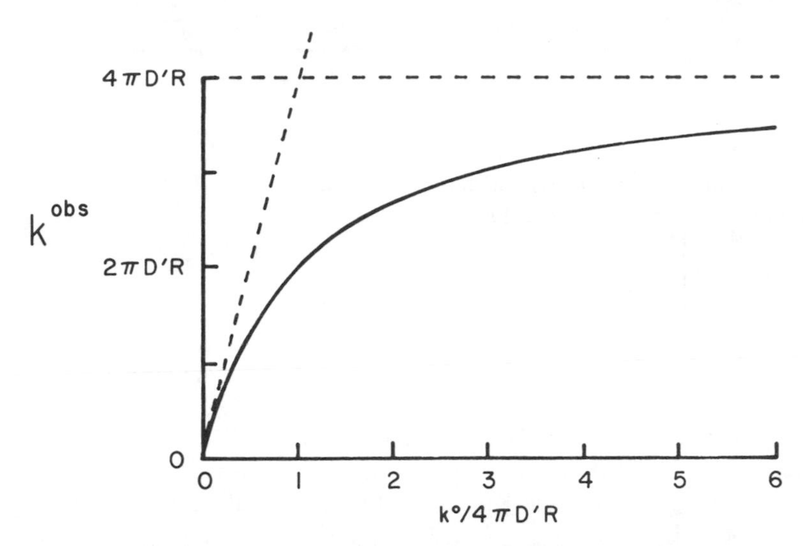

FIGURE 1.. The observed bimolecular rate constant as a function of the intrinsic reactivity, k^o, as determined by the Collins-Kimball theory. The dashed lines represent $k^{obs} = k^o$ and $k^{obs} = 4\pi D'R$.

slow reaction, and Equation 18 reduces to $k^{obs} = k^o$. This is sometimes called reaction control. Intermediate cases are possible and, for the association of atomic iodine, it has been estimated[22] that $4\pi D'R = 0.35k^o$. Because of the form of Equation 18, k^{obs} is always less than both k^o and $4\pi D'R$. This is shown schematically in Figure 1. Thus, for any reaction, the Collins-Kimball result predicts that $4\pi D'R$ is the maximum value for an observed rate constant. By rewriting Equation 18 in the form:

$$k^{obs}/4\pi D'R = 1/(1 + 4\pi D'R/k^o) \tag{19}$$

we can use the right-hand side to express the fractional effect of diffusion on the reaction. For the bimolecular association of iodine atoms mentioned above, we can then say that its rate constant is 74% diffusion controlled. The right-hand side of Equation 19, incidentally, is the same as the expression that can be derived for $\rho_B(R)/\bar{\rho}_B$ using Equations 17 and 18. Thus, the percentage of diffusion control is proportional to the fractional decrease in density of B at the encounter radius compared to its value in bulk. The density profile, $\rho_B(r)$, as given by Equation 17, is shown in Figure 2 for a reaction which is 74% diffusion controlled.

2. Interacting Molecules

Another improvement of Smoluchowski's original theory was made by Debye,[23] who considered the effect of pairwise interactions between reactants. This is done by using the so-called potential of mean force,[20] w(r). The potential of mean force is the average interaction potential between an A molecule and a B molecule which is separated by the distance r. The average is taken over the positions and momenta of all other A and B molecules and the solvent. The negative of the gradient of the potential of mean force gives the average force which an A molecule exerts on a B molecule at that distance and depends explicitly on solvent characteristics, such as the dielectric constant. For ions in dilute solution, the potential of mean force can be shown using Debye-Hückel theory[23] to be

$$w(r) = \frac{q_A q_B \exp[-\kappa(r - R)]}{\epsilon r(1 + \kappa R)} \tag{20}$$

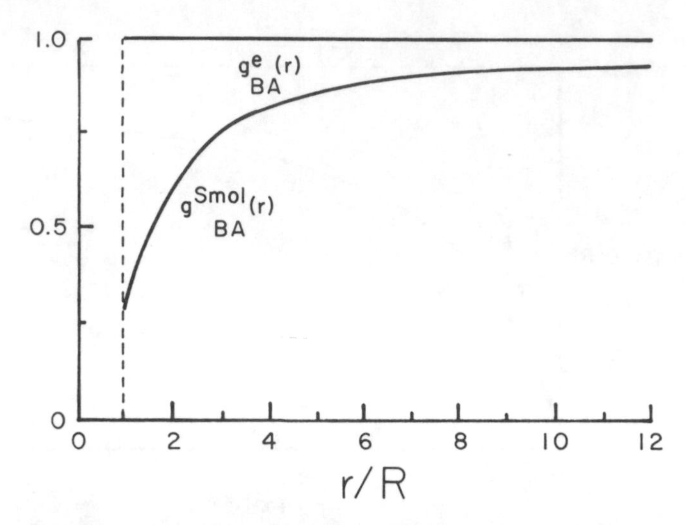

FIGURE 2. The distance dependence of the radial distribution function of B molecules as one moves out from a central A molecule located at r = O. The curve is drawn for the Smoluchowski theory for a reaction which is 74% diffusion controlled.

where $r > R$, q_A and q_B are the charges on A and B, ϵ is the dielectric constant of the solvent, and κ^{-1} is the Debye-Hückel length.

To incorporate the effects of interaction between A and B, it is necessary to modify the diffusion equation. In addition to the relative diffusion flux of B molecules towards A, there is a streaming flux caused by the attraction or repulsion of B molecules toward the central A molecule. This term can be written in terms of the friction constant, ζ, and the mean force **F** as

$$(\mathbf{F}/\zeta)\ \rho_B = -(\nabla w/\zeta)\ \rho_B = -D'(\nabla w/k_BT)\ \rho_B \tag{21}$$

since $\mathbf{F} = -\nabla w$ and in the second equality the Nernst-Einstein relationship $D' = k_BT/\zeta$ was used. Thus, the total flux of B molecules toward a central A is given by

$$\mathbf{j} = -D'[\nabla \rho_B + (\nabla w/k_BT)\ \rho_B] \tag{22}$$

The magnitude of total flux of B through a sphere of radius r around the A molecule then is

$$j_r' = 4\pi r^2 D'[\partial \rho_B/\partial r + \beta(\partial w/\partial r)\ \rho_B] \tag{23}$$

where $\beta \equiv 1/k_BT$. Following the logic used by Smoluchowski to obtain Equation 4, the rate expression for interacting molecules becomes

$$d\overline{\rho}_A/dt = -j_R'\overline{\rho}_A \tag{24}$$

where j'_R is obtained by evaluating Equation 23 at the encounter radius, R.

It is easy[2] to obtain an expression for the observed rate constant at steady state using Equations 23 and 24 and the Collins-Kimball boundary conditions in Equations 14 and 15. Since at steady state $j_r' = j_R'$, one obtains the differential equation:

$$d\rho_B/dr + \beta\rho_B dw/dr = k^{obs}\overline{\rho}_B/4\pi r^2 D' \tag{25}$$

From equation 25, one can deduce that $X = \exp(\beta w)\rho_B$ solves the equation:

$$dX/dr = k^{obs}\bar{\rho}_B \exp(\beta w)/4\pi r^2 D' \tag{26}$$

Equation 26 can be integrated directly and yields the solution:

$$\bar{\rho}_B - X(r) = \frac{k^{obs}\bar{\rho}_B}{4\pi D'} \int_r^{\infty} \frac{\exp(\beta w)}{r'^2} dr' \tag{27}$$

where we have used the fact that $X(\infty) = \bar{\rho}_B$, since the potential of mean force vanishes at infinity. Substituting the definition of X in Equation 27 and rearranging then gives:

$$\rho_B(r) = \bar{\rho}_B \exp[-\beta w(r)]\left[1 - \frac{k^{obs}}{4\pi D'} \int_r^{\infty} \frac{\exp(\beta w)}{r'^2} dr'\right] \tag{28}$$

It is convenient to define an average encounter radius, $\bar{R}$, by

$$\int_R^{\infty} \frac{\exp(\beta w)}{r'^2} dr' \equiv 1/\bar{R} \tag{29}$$

Using the Collins-Kimball boundary condition at R in Equation 15, it is then easy to show that

$$k^{obs} = 4\pi D'\bar{R}\bar{k}^o/(\bar{k}^o + 4\pi D'\bar{R}) \tag{30}$$

where

$$\bar{k}^o \equiv \exp[-\beta w(R)]\, k^o \tag{31}$$

Comparing Equation 30 to Equation 18, we see that the effect of interactions is simply to modify the encounter radius, $\bar{R}$, and the intrinsic rate constant, $\bar{k}^o$, as indicated in Equations 29 and 31. Thus, the observed rate constant for rapid reactions, i.e., those which satisfy the condition $\bar{k}^o >> 4\pi D'\bar{R}$, takes the form

$$k^{obs} = 4\pi D' \Big/ \int_R^{\infty} [\exp(\beta w)/r'^2]\, dr' \tag{32}$$

An estimate of the size of the effect of charge interactions on the rapid reactions of the rate constants can be obtained from Equations 20 and 32. At concentrations below 10^{-2} *M*, the Debye-Hückel length, κ^{-1}, is much larger than molecular dimensions.[24] Thus, to a good approximation, we may expand Equation 20 in the ratio $R\kappa$ and write the potential of mean force as

$$w(r) = (z_A z_B e^2/\epsilon)\left(\frac{1}{r} - \kappa\right) \tag{33}$$

where e is the electron charge. Substituting this expression into Equation 32 and integrating gives:

$$k^{obs} = 4\pi D' z_A z_B r_o e^{z} A^{z} B^{r} o^{\kappa}/(e^{z_A z_B r_o/R} - 1) \tag{34}$$

For charged ions of unit charge at infinite dilution, $\kappa^{-1} \rightarrow \infty$, and Equation 34 becomes

$$k^{obs} = 4\pi D' r_o/(1 - e^{\pm r_o/R}) \quad (35)$$

with the lower sign for opposite charges and the upper sign for like charges. For small molecules, R is also of the order of 7 Å, and for $r_o = R$, Equation 35 gives:

$$k^{obs} = 4\pi D' R \times (1.58)^{\mp 1} \quad (36)$$

Thus, the effect of attractive charge interactions between reactants is to increase the effective encounter radius by a factor of about two, while repulsive interactions decrease the effective radius by the same factor. Since the definition of the encounter radius is somewhat arbitrary in the Smoluchowski theory, these small changes do not appreciably modify the theory. At higher concentrations of ionic reactants, Equation 34 can be written:

$$\ln k^{obs} = \ln k_o^{obs} + z_A z_B r_o \kappa \quad (37)$$

where k_o^{obs} is the value at infinite dilution. According to the Debye-Hückel theory, κ is proportional to the square root of the ionic strength. Thus, Equation 37 predicts that the effect of increasing the ionic strength is to decrease the observed rate constant for ions of opposite charge and to increase it for ions of the same charge. This is an example of the kinetic salt effect and has been verified to hold for the diffusion-controlled reduction of NO_2^- by hydrated electrons.[26]

B. Extensions of the Basic Theory

1. Unimolecular Lifetime Effects

The Smoluchowski theory outlined in the previous section treats the bimolecular reaction as the only reactive channel available to the reactants. However, for rapid bimolecular reactions, it is often the case that one of the reactants decays by a unimolecular process. This is obviously the case for a reactant which is activated photochemically to an excited state,[14] and so may be important for quenching reactions. Thus, it is natural to consider reactants that have a finite unimolecular lifetime, τ_u, associated with the additional reaction:

$$B \xrightarrow{k_u} \text{product} \quad (38)$$

where $\tau_u = 1/k_u$. Equation 38 could also represent pseudounimolecular decay by reaction with the solvent which might occur, e.g., if B is a free radical. An estimate of the importance of the unimolecular lifetime can be made using the expression:

$$\langle r^2 \rangle^{1/2} = (D' \tau_u)^{1/2} \quad (39)$$

which gives the change in the root mean square radial separation[20] between A and B during the lifetime of B. If this distance is the order of molecular dimensions, then there will be insufficient time for diffusion to effect the reaction rate. In that case, bimolecular reaction would occur only between pairs which started out near the encounter radius. Taking $D' = 10^{-5}$ cm^2 sec^{-1}, it follows from Equation 39 that if τ_u is of the order of a nanosecond, $\langle r^2 \rangle^{1/2}$ will be about 10 Å. Thus, this effect should be important when $\tau_u \sim 10^{-9}$ sec.

One might think to approach this problem using a simple modification of the Smoluchowski Equation 6. Again letting ρ_B represent the average number concentration of B around a central A, it seems natural to write:[27]

$$\partial \rho_B/\partial t = D'\nabla^2\rho_B - \tau_u^{-1}\rho_B \tag{40}$$

For radial symmetry, we can use the substitution $\rho = \rho_B r$ to obtain the simpler equation:

$$\partial \rho/\partial t = D'\partial^2\rho/\partial r^2 - \tau_u^{-1}\rho \tag{41}$$

If B molecules are added to the solution at the rate they are being removed by reaction, Equation 41 implies a steady state solving:

$$\frac{d^2\rho}{dr^2} = \xi^2\rho \tag{42}$$

where $\xi^{-1} = (D'\tau_u)^{1/2}$. The general solution to Equation 42 is

$$\rho = Ce^{-\xi r} + C'e^{\xi r}$$

where C and C′ are constants determined by the boundary conditions. Thus, at steady state:

$$\rho_B = (Ce^{-\xi r} + C'e^{\xi r})/r \tag{43}$$

However, the Smoluchowski radiation boundary conditions are

$$\lim_{r\to\infty} \rho_B(r) = \bar{\rho}_B,\ k^o\rho_B(R) = k^{obs}\bar{\rho}_B = 4\pi R^2 D(\partial\rho_B/\partial r)_R \tag{44}$$

These conditions cannot be met since, for finite C, the first term in Equation 43 vanishes at $r = \infty$ and, unless $C' = 0$, the second term becomes infinite. Thus, the Smoluchowski approach, which works so well in the absence of lifetime effects, cannot describe the steady state which develops in this system. Although Equation 40 has been applied for transient kinetics[27] involving unimolecular effects, the fact that it does not apply to steady states brings these applications into doubt.

A slight modification of the interpretation of Equation 40 can be introduced which allows Smoluchowski's approach to be applied to this problem. Notice that Equation 6 involves derivatives on both sides so that it can also be written as

$$\partial \Delta\rho/\partial t = D'\nabla^2\Delta\rho \tag{45}$$

where $\Delta\rho \equiv \rho_B - \bar{\rho}_B$. Thus, another possible generalization of the Smoluchowski theory to the lifetime problem could be

$$\partial \Delta\rho/\partial t = D'\nabla^2\Delta\rho - \tau_u^{-1}\Delta\rho \tag{46}$$

The boundary conditions, of course, are the same as before, except that they need to be applied to $\rho_B(r) = \Delta\rho + \bar{\rho}_B$. Using this interpretation at steady state, the function $\hat{\rho} \equiv r\Delta\rho$ is found to satisfy:

$$\frac{d^2\hat{\rho}}{dr^2} = \xi^2\hat{\rho}$$

This equation is the same as Equation 42 and has the same general solution given in Equation 43. Now, however, the boundary condition at infinity can be satisfied and one finds that

$$\Delta\rho = \rho_B(r) - \overline{\rho}_B = Ce^{-\xi r}/r \tag{47}$$

Using the other boundary condition in Equation 44, C and k^{obs} can be found. One obtains:

$$\rho_B(r) = \overline{\rho}_B(1 - k^{obs}e^{-\xi(r-R)}/4\pi D'r[1 + \xi R]) \tag{48}$$

$$k^{obs} = \frac{4\pi D'R[1 + \xi R]k^o}{4\pi D'R[1 + \xi R] + k^o}, \quad \xi^{-1} = (D'\tau_u)^{1/2} \tag{49}$$

Thus, the rate constant is modified from the Collins-Kimball formula in Equation 18 by the factor $1 + \xi R = 1 + R/(D'\tau_u)^{1/2}$. Equation 49 satisfies our intuition that at the root mean square distance an A-B pair moves by diffusion during its lifetime and must be comparable to R for the lifetime effects to be important. In fact, in the limit that τ_u approaches zero, ξ becomes infinite and Equation 49 reduces to $k^{obs} = k^o$. In this limit, diffusion has no effect on the rate constant since the reactant B does not live long enough to diffuse. For a rapid reaction, on the other hand, one obtains from Equation 49:

$$k^{obs} = 4\pi D'R[1 + R/(D'\tau_u)^{1/2}] \tag{50}$$

An alternative expression for k^{obs} can be written in terms of the viscosity coefficient, η, of the solvent by invoking the Stokes-Einstein relationship for the diffusion constants:

$$D_i = k_BT/6\pi\eta R_i \tag{51}$$

where k_B is the Boltzmann constant and T is the Kelvin temperature. Applying Equation 51 to A and B with $R_A = R_B = R/2$, Equation 50 becomes

$$k^{obs} = (8k_BT/3\eta)(1 + [3\pi\eta R^3/2k_BT\tau_u]^{1/2}) \tag{52}$$

Notice that, for finite lifetimes, Equation 52 depends on both the encounter radius and the viscosity coefficient. When τ_u is infinite, k^{obs} is independent of R and one recovers the usual Smoluchowski expression:

$$k^{obs} = 8k_BT/3\eta \tag{53}$$

Clearly, the generalization of the Smoluchowski approach to describe the lifetime based on Equations 45 and 46 is to be preferred over Equation 40. The reasons for this will become clear when the statistical approach to these problems is presented in the next major section.

2. Concentration Effects

Even if molecules A and B are stable to unimolecular decay, there is an important bimolecular lifetime effect missing in the basic Smoluchowski theory. Indeed, the lifetime which characterizes the disappearance of B caused by bimolecular reaction with A is

$$\tau_B = 1/k^{obs}\overline{\rho}_A \tag{54}$$

This lifetime is infinite only in the limit that the density of A vanishes. Thus, one expects corrections to the Smoluchowski theory as one increases the concentration of A. Similarly, the bimolecular lifetime of A is

$$\tau_A = 1/k^{obs}\overline{\rho}_B \tag{55}$$

and, except in the limit that B is infinitely dilute, there should be effects due to the disappearance of A.

It is difficult to incorporate both of these lifetime effects into the Smoluchowski theory. The asymmetric case, however, in which the density of B is kept small and the density of A is increased, can be treated in several ways.[28] The simplest is to follow the same method used in the preceding section to describe the unimolecular lifetime effect.[29] Since the bimolecular lifetime of B in Equation 54 is independent of B, we can write an equation for the diffusion of B toward A which is analogous to Equation 46, i.e.

$$\partial\Delta\rho/\partial t = D'\nabla^2\Delta\rho - \tau_B^{-1}\nabla\rho \quad (56)$$

Except for the substitution of τ_B for τ_u, this equation is identical to Equation 46. Thus, for the radiation boundary condition, the steady-state solution to Equation 56 is

$$\rho_B(r) = \bar{\rho}_B(1 - k^{obs}e^{-\xi(r-R)}/4\pi D'r[1 + \xi R]) \quad (57)$$

$$k^{obs} = \frac{4\pi D'R[1 + \xi R]\, k^o}{4\pi D'R[1 + \xi R] + k^o},\ \xi^{-1} = (D'/k^{obs}\bar{\rho}_A)^{1/2} \quad (58)$$

In case there are both uni- and bimolecular lifetime effects, one obtains the same results, with $\xi^{-1} = [D'\tau_u\tau_B/(\tau_u + \tau_B)]^{1/2}$. Under conditions of diffusion control, Equation 58 becomes

$$k^{obs} = 4\pi D'R[1 + (R^2\bar{\rho}_A k^{obs}/D')^{1/2}] \quad (59)$$

This equation is somewhat different from the comparable Equation 50 for the unimolecular lifetime effect since both sides of Equation 59 depend on k^{obs}. To solve Equation 59, we write it in the form:

$$\frac{k^{obs}}{4\pi D'R} = [1 + (3\phi k^{obs}/4\pi D'R)^{1/2}] \quad (60)$$

where

$$\phi \equiv 4\pi R^3\bar{\rho}_A/3 \quad (61)$$

is the ratio of the volume taken up by the A molecues at density $\bar{\rho}_A$ compared to the close packing volume of A and B. Squaring both sides of Equation 60 and solving the resulting quadratic equation for $k^{obs}/4\pi D'R$ gives:

$$k^{obs} = 4\pi D'R[1 + 3\phi/2 + (12\phi + 9\phi^2)^{1/2}/2] \quad (62)$$

where we have used the plus root in the quadratic formula since, as we saw in the previous section, lifetime effects increase k^{obs} by lessening the effect of diffusion. Thus, to lowest order in ø, Equation 62 predicts that

$$k^{obs} = 4\pi D'R[1 + (3\phi)^{1/2}] \quad (63)$$

Effects of increased density on rapid reactions have been observed in several different systems, including fluorescence quenching experiments.[30-32] The lowest order correction term $(3\phi)^{1/2}$ is about 0.023. A more thorough discussion of these effects is given in the next section.

Another way to model the bimolecular lifetime effect is to add more A molecules to the solution. In other words, instead of considering a single, central A molecule with B molecules clustered around it, one considers an ensemble of A molecules, some of which might be close to one another.[33] In this extension of the Smoluchowski theory, it is not possible to take the diffusive motion of the A molecules explicitly into account. The effect of diffusion is assumed only to disperse the A molecules randomly throughout the medium. Since the A molecules do not move, they are thought of as static sinks. Thus, the problem of calculating the diffusion-controlled rate constant is reduced to solving Fick's diffusion Equation 6 with appropriate boundary conditions on the spherical surfaces of randomly distributed spherical sinks.[34] Although this is a well-defined mathematical problem, it is of limited interest for chemical kinetics and little progress has been made in solving it. Different approximations to the solution, which have been reviewed[35] recently, give different results even at low volume fractions, ϕ. The consensus, evidently, is that to lowest order in ϕ, this extension of the Smoluchowski theory agrees with Equation 63.

3. Membranes and Two-Dimensional Systems

The straightforward extension of the Smoluchowski theory to rapid reactions constrained on a surface does not work. The problem is similar to that encountered in the simple extension of the theory to describe unimolecular lifetime effects at steady state, namely, the boundary condition at infinity cannot be satisfied. In two dimensions,[36] the steady-state diffusion equation for radial symmetry becomes

$$(D'/r)\frac{d}{dr}(r d\rho_B/dr) = 0 \tag{64}$$

Integrating this twice, the most general solution is seen to have the form:

$$\rho_B(r) = C\ln(r/r^*) \tag{65}$$

where C and r* are constants. Taking, as usual, the value of ρ_B at infinity to be $\overline{\rho_B}$, we see that none of the solutions to Equation 64 can satisfy this condition. Unlike the unimolecular lifetime extension, in this case it does help to say that Equation 64 is satisfied by $\Delta\rho_B = \rho_B - \overline{\rho_B}$, since the general solution still blows up at infinity.

Several modifications of the Smoluchowski theory which circumvent this problem have been suggested.[37,38] The basic idea in these modifications of the Smoluchowski theory is to substitute for the boundary condition at infinity a plausible condition on a boundary nearby the sink. For example, taking the A molecule to be a circular disk with an encounter radius R, we can locate a second circular boundary at a radius $b = (1/\pi\overline{\rho_A})^2 > R$. This corresponds to associating with each A molecule an area πb^2 equal to its average fraction of the surface area. The steady-state solution to the diffusion Equation 64 is still given by Equation 65. Adopting the Smoluchowski boundary condition $\rho_B(R) = 0$, then the solution must have the form:

$$\rho_B(r) = C\ln(r/R) \tag{66}$$

To determine C, we can assume, e.g., a constant flux, j_B, of B at the boundary $r = b$. Differentiating Equation 66, we find that

$$j_B \equiv 2\pi D' b (d\rho_B/dr)_b = 2\pi D' A \tag{67}$$

from which we deduce that $C = j_B/2\pi D'$. Finally, to find k^{obs}, we use the analog of Equation 9 to write:

$$\begin{aligned} k^{obs} &= 2\pi D' R (d\rho_B/dr)_R/\bar{\rho}_B \\ &= 2\pi D' C/\bar{\rho}_B \end{aligned} \tag{68}$$

In this expression, the constant, $\bar{\rho}_B$, is taken as the average density of B in the annulus extending between $r = R$ and $r = b$. Thus,

$$\begin{aligned} \bar{\rho}_B &= \int_R^b \rho_B(r)\, r dr \Big/ \int_R^b r dr \\ &= C \frac{[b^2 \ln(b/R) - b^2/2 + R^2/2]}{[b^2 - R^2]} \end{aligned} \tag{69}$$

Combining this expression for $\bar{\rho}_B$ with Equation 68 gives:

$$k^{obs} = 2\pi D'[1 - (R/b)^2]/\{\ln(b/R) - {}^1/_2[1 - (R/b)^2]\} \tag{70}$$

In the limit of low density of A, one has $b = (1/\pi\ \bar{\rho}_A)^{1/2} \gg R$ and the asymptotic formulae:

$$k^{obs} = 2\pi D'/[\ln(b/R) - {}^1/_2] \tag{71}$$

This result was first obtained, using other methods, by Adam and Delbrück.[37] Using a slightly different modification of the Smoluchowski theory, Berg and Purcell[38] find a result similar to Equation 71 with the term $^1/_2$ replaced by $^3/_4$. We will argue in the next section that the correct term is actually $\gamma - \ln\sqrt{2} = 0.2306\ldots$, where γ is Euler's constant.

4. Rotational Effects

Many reactive molecular interactions are localized at specific regions of the reacting molecules. This implies that reactivity depends on the relative orientation of the two molecules in addition to the distance between their centers of mass. The effect of such an orientational requirement for reaction can be important for rapid bimolecular reactions. This is certainly clear for reactions in the gas phase, since many reactions can occur only when a collision brings appropriate atoms close enough together. In condensed phases, the situation is somewhat more complicated since once molecules get close to each other they tend to be caged by the solvent. If they stay close together long enough for significant molecular reorientation to occur, then reaction will be able to occur at every encounter. A measure of the time that a pair remains caged by solvent can be obtained using Equation 39, which gives the root mean square distance that the pair moves apart by diffusion. Accordingly, the caging lifetime should be about

$$\tau_c = R^2/6D' \tag{72}$$

On the other hand, the time required for the reorientation of one of the molecules, τ_R, is given by the inverse of rotational diffusion constant,[39] D^r. Hence,

$$\tau_R/\tau_c = 6D'/R^2 D^r \tag{73}$$

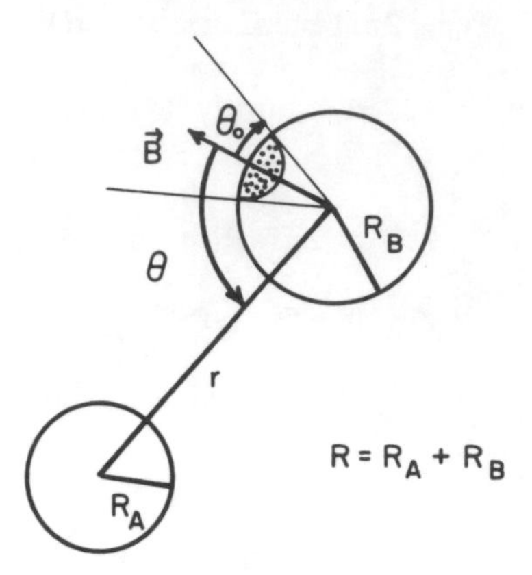

FIGURE 3. The geometry for assessing rotational effects on reactions in the Ŝolc-Stockmayer theory. The spherical molecules of radii, R_A and R_B, are separated by a distance, r. The stippled area represents the reactive region on the B molecule defined by the angle θ_o, which is measured from the orientation vector $\vec{B}$. The vector $\vec{B}$ and the line of centers form the angle θ.

To estimate this ratio, we can use the Stokes-Einstein expression[39] for D^r, i.e.,

$$D_i^r = k_BT/8\pi\eta R_i^3 \tag{74}$$

and the comparable expression for D′ based on Equation 51, i.e.

$$D' = k_BTR/6\pi\eta R_AR_B \tag{75}$$

Taking $R_A = R_B = R/2$ and substituting Equations 74 and 75 into Equation 73 gives $\tau_R/\tau_c = 4$. Since these two times are comparable, orientational effects cannot be neglected for rapid reactions. Notice that this estimate is independent of the magnitude of the encounter radius, and so orientational effects can be significant even for small molecules of similar size.

An extension of the Smoluchowski theory to include orientational effects has been made by Ŝolc and Stockmayer.[40] Their treatment adds the effect of rotational diffusion of the reactants to translational diffusion and leads, in the general case, to rather complicated differential equations. A variety of simplifications have been proposed which are appropriate for molecules of disparate size,[41,42] e.g., ligands binding to proteins. For small molecules, the example in Figure 3 is illustrative of the kind of problem that can be attacked with the Ŝolc-Stockmayer theory. The spherical molecule B in that figure is assumed to have a limited reactivity region on its surface, while the A molecule is taken to be uniformly reactive. The reactive region is specified by the polar angle measured from the orientation vector, **B**. Reaction can occur if and only if the distance r separating A and B equals the encounter radius R, and, simultaneously, the angle θ, which corresponds to the point of contact between A and B, satisfies $0 \leq \theta \leq \theta_o$. Taking the A molecule as fixed at the origin, Ŝolc and Stockmayer keep track of the density of B molecules that are at the distance r and that have the orientation vector **B** at the angle θ. Writing this density ρ(r,θ,t), it follows that at steady state, ρ solves the analog of the Smoluchowski equation, namely,[40]

$$\frac{D'}{r^2}\frac{\partial}{\partial r}(r^2\partial\rho/\partial r) + \left(D_B^r + \frac{D'}{r^2}\right)\frac{1}{\sin\theta}\frac{\partial}{\partial\theta}(\sin\theta\partial\rho/\partial\theta) = 0 \tag{76}$$

where D_B^r is the rotational diffusion constant of B.

To solve Equation 76, boundary conditions need to be specified. At the encounter radius, a generalization of the radiation boundary condition of the form:[40]

$$(\partial\rho/\partial r)_R = k^o(\theta)\,\rho(R, \theta) \tag{77}$$

where

$$k^o(\theta) = \begin{cases} k^o & 0 \leq \theta \leq \theta_o \\ 0 & \theta_o \leq \theta \leq \pi \end{cases} \tag{78}$$

expresses the fact that only B molecules of the right orientation can react with the intrinsic rate constant, k^o. By analogy to the usual Smoluchowski theory, at $r = \infty$ the density should satisfy:

$$\lim_{r\to\infty} \rho(r, \theta) = \bar{\rho}_B/2 \tag{79}$$

so that

$$\lim_{r\to\infty} \int_0^{\pi} \rho(r, \theta)\, \sin\theta d\theta = \bar{\rho}_B \tag{80}$$

Solution of Equation 76 with these boundary conditions is difficult. Slightly modified boundary conditions which allow Equation 76 to be solved analytically have been proposed.[43] These boundary conditions change Equation 77 to a constant flux condition and evaluate this constant by requiring that Equation 77 hold after integrating over the angle θ. An analytical solution of Equation 76 with these boundary conditions has been given in terms of a series of Legendre polynomials and modified Bessel functions.[43] The resulting expression for the diffusion-controlled rate constant can then be written as

$$k^{obs} = 4\pi D'Rk^* \tag{81}$$

where k^* involves corrections due to rotational reorientation. The term k^* depends explicitly on the ratio τ_R/τ_c given in Equation 73, as well as the size of θ_o, R, and D'. If the ratio $\tau_R/\tau_c = 6D'/R^2D^r$ is very small, so that there is ample time for reorientational motion, it is found that $k^* \to 1$ and $k^{obs} \to 4\pi D'R$. On the other hand, if τ_R/τ_c is close to the value of 4, expected on the basis of the Stokes-Einstein relationship for molecules of equal size, then k^* is found to be a monotone-increasing function of θ_o which vanishes at $\theta_o = 0$ and is unity at $\theta_o = \pi$. In this case, if the reaction region angle θ_o is less than about 30°, k^* is less than about $^1/_4$. Thus, even for small molecules, reorientational effects on the diffusion-controlled rate constant are predicted to be significant.

III. STATISTICAL NONEQUILIBRIUM THERMODYNAMICS AND RAPID CHEMICAL REACTIONS

The Smoluchowski theory of rapid chemical reactions is based on a calculation of the density of B molecules around a central A molecule. Although the ideas in the Smoluchowski calculation seem intuitively correct, we have seen that this calculation is not always easy

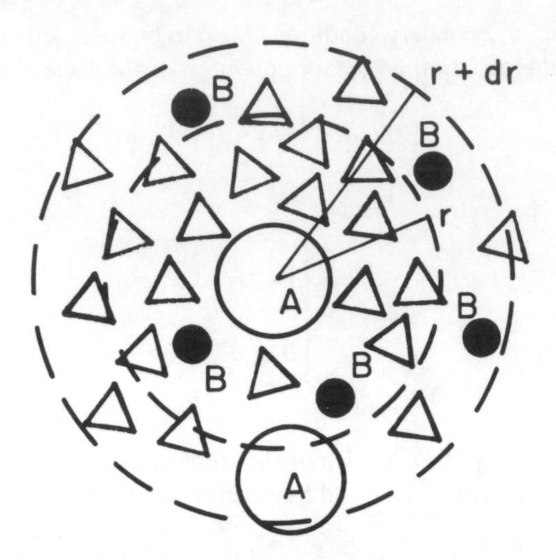

FIGURE 4. A schematic representation of a single configuration for a solution of A and B molecules (circles) in a liquid solvent (triangles). The radial distribution function is determined by the number of B molecules whose centers lie in the region between the spheres of radius r and r + dr.

to generalize. For example, in Section II we found that the method of calculation had to be seriously modified in order to include lifetime effects and the effect of diffusion in two dimensions. These ad hoc modifications make the Smoluchowski theory rather unsystematic and tend to cast doubt on the basis of the entire theory. In this section, we present an alternative theoretical approach for studying rapid reactions. This approach is based on the statistical theory of nonequilibrium thermodynamics, and permits a systematic calculation of the average density distributions of one reactant around another.

A. Basic Ideas

1. Radial Distribution Function

The statistical quantity which measures the average density of one kind of molecule around a central molecule is called the radial distribution function.[44] Figure 4 illustrates this with a schematic diagram of a molecular configuration which involves a solvent (triangles) and two kinds of solutes (circles, A and B). This snapshot represents one of an uncountable number of different configurations in the solution, which for simplicity we take to be in a steady state. For each such configuration, we pick our coordinate system at the center of an A molecule and count the number of B molecules, $N_B(r)$, in the spherical shell in between r and r + dr as indicated by the dotted circles. The average density of B in this shell then is

$$\bar{\rho}_B(r) \equiv \langle N_B(r) \rangle / 4\pi r^2 dr \tag{82}$$

where the brackets indicate an average overall possible configuration like that in Figure 4. In terms of this average density, the radial distribution function is defined as[44]

$$g_{BA}(r) \equiv \bar{\rho}_B(r)/\bar{\rho}_B \tag{83}$$

In this expression, $\bar{\rho}_B$ is the average density of B molecules in the overall solution, and the subscripts remind us that there is an A molecule located at r = 0. Because the solution is uniform on the average, $\bar{\rho}_B$ is the same as the bulk density of B.

Based on the definition in Equation 83, it is obvious that the radial distribution function is the central quantity in the Smoluchowski theory of diffusion-controlled reactions. Thought of as a method for calculating the radial distribution function, the Smoluchowski theory is physically appealing even though it is ad hoc. At equilibrium, of course, statistical mechanics provide rigorous expressions for the radial distribution function,[44] as well as a number of approximation schemes for its calculation.[45] However, the statistical ensemble corresponding to the steady state supported by a rapid reaction is a nonequilibrium ensemble. Thus, equilibrium statistical mechanics cannot be used to calculate the radial distribution function. This is illustrated graphically in Figure 2 where the Smoluchowski radial distribution function for a diffusion-controlled reaction is plotted along with the equilibrium radial distribution function:

$$g^{e}_{BA}(r) = \begin{cases} 0 & 0 \leqslant r \leqslant R \\ 1 & R < r \end{cases} \tag{84}$$

for dilute hard spheres. Comparison of the two graphs makes it clear that nonequilibrium effects are large and cannot be ignored.

Clearly, what is needed to describe rapid reactions is a systematic method of calculating nonequilibrium radial distribution functions. The approach considered in this section, which does just that, relies on the statistical description of density fluctuations.[46-48] Density fluctuations measure the disparity between the instantaneous density of molecules at position $\mathbf{r}$ and their ensemble average value at the same position. For the reactant molecules A and B, density fluctuations are defined by $\delta\rho_A(\mathbf{r},t) \equiv \rho_A(\mathbf{r},t) - \overline{\rho_A}(\mathbf{r},t)$ and $\delta\rho_B(\mathbf{r},t) \equiv \overline{\rho}_B(\mathbf{r},t) - \overline{\rho}_B(\mathbf{r},t)$. Referring to the instantaneous configuration in Figure 4, the density fluctuation of A in that small region is the difference in the actual number of A there (two) minus the average number, all divided by the volume of the region. Although in calculating $\delta\rho_A$ there is no requirement that an A molecule be located at the center of the volume element, there is a relationship between the density-density correlation function and the radial distribution function. For molecules like A and B which are distinguishable, one has the relationship:[49,50]

$$g_{BA}(\mathbf{r}, \mathbf{r}', t) = 1 + \langle\delta\rho_A(\mathbf{r}, t)\, \delta\rho_B(\mathbf{r}', t)\rangle/\overline{\rho}_A(\mathbf{r}, t)\, \overline{\rho}_B(\mathbf{r}', t) \tag{85}$$

The quantity in square brackets is called the density-density correlation function. It measures the correlation between density fluctuations of A molecules at $\mathbf{r}$ and B molecules at $\mathbf{r}'$. If the solution has density gradients, the density-density correlation function may depend on the exact location of $\mathbf{r}$ and $\mathbf{r}'$. In a homogeneous, isotropic solution at steady state, it no longer depends on time and is a function of only the absolute value of the difference in positions. Thus, one can write:

$$g_{BA}(|\mathbf{r} - \mathbf{r}'|) = 1 + \langle\delta\rho_A(\mathbf{r}, t)\, \delta\rho_B(\mathbf{r}', t)\rangle/\overline{\rho}_A\overline{\rho}_B \tag{86}$$

where $\overline{\rho_A}$ and $\overline{\rho}_B$ are the bulk steady-state values of the densities. If the molecules are identical (say, both A molecules), then the relationship between the radial distribution function and the density-density correlation is modified slightly. In that case, one has:[50]

$$g_{AA}(|\mathbf{r} - \mathbf{r}'|) = 1 - \delta(\mathbf{r} - \mathbf{r}')/\overline{\rho}_A + \langle\delta\rho_A(\mathbf{r}, t)\, \delta\rho_A(\mathbf{r}', t)\rangle/\overline{\rho}_A^2 \tag{87}$$

The expressions in Equations 86 and 87 show how to calculate the radial distribution function in terms of the density-density correlation function. Later in this section, we show

how the statistical theory of nonequilibrium thermodynamics can be used to calculate the density-density correlation function in a systematic fashion.

2. Reactivity Functions

In the original Smoluchowski theory, reaction between a bimolecular pair is assumed to occur only at a well-defined encounter radius, R. Although the addition of rotational effects to the calculation allows an angular dependence to be added to the intrinsic reactivity, reaction remains restricted to a fixed radius, R. This is a reasonable first approximation for rapid reactions which require overlap of electron density. However, reactions like energy transfer can involve dipole-dipole interactions which permit reaction to occur even when the molecules are rather far apart. According to the Förster theory[51] of dipolar energy transfer, the intrinsic reaction rate constant depends on the inverse of the separation of the centers of mass to the sixth power, i.e.

$$k^o(r) = \tau_o^{-1}(r_o/r)^6 \tag{88}$$

where τ_o is the fluorescence lifetime and r_o is an effective length that can be as large as 50 Å.

As a generalization of these considerations, it makes sense to introduce an intrinsic reaction rate constant[49,50] that depends explicitly on the location of A and B. This is termed the intrinsic reactivity function and is written as $k^o(\mathbf{r},\mathbf{r}')$. Since bimolecular reactions involve pairs of molecules, the local reaction rate at $\mathbf{r}$ can be written:

$$V_{AB}^{+}(\mathbf{r}, t) = \int k^o(\mathbf{r}, \mathbf{r}')\, \rho_{AB}(\mathbf{r}, \mathbf{r}', t)\, d\mathbf{r} \tag{89}$$

where $\rho_{AB}(\mathbf{r},\mathbf{r}',t)$ is the density of molecular pairs, with the A at $\mathbf{r}$ and the B at $\mathbf{r}'$. At a homogeneous, isotropic steady state, Equation 89 gives rise to the expression for the observed rate constant:[50]

$$k^{obs} = 4\pi \int_0^\infty k^o(r)\, g_{BA}(r)\, r^2 dr \tag{90}$$

This expression can be generalized without difficulty to take into account orientational requirements for reaction and internal states.

The simplest reactivity function is

$$k^o(r) = k^o \delta(r - R)/4\pi r^2 \tag{91}$$

which we call the Smoluchowski reactivity.[49,50] For atomic recombination one might use, instead, a function of the form:

$$k^o(r) = k^o(T) \exp(-r/r_o) \tag{92}$$

where r_o measures the distance over which the electron density falls off in the appropriate quantum states. The reactivity function is an intrinsic property of the reactants in a particular solvent, and its calculation requires quantum or semiclassical considerations. Whatever the form of the intrinsic reactivity, Equation 90 shows that the radial distribution function is required in order to calculate the observed rate constant.

B. Calculation of Rate Constants

1. Calculation of the Density-Density Correlation Function

The mechanistic statistical theory of nonequilibrium thermodynamics can be used to describe molecular processes at various levels.[46-48] The simplest description which includes the effect of molecular diffusion on rapid chemical reactions is the hydrodynamic level.[52] As a first approximation, the only molecular processes which need to be included at the hydrodynamic level are diffusion of the reactants and the chemical reaction. To illustrate how the theory works, we consider first the atomic recombination reaction:[50]

$$A + A \xrightarrow{k^{obs}} A_2 \tag{93}$$

Although it is not necessary, we neglect effects due to the presence of A_2 and assume that A is an ideal solute. In this case, the mechanistic statistical theory[46-50] predicts that, on the average, the density of A changes according to the equation:

$$\partial\bar{\rho}_A/\partial t = D\nabla^2\bar{\rho}_A - 2k^{obs}\bar{\rho}_A^2 + 2K \tag{94}$$

The term 2K in this equation represents an external process which generates the atoms at a constant rate. At steady state in a system with no gradients, Equation 94 has the solution $\bar{\rho}_A = (K/k^{obs})^{1/2}$. According to the mechanistic statistical theory, fluctuations in the density of A around this steady-state value satisfy a linearized version of Equation 94, i.e.

$$\partial\delta\rho_A/\partial t = D\nabla^2\delta\rho_A - 4k^{obs}\bar{\rho}_A\delta\rho_A + \tilde{f} \tag{95}$$

The final term in this equation, $\tilde{\mathbf{f}}$, reflects the random nature of molecular diffusion and reaction. It is a Gaussian random process[46] which vanishes on the average and has the correlation function:

$$\langle\tilde{f}(\mathbf{r}, t)\,\tilde{f}(\mathbf{r}', t')\rangle = (4k^{obs}\bar{\rho}_A^2 - 2D\bar{\rho}_A\nabla_r^2)\,\delta(\mathbf{r} - \mathbf{r}') \tag{96}$$

Equations 95 and 96 are the stochastic differential equations which allow the density-density correlation function to be calculated. Notice that they contain only kinetic constants and the bulk density of A. There are no adjustable parameters.

For ease in writing, we will denote the density-density correlation function by

$$\sigma(\mathbf{r}, \mathbf{r}', t) \equiv \langle\delta\rho_A(\mathbf{r}, t)\,\delta\rho_A(\mathbf{r}', t)\rangle \tag{97}$$

Equations 95 and 96 can be used to obtain an equation for the time evolution of σ. To compare with the Smoluchowski theory, it is simpler to consider the function:

$$\Delta\rho \equiv \bar{\rho}_A g_{AA}(|\mathbf{r} - \mathbf{r}'|, t) - \bar{\rho}_A \tag{98}$$

According to the relationship between g_{AA} and σ in Equation 87, Equation 98 can be written:

$$\Delta\rho = -\delta(\mathbf{r} - \mathbf{r}') + \sigma/\bar{\rho}_A \tag{99}$$

To obtain an equation for $\boldsymbol{\Delta\rho}$, we evaluate Equation 95 at $\mathbf{r}'$ and t, multiply on the left by $\delta\rho_A(\mathbf{r},t)$, and then average. Adding this result to the comparable result obtained by evaluating Equation 95 at $\mathbf{r}$ and t and then multiplying by $\delta\rho_A(\mathbf{r}',t)$, followed by some straightforward manipulations, gives the equation:

$$\partial\Delta\rho/\partial t = 2(D\nabla^2\Delta\rho - 4k^{obs}\bar{\rho}_A\Delta\rho) - 4k^{obs}\bar{\rho}_A\delta(\mathbf{r} - \mathbf{r}') \tag{100}$$

The limit $\bar{\rho}_A \rightarrow 0$ reduces Equation 100 to the simpler equation:

$$\partial\Delta\rho/\partial t = D'\nabla^2\Delta\rho \tag{101}$$

where we have written $D' = 2D$ for the relative diffusion constant of the two A. Recall that this is precisely the differential equation used by Smoluchowski, as we noted in Equation 45. There are, however, distinct differences between Equations 100 and 45. Equation 100, for instance, requires no boundary condition except at $|\mathbf{r} - \mathbf{r}'| \rightarrow \infty$. Since for large separations the density-density correlation function vanishes, Equation 99 implies that, in this limit, $\Delta\rho = 0$.

At steady state, the left-hand side of Equation 100 vanishes. Thus, $\Delta\rho$ solves the equation:

$$(D\nabla^2\Delta\rho - 4k^{obs}\Delta\rho) = 2k^{obs}\bar{\rho}_A\delta(\mathbf{r} - \mathbf{r}') \tag{102}$$

This is a standard differential equation whose solution is

$$\Delta\rho = \frac{-k^{obs}\bar{\rho}_A\exp(-\xi r)}{2\pi D r} \tag{103}$$

where

$$\xi^{-1} = (D/4k^{obs}\bar{\rho}_A)^{1/2} \tag{104}$$

is the root mean square distance that an A moves in its bimolecular lifetime. Using the definition of $\Delta\rho$ in Equation 98, the radial distribution function of A is found to be

$$g_{AA}(r) = \left(1 - \frac{k^{obs}\exp(-\xi r)}{2\pi D r}\right) \tag{105}$$

It is instructive to compare this expression with the radial distribution function obtained using the Smoluchowski theory for this reaction,[22] i.e.

$$g_{AA}^{Smol}(r) = \left(1 - \frac{k^{obs}}{2\pi D' r}\right) \tag{106}$$

The Smoluchowski expression is missing the exponential factor which is due to the bimolecular lifetime of A and involves $D' = 2D$, rather than D, in the denominator. Neither of these differences can be removed by modifying the Smoluchowski theory. This can be seen by comparing the form of Equation 100 with the Smoluchowski equation, modified as in Equation 56 to include the bimolecular lifetime effect.

The same type of techniques can be used to calculate the radial distribution function for the generic bimolecular reaction:

$$A + B \xrightarrow{k^{obs}} \text{products} \tag{107}$$

The calculation, however, is complicated by the fact that fluctuations in the density of A and B are coupled. In an ideal solution, the average equations for the densities, which again include only the effects of diffusion and reaction, are

$$\partial\bar{\rho}_A/\partial t = D_A\nabla^2\bar{\rho}_A - k^{obs}\bar{\rho}_A\bar{\rho}_B + K$$

$$\partial\bar{\rho}_B/\partial t = D_B\nabla^2\bar{\rho}_B - k^{obs}\bar{\rho}_A\bar{\rho}_B + K \quad (108)$$

The constant input term guarantees a uniform steady state at which $K = K^{obs}\bar{\rho}_A\bar{\rho}_B$. The density fluctuations around this steady state satisfy the stochastic differential equations:[46,47]

$$\partial\delta\rho_A/\partial t = D_A\nabla^2\delta\rho_A - k^{obs}\bar{\rho}_B\delta\rho_A - k^{obs}\bar{\rho}_A\delta\rho_B + \tilde{f}_A$$

$$\partial\delta\rho_B/\partial t = D_B\nabla^2\delta\rho_B - k^{obs}\bar{\rho}_A\delta\rho_B - k^{obs}\bar{\rho}_B\delta\rho_A + \tilde{f}_B \quad (109)$$

The two random terms, $\tilde{\mathbf{f}}_A$ and $\tilde{\mathbf{f}}_b$, are due to the random diffusion of A and B separately and their mutual reaction. Since there are two different $\tilde{\mathbf{f}}$, their correlation function is now a matrix. Its form is given by the mechanistic statistical theory and is[50]

$$\langle\tilde{f}_A(\mathbf{r}, t)\,\tilde{f}_A(\mathbf{r}', t')\rangle = [k^{obs}\bar{\rho}_A\bar{\rho}_B - 2D_A\bar{\rho}_A\nabla^2\mathbf{r}]\,\delta(\mathbf{r} - \mathbf{r}')\,\delta(t - t')$$

$$\langle\tilde{f}_A(\mathbf{r}, t)\,\tilde{f}_B(\mathbf{r}', t')\rangle = \langle\tilde{f}_B(\mathbf{r}, t)\,\tilde{f}_A(\mathbf{r}', t')\rangle = k^{obs}\bar{\rho}_A\bar{\rho}_B\delta(\mathbf{r} - \mathbf{r}')\,\delta(t - t')$$

$$\langle\tilde{f}_B(\mathbf{r}, t')\,\tilde{f}_B(\mathbf{r}', t')\rangle = [k^{obs}\bar{\rho}_A\bar{\rho}_B - 2D_B\bar{\rho}_B\nabla_r^2]\,\delta(\mathbf{r} - \mathbf{r}')\,\delta(t't') \quad (110)$$

The fact that the random terms and the density fluctuations themselves are coupled by Equations 109 and 110 means that it is not possible to obtain a single differential equation, like Equation 100, for the radial distribution function, $g_{BA}(r,t)$. One obtains, instead, three coupled inhomogeneous partial differential equations for $g_{AA}(r,t)$, $g_{BA}(r,t)$, and $g_{BB}(r,t)$. The coupling comes through the chemical reaction, and it can be neglected only in the limit that $\bar{\rho}_A$ and $\bar{\rho}_B$ approach zero. In this limit of dilute solution, one finds that

$$\Delta\rho \equiv \bar{\rho}_B g_{BA}(|\mathbf{r} - \mathbf{r}'|, t) - \bar{\rho}_B \equiv \sigma_{BA}/\bar{\rho}_A \quad (111)$$

satisfies the modified Smoluchowski equation in Equation 101. Thus, in dilute solution, we recover the Smoluchowski theory. If this limit is not taken, the solution of the stochastic equations is carried out most easily using the technique of Fourier transformation. The details of this have been explained elsewhere,[50] and we simply quote the end result for the radial distribution function. At steady state, one finds:

$$g_{BA}(r) = 1 - \frac{k^{obs}}{4\pi D'r}\left[\frac{(\alpha - 2\beta)}{2(\alpha - \beta)}\exp(-r\beta^{1/2}) + \frac{\alpha}{2(\alpha - \beta)}\exp(-r\alpha^{1/2})\right] \quad (112)$$

where

$$\alpha^{-1/2} = [D_AD_B/k^{obs}(\bar{\rho}_BD_B + \bar{\rho}_AD_A)]^{1/2}$$

$$\beta^{-1/2} = [D'/k^{obs}(\bar{\rho}_A + \bar{\rho}_B)]^{1/2} \quad (113)$$

The bracketed term in Equation 112 is a correction factor to the usual Smoluchowski result. Notice that, in the limit of dilute solution, the characteristic lengths $\alpha^{-1/2}$ and $\beta^{-1/2}$ in Equation 113 become very large. In this limit, the exponentials in the correction factor in Equation 112 asymptotically approach one. Thus, at low density, the correction factor becomes one and we recover the Smoluchowski radial distribution function. At higher densities of reactants, the correlation lengths may cause significant corrections to the Smoluchowski theory. This will occur when $\alpha^{-1/2}$ or $\beta^{-1/2}$ is comparable to a molecular dimension, R. For example, if $\overline{\rho_B} = 0$, then:

$$R\beta^{1/2} = (R^2k^{obs}\overline{\rho}_A/D')^{1/2} \approx (4\pi R^3\overline{\rho}_A)^{1/2} = (3\phi)^{1/2} \tag{114}$$

where we have used the estimate $k^{obs} \approx 4\pi D'R$, and ϕ is the volume fraction defined in Equation 61. Thus, these correlation effects, which are due to the finite bimolecular lifetime, become appreciable at concentrations above a few hundredths molar. It is worth noting that the coupling between A and B leads to a dependence of $g_{BA}(r)$ on both the relative diffusion constant (D') and the individual diffusion constants of A and B. This feature is completely missing in the Smoluchowski theory.

2. Calculation of Rate Constants in Solution

The fundamental connection between the radial distribution function and the bimolecular rate constant at steady state is given in Equation 90. For a given bimolecular reaction, the radial distribution function can be obtained using statistical nonequilibrium thermodynamics as described above. To evaluate the observed rate constant one must know, in addition, the intrinsic rate constant, $k^o(r)$. For the sake of illustration, we shall use the Smoluchowski reactivity in Equation 91. Combining this expression with Equation 90 gives:[50]

$$k^{obs} = k^o g_{BA}(R) \tag{115}$$

On first glance, Equation 115 appears to provide a simple expression for k^{obs}. However, examining the form of the radial distribution function in Equation 112, we notice that it depends explicitly on k^{obs}. Indeed, for the radial distribution in Equation 112, we find that Equation 115 becomes

$$k^{obs} = k^o\left(1 - \frac{k^{obs}}{4\pi D'R}\left[\frac{(\alpha - 2\beta)}{2(\alpha - \beta)}\exp(-R\beta^{1/2}) + \frac{\alpha}{2(\alpha - \beta)}\exp(-R\alpha^{1/2})\right]\right) \tag{116}$$

Since α and β both depend on k^{obs} as indicated in Equation 113, this expression turns out to be a transcendental equation which must be solved for k^{obs}. If the solution is dilute in both A and B, the expression in square brackets goes to one, as we have seen. Thus, in dilute solution, we find that

$$k^{obs} = k^o\left(1 - \frac{k^{obs}}{4\pi D'R}\right) \tag{117}$$

This is easily solved for k^{obs} and gives:

$$k^{obs} = 4\pi D'Rk^o/(4\pi D'R + k^o) \tag{118}$$

which is the Collins-Kimball expression for a rate constant in solution.

If the solution is not dilute in A and B, then Equation 116 must be solved numerically. As a practical matter, this can be done quite rapidly using iteration even on a hand-held calculator. To see what is required, we consider the diffusion-controlled limit of Equation 116. This implies that $k^o \gg k^{obs}$, which means that the term in the large parentheses in Equation 116 can be set equal to zero. Hence,

$$k^{obs} = 4\pi D'R \Big/ \left[\frac{(\alpha - 2\beta n)}{2(\alpha - \beta)} \exp(-R\beta^{1/2}) + \frac{\alpha}{2(\alpha - \beta)} \exp(-R\alpha^{1/2})\right] \quad (119)$$

To begin the iteration, we approximate k^{obs} by $k_o^{obs} = 4\ \pi D'R$. Substituting this value into the expressions for α and β on the right-hand side of Equation 119 gives a new approximation for k^{obs}, k^{obs}. This, in turn, is substituted into the right-hand side of Equation (119) leading to k_2^{obs}, etc. As long as the densities are well below close packing, this procedure rapidly converges to the solution of Equation 119. In this way, the size of the correction term in Equation 119 can be ascertained.

A simple special case is instructive as to the size of these effects. Let us assume that D_A equals zero and that B is dilute. This implies that the A act as static traps[53] for low density B, as might correspond to exciton trapping in solids. Equation 119 then reduces to

$$k^{obs} = 4\pi RD'\exp[R/(D'k^{obs}\bar{\rho}_A)^{1/2}] \quad (120)$$

where $D' = D_B$. If we take $D_B = 1.0 \times 10^{-5}$ cm^2 sec^{-1} and R = 4 Å, then at a density of A corresponding to a concentration of 1.7×10^{-4} *M*, the correlation length obtained by solving Equation 120 iteratively is $(D'/k^{obs}\bar{\rho}_A)^{1/2} = 400$ Å. The value of k^{obs} at this concentration is $k^{obs} = 3.0 \times 10^9\ M^{-1}\ sec^{-1}$, which is just that expected on the basis of the Smoluchowski theory. At a concentration of A equivalent to 0.52 *M*, on the other hand, the correlation length is $(D'/k^{obs}\bar{\rho}_A)^{1/2} = 5.6$ Å and $k^{obs} = 6.0 \times 10^9\ M^{-1}\ sec^{-1}$. This increase is caused by the decrease in the bimolecular lifetime of the B molecules at higher concentrations of A. Thus, diffusion has less time to effect the radial distribution of B molecules around the A molecules.

If the intrinsic reactivity is different from the Smoluchowski reactivity, then the transcendental expressions for the rate constant based on Equation 90 are more complicated. Let us assume that the reactivity function has the form:

$$k^o(r) = \begin{cases} (1/\tau_o)\exp(-r/r_o) & r > R \\ 0 & r \leq R \end{cases}$$

When this truncated exponential reactivity is substituted into Equation 90 along with the radial distribution function in Equation 112 for the reaction A + B → products, one finds that[50]

$$k^{obs} = \frac{4\pi r_o^3}{\tau_o}\{[(R/r_o) + 1]^2 + 1\}\exp(-R/r_o)$$
$$\frac{-k^{obs}}{D'\tau_o}\left\{\frac{\alpha[R(\alpha^{1/2} + 1/r_o) + 1]\exp(-R\alpha^{1/2})}{2(\alpha - \beta)(\alpha^{1/2} + 1/r_o)^2}\right.$$
$$\left. + \frac{(\alpha - 2\beta)[R(\beta^{1/2} + 1/r_o) + 1]\exp(-R\beta^{1/2})}{2(\alpha - \beta)(\beta^{1/2} + 1/r_o)^2}\right\} \quad (121)$$

Although this expression is considerably more complicated than that in Equation 116, it can also be solved interatively on a hand calculator. Results of such a calculation,[50] which begins the iteration at the dilute solution value of k^{obs}, are shown in Figure 5. The values used in the calculation are $D'/D_B = 1.4$, $D_A = 2 \times 10^{-5}$ cm^2 sec^{-1}, R = 6 Å, $r_o = 3$

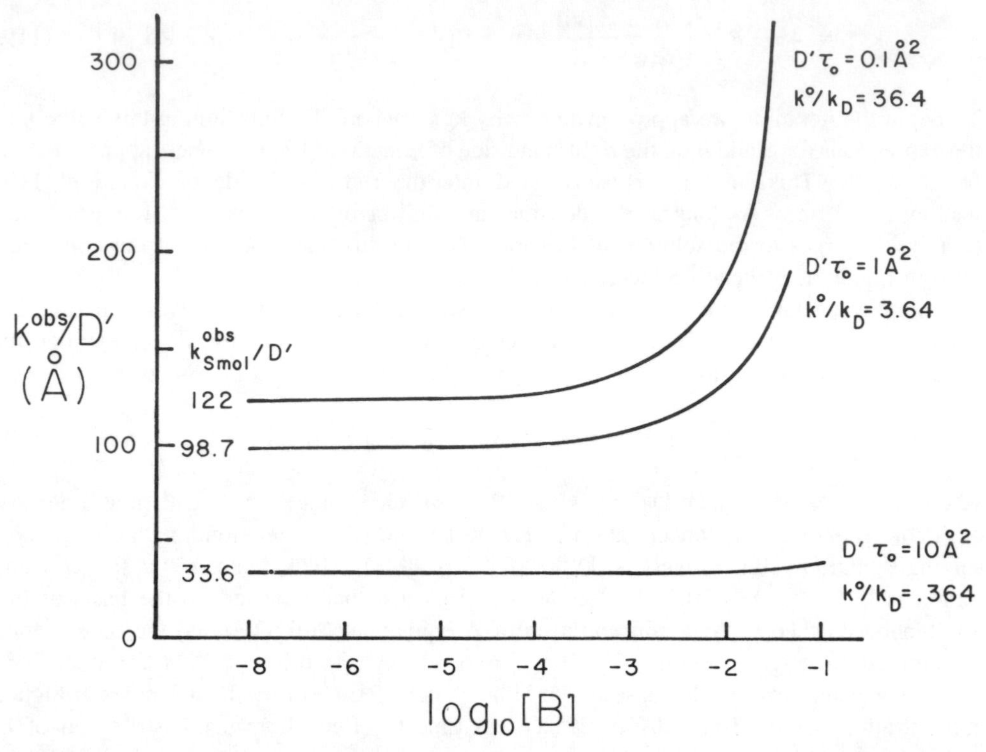

FIGURE 5. Graph of the observed rate constant as a function of the logarithm of the molar density of B for the reactivity function described by Equation 120 in the text. The parameter values used are summarized at the right of each curve, and the limiting low concentration values of k^{obs}/D' are given at the left of each curve.

Å, and the three values of $D' \tau_o$ indicated. Also shown there are the effective values of $k^o/4\pi D'R_o$ which occur in dilute solution if Equation 121 is cast into the Collins-Kimball form:

$$k^{obs} = 4\pi D'R_o k_o/(4\pi D'R_o + k_o) \qquad (122)$$

The curves in Figure 5 are all asymptotic to Equation 122 at low density of B. The lower curve corresponds to a slow intrinsic reaction, which is about 25% diffusion controlled. Changing the molar concentration of B even by seven orders of magnitude has little effect on k^{obs} under this condition. The upper curve, on the other hand, represents a reaction which is about 97% diffusion controlled. In this case, the effect of changing the concentration of B becomes significant above 10^{-4} *M*. In the absence of other effects, experimental measurement of bimolecular rate constants as a function of concentration should provide a method of assessing the amount of diffusion control without having to change solvent.

3. *Calculation of Rate Constants in Two-Dimensional Systems*

In the two preceding subsections, we showed how to use the statistical theory of nonequilibrium thermodynamics to calculate reaction rate constants in solutions of three spatial dimensions. Rapid reactions, however, also occur in two-dimensional environments such as biological membranes in vivo or in isolated phospholipid bilayers in vitro. It is straightforward to extend the statistical theory to two-dimensional, or even one-dimensional, systems.[50] All that is required is to restrict the coordinates to a two-dimensional manifold and to use the proper definition of the Laplacian operator, ∇^2, on the manifold. The simplest situation to consider is a flat plane. In that case, instead of spherical polar coordinates, one can use the plane polar coordinates r and θ, or the Cartesian coordinates x and y. In this case, the relationship in Equation 90 between the intrinsic reactivity and the radial distribution function becomes

$$k^{obs} = 2\pi \int_0^\infty k^o(r)\, g_{BA}^{(2)}(r)\, rdr \tag{123}$$

where the superscript emphasizes that the radial distribution function is calculated in two dimensions.

The way in which the calculations are modified is seen most easily for the atomic recombination reaction, $A + A \rightarrow A_2$. As we noted, only the form of the Laplacian which appears in Equations 94 to 96 is changed. It follows that Equation 100 is still valid if ∇^2 is taken to be the two-dimensional Laplacian. Consequently, at steady state, we need to solve the two-dimensional version of Equation 102, i.e.

$$\frac{D}{r}\frac{d}{dr}\, r(d\Delta\rho/dr) - 4k^{obs}\bar{\rho}_A\Delta\rho = 2k^{obs}\bar{\rho}_A \frac{\delta(r)}{2\pi r} \tag{124}$$

where we have introduced plane polar coordinates. This equation is easy to solve by Fourier transformation,[50] which leads to the solution:

$$\Delta\rho(r) = -(k^{obs}/\pi D)\, K_o(\xi r) \tag{125}$$

where K_o is a Bessel function of imaginary argument (also called the McDonald function of order zero) and ξ^{-1} is the correlation length in Equation 104. Consequently, the radial distribution function is

$$g_{AA}^{(2)}(r) = 1 - (k^{obs}/\pi D)\, K_o(\xi r) \tag{126}$$

In a similar fashion, the radial distribution function for the generic bimolecular reaction $A + B \rightarrow$ products is found to be

$$g_{BA}^{(2)}(r) = 1 - \frac{k^{obs}}{2\pi D'}\left[\frac{(\alpha - 2\beta)}{2(\alpha - \beta)} K_o(\beta^{1/2} r) + \frac{\alpha}{2(\alpha - \beta)} K_o(\alpha^{1/2} r)\right] \tag{127}$$

with α and β defined in Equation 113. Indeed, these results illustrate a rather general difference between two- and three-dimensional radial distribution functions. Terms like $\exp(-\xi r)/2r$ in the three dimensions are replaced by $k_o(\xi r)$ in two dimensions. This can be checked for the two reactions already treated by comparing Equations 126 and 127 with Equations 105 and 112.

We can use this correspondence to compare our results in two dimensions to the modifications of the Smoluchowski theory discussed previously. For the special case of static A molecules ($D_A = 0$) and dilute B molecules ($\rho_B \rightarrow 0$), Equation 112 reduces to

$$g_{BA}(r) = 1 - \frac{k^{obs}}{4\pi D' r} \exp[-(\bar{\rho}_A k^{obs}/D_B)^{1/2}\, r] \tag{128}$$

Using the correspondence with two dimensions, we conclude that

$$g_{BA}^{(2)}(r) = 1 - \frac{k^{obs}}{2\pi D'} K_o([\bar{\rho}_A k^{obs}/D']^{1/2}\, r) \tag{129}$$

This is the expression for the radial distribution of dilute molecules which are trapped at static sinks,[54] which is the problem considered by Adam and Dëlbruck[37] and Berg and Purcell.[38] Using the two-dimensional Smoluchowski reactivity, i.e.

$$k^{o}(r) = k^{o}\delta(r - R)/2\pi r \tag{130}$$

and the expression for $g_{BA}^{(2)}(r)$ in Equation 129, we obtain:

$$k^{obs} = k^{o}\left(1 - \frac{k^{obs}}{2\pi D'} K^{o}([\bar{\rho}_A k^{obs}/D']^{1/2} r)\right) \tag{131}$$

Since the reaction at static sinks is assumed to be diffusion controlled, we have $k^{o} \gg k^{obs}$. Thus, Equation 131 gives:

$$k^{obs} = 2\pi D'/K_o([\bar{\rho}_A k^{obs}/D']^{1/2} R) \tag{132}$$

as we could have anticipated from the corresponding three-dimensional result in Equation 120. Using this expression, the radial distribution function in Equation 129 can be written:

$$g_{BA}^{(2)}(r) = \frac{K_o([\bar{\rho}_A k^{obs}/D']^{1/2} R) - K_o([\bar{\rho}_A k^{obs}/D']^{1/2} r)}{K_o([\bar{\rho}_A k^{obs}/D']^{1/2} R)} \tag{133}$$

When the density of traps, $\bar{\rho}_A$, is small, we can simplify Equation 133 using the asymptotic formula:[55]

$$K_o(x) \sim -[\ln(x/2) + \gamma] \tag{134}$$

where $\gamma = 0.5772 \ldots$ is Euler's constant. Thus,

$$g_{BA}^{(2)}(r) = \ln(r/R)/K_o[\bar{\rho}_A k^{obs}/D']^{1/2} R) \tag{135}$$

Compared to Equation 66, this result is seen to be identical in form to the expression obtained using the modified Smoluchowski theory. Indeed, if the comparison is taken a step further, one can also verify for $\bar{\rho}_A \rightarrow 0$ that[54]

$$k^{obs} = 2\pi D'/[\ln(b/R) - \gamma + \ln\sqrt{2}] \tag{136}$$

where $b^2 = \bar{\rho}_A/\pi$. Except for the term $-\gamma + \ln\sqrt{2} = 0.231 \ldots$; this is identical to the limiting forms obtained by the modified Smoluchowski theory in Equation 71. Equation 132, of course, provides a more general expression for the rate constant, which must be solved by iteration.

4. Effect of Molecular Interactions

Molecular interactions appear naturally in the mechanistic statistical theory of fluctuations.[56,57] In the preceding subsections, we have restricted attention to reactions that occurred in ideal solutions. Only in this limiting case do the fluctuations satisfy simple partial differential equations like Equations 95 and 109. In the more general case, fluctuations in the density at one point in space are coupled to those at nearby points by the local equilibrium direct correlation function.[58] The direct correlation function, in turn, is related to the local equilibrium radial distribution function. Finally, the local equilibrium radial distribution function is related to the potential of mean force by the definition:[22]

$$w(r) = -k_BT\ln g^{le}(r) \tag{137}$$

The superscripts here imply the local equilibrium form, e.g., such as found in the Debye-Hückel expression in Equation 20. In this way, the fluctuations are affected by molecular interactions in a way comparable to the Debye modification of the Smoluchowski theory. Detailed results and calculations, however, differ somewhat.

We again rely on the recombination reaction A + A → products to highlight the main points. In general, the mechanistic statistical theory implies that the average equations for this reaction are[50,56,57]

$$\partial\bar{\rho}_A/\partial t = -2k_*^{obs}\bar{z}_A^2 + D^o\nabla^2\bar{z}_A + 2K \tag{138}$$

In this equation, z_A is the activity, defined in terms of the deviation of the chemical potential, μ, from its standard state value, μ^o, by

$$z_A = \exp[(\mu - \mu^o)/k_BT] \tag{139}$$

The rate constant, k_*^{obs}, is now based on activity[59] rather than density, so that

$$k_*^{obs} = k^{obs}/\gamma_A^2 \tag{140}$$

where γ_A is the activity coefficient of A. Furthermore, the transport coefficient, D_o, is activity based and defined by

$$D^o = D(\partial\bar{\rho}_A/\partial\bar{z}_A)_T \tag{141}$$

In dilute solution, of course, one has $\mu - \mu^o = k_BT\ln\rho_A$ and $\gamma_A = 1$. In this case, it is easy to see that $\bar{z}_A = \bar{\rho}_A$, and Equation 138 reduces to the earlier equation 94, which we used to describe this reaction.

The equation which governs the density fluctuations, $\delta\rho_A$, is obtained by linearizing Equation 138 and adding the appropriate term to account for random molecular changes. One finds that[50]

$$\partial\delta\rho_A/\partial t = -4k_*^{obs}\bar{z}_A\delta z_A + D^o\nabla^2\delta z_A + \tilde{f} \tag{142}$$

where

$$\langle\tilde{f}(\mathbf{r}, t)\,\tilde{f}(\mathbf{r}', t')\rangle = (4k_*^{obs}\bar{z}_A^2 - 2D^o\bar{z}_A\nabla_r^2)\,\delta(\mathbf{r} - \mathbf{r}')\,\delta(t - t') \tag{143}$$

The most significant difference between these equations and their dilute solution analogs, Equations 95 and 96, is that fluctuations in the activity at point **r** depend on fluctuations in the density at neighboring points. The relationship has the form:[56]

$$\delta z_A(\mathbf{r}) = (\bar{z}_A/\bar{\rho}_A)\,\delta\rho_A(\mathbf{r}) - \bar{z}_A \int c_{AA}(|\mathbf{r} - \mathbf{r}'|)\,\delta\rho_A(\mathbf{r}')\,d\mathbf{r}' \tag{144}$$

where $C_{AA}(|\mathbf{r}-\mathbf{r}'|)$ is the local equilibrium direct correlation function for an isotropic solution. The coefficient $\bar{z}_A$ of the integral insures that the effect of this term vanished in dilute solution. The direct correlation function has the range of the molecular interaction. In Debye-Hückel theory, e.g., one has:[56]

$$c_{AA}(r) = -q_A^2/k_BT\epsilon r \tag{145}$$

The nonlocal effect of molecular interactions is taken into account by substituting Equation 144 into Equation 142 and solving for the density-density correlation function. Since the nonlocal term in Equation 144 is linear, it is easy to do this using Fourier transforms. One finds that[50]

$$g_{AA}(|\mathbf{r} - \mathbf{r}'|) = g_{AA}^{le}(|\mathbf{r} - \mathbf{r}'|) - k_*^{obs} \frac{(\bar{z}_A/\bar{\rho}_A) \exp(-\xi_*|\mathbf{r} - \mathbf{r}'|)}{2\pi D^o|\mathbf{r} - \mathbf{r}'|}$$

$$- \int \frac{[g_{AA}^{le}(|\mathbf{r} - \mathbf{r}''| - 1)\, k_*^{obs}\bar{z}_A \exp(-\xi_*|\mathbf{r}'' - \mathbf{r}'|)\, d\mathbf{r}''}{2\pi D^o|\mathbf{r}'' - \mathbf{r}'|} \tag{146}$$

where $\xi_*^{-1} \equiv (D/4k_*^{obs}\, \bar{z}_A)^{1/2}$. In an ideal solution, $g_{AA}^{le} \equiv 1$ and it is easy to see that Equation 146 correctly reduces to our previous result in Equation 105. We can compare this result to that obtained by Debye for the recombination reaction by referring to Equation 28. Recalling the definition of w(r) in Equation 137 and the fact that identical particles change the 4π to 2π in Equation 128, we obtain:

$$g_{AA}^{Debye}(|\mathbf{r} - \mathbf{r}'|) = g_{AA}^{le}(|\mathbf{r} - \mathbf{r}'|) - \frac{k^{obs}\Delta}{2\pi D(|\mathbf{r} - \mathbf{r}'|)} \tag{147}$$

with Δ defined by

$$\Delta = (|\mathbf{r} - \mathbf{r}'|/2) \int_{|\mathbf{r}-\mathbf{r}'|}^{\infty} [g_{AA}^{le}(R)/g_{AA}^{le}(x)\, x^2]\, dx \tag{148}$$

Comparing this with Equation 146, we see that the Debye modification of the Smoluchowski theory yields expressions that are comparable to only the first two terms in Equation 146.

The size of the integral term in Equation 146 can be estimated by assuming explicit expressions for $g_{AA}^{le}(r)$. A choice which leads to analytical results is

$$g_{AA}^{le}(r) = 1 - \exp(-\kappa r)/4\pi\bar{\kappa} r \tag{149}$$

where κ is a molecular correlation length. For this local equilibrium radial distribution function, we find:[50]

$$g_{AA}(r) = 1 + \frac{\exp(-\kappa r)}{4\pi\kappa r} - \frac{k_*^{obs}(\bar{z}_A/\bar{\rho}_A)}{2\pi D^o r} [\exp(-\xi_* r)$$

$$- (\bar{\rho}_A/\bar{\kappa}(\xi_*^2 - \kappa^2))\, (\exp(-\xi_* r) - \exp(-\kappa r))] \tag{150}$$

The nonlocal term in Equation 146 leads to the second term in square brackets. By expanding the exponentials, we see that it will be nonnegligible if r, ξ_*^{-1}, and κ^{-1} are of comparable size and the number of molecules in a volume of magnitude $r\bar{\kappa}^{-1}\xi_*^{-1}$, i.e., $\bar{\rho}_A r\bar{\kappa}^{-1}\xi_*^{-1}$, is of the order of 0.1. If the molecules A have a single negative charge, then Equation 149 represents the local equilibrium Debye-Hückel radial distribution function if κ^{-1} is the Debye-Hückel length and $\bar{\kappa}^{-1} = 2r_o e^{\kappa a}/(1 + \kappa a)$, with r_o the Onsager length. The quantity $\bar{\kappa}^{-1}$ is of the order of 14 Å in water at room temperature. The length κ^{-1} is of the order[24] of 10 Å for a one-one electrolyte at 0.1 *M* and ξ_*^{-1}, as we have seen, will also be of the order of 10 Å at that concentration. Thus, at this concentration and r = 10 Å, we estimate that $\bar{\rho}_A r\bar{\kappa}^{-1}\xi_1^{-*} \sim 0.1$. Consequently, this term can be appreciable in the range of 0.1 *M* concentrations.

C. Complex Systems

Thus far, we have focused our attention on a single bimolecular reaction step, either the recombination reaction $A + A \rightarrow A_2$ or the generic bimolecular reaction $A + B \rightarrow$ products. The statistical theory that we have outlined, however, is applicable to more general chemical reactions. Here, we examine a variety of more complex schemes. The first is a coupled unimolecular/bimolecular reaction mechanism which can be used to describe fluorescence quenching. Second, we examine the influence of the back reaction on the recombination reaction. As a third example, we discuss a coupled mechanism involving three reactants which has been used to model the trapping of membrane-bound receptors by regions called coated pits. We complete our discussion of complex systems by outlining how the effects of other transport processes, in particular, rotational diffusion and heat conduction, can be included in the theory.

1. Fluorescence Quenching: Uni- and Bimolecular Lifetime Effects

Two examples of fluorescence quenching reactions are listed in Table 1. This sort of reaction involves a fluorescent molecule, say, A, which is called the fluorophore. The fluorophore is in an excited electronic state and reacts with the quencher, say, B, which removes the excitation energy from A. Three molecular processes are involved in fluorescence quenching:[14] excitation of A, fluorescence by A, and the quenching itself. Symbolically,

$$\overline{A} + h\nu \rightarrow A$$

$$A \xrightarrow{k_u} \overline{A} + h\nu'$$

$$A + B \xrightarrow{k^{obs}} \overline{A} + B' \tag{151}$$

where $\bar{A}$ represents the ground state of A. As long as $\bar{A}$ is in great excess, we can treat its density as constant. Thus, the average differential equations for this reactions, including diffusion, are[49]

$$\partial\overline{\rho}_A/\partial t = D_A\nabla^2\overline{\rho}_A - k^{obs}\overline{\rho}_A\overline{\rho}_B - k_u\overline{\rho}_A + K$$

$$\partial\overline{\rho}_B/\partial t = D_B\nabla^2\overline{\rho}_B - k^{obs}\overline{\rho}_A\overline{\rho}_B + K' \tag{152}$$

Here, K is proportional to the density of $\bar{A}$ and the intensity of the excitation light, K′, is a constant rate which accounts for reappearance of B from the state B′; in addition, $k_u^{-1} = \tau_o$, the fluorescence lifetime in the absence of quencher. A more complete scheme would also include the kinetic step $B' \rightarrow B$. This, however, would require the use of another density and, as long as the concentration of B′ is small, it can be neglected. During steady illumination in a uniform system, Equation 152 shows that A achieves the steady density $\bar{\rho}_A = K/(k^{obs}\bar{\rho}_B + k_u)$.

To obtain the radial distribution for A and B at steady state, we linearize these equations around the steady state and add the random terms dictated by the mechanistic statistical theory. The equations for the fluctuations are[49,50]

$$\partial\delta\rho_A/\partial t = D_A\nabla^2\delta\rho_A - k^{obs}\overline{\rho}_B\delta\rho_A - k^{obs}\overline{\rho}_A\delta\rho_B - k_u\delta\rho_A + \tilde{f}_A$$

$$\partial\delta\rho_B/\partial t = D_B\nabla^2\delta\rho_B - k^{obs}\overline{\rho}_A\delta\rho_B - k^{obs}\overline{\rho}_B\delta\rho_A + \tilde{f}_B \tag{153}$$

Except for the term $-k_u\delta\rho_A$ in the first equation, these are identical to those in Equation

109 for the reaction A + B → products. The correlations of the random terms are also the same as those in Equation 110, except now

$$\langle \tilde{f}_A(\mathbf{r}, t)\, \tilde{f}_A(\mathbf{r}', t) \rangle = [k^{obs}\bar{\rho}_A\bar{\rho}_B + k_u\bar{\rho}_A - 2D_A\bar{\rho}_A\nabla_r^2]\, \delta(\mathbf{r} - \mathbf{r}')\, \delta(t - t') \quad (154)$$

As before, these equations can be solved using Fourier transforms to obtain the density-density correlation function. Using this result and Equation 86, we find that the radial distribution function is[49]

$$g_{BA}(r) = 1 - \frac{k^{obs}}{4\pi D'r}\left\{\left[\frac{\alpha(\beta - \gamma)}{2(\beta - \gamma_1)(\beta - \gamma_2)} + 1\right] \exp(-r\beta^{1/2}) + \left[\frac{\alpha(\lambda_1 - \gamma)}{2(\lambda_1 - \beta)(\lambda_1 - \lambda_2)}\right] \exp(-r\lambda_1^{1/2}) + \left[\frac{\alpha(\lambda_2 - \gamma)}{2(\lambda_2 - \beta)(\lambda_2 - \lambda_1)}\right] \exp(-r\lambda_2^{1/2})\right\} \quad (155)$$

Three correlation lengths, $\beta^{-1/2}$, $\lambda_1^{-1/2}$, and $\lambda_2^{-1/2}$, appear in this expression. Along with α and γ, they are defined by

$$\gamma = 2k_u k^{obs}\bar{\rho}_A/[(k_u + k^{obs}\bar{\rho}_B)\, D_B + k^{obs}\bar{\rho}_A D_A]$$

$$\beta = [k_u + k^{obs}(\bar{\rho}_A + \bar{\rho}_B)]/D'$$

$$\lambda_1 = (\alpha/2)(1 + [1 - 2\gamma/2\alpha]^{1/2})$$

$$\lambda_2 = (\alpha/2)(1 - [1 - 2\gamma/\alpha]^{1/2})$$

$$\alpha = [(k_u + k^{obs}\bar{\rho}_B)\, D_B + k^{obs}\bar{\rho}_A D_A]/D_A D_B \quad (156)$$

The complicated term in curly brackets in Equation 155 gives a correction to the Smoluchowski radial distribution function. Part of this correction is caused by the bimolecular lifetime effect and part by the lifetime of the fluorophore. As we have seen, keeping the concentration of $\bar{A}$ and B below about 10^{-4} *M* eliminates the bimolecular lifetime effect (cf. Figure 5). Indeed, setting $\bar{\rho}_A = \bar{\rho}_B = 0$ and for simplicity taking $D_A = D_B$, Equation 156 gives $\gamma = \gamma_2 = 0$, $2\beta = \alpha = k_u/D_A$. Thus, Equation 155 reduces to

$$g_{BA}(r) = 1 - \frac{k^{obs}}{4\pi D'r} \exp(-r\alpha^{1/2}) \quad (157)$$

The correlation length is $\alpha^{-1/2} = (D_A\tau_o)^{1/2}$, and depends on only the fluorescence lifetime. Using the Smoluchowski reactivity, the corresponding diffusion-controlled rate constant is

$$k^{obs} = 4\pi D'R\exp(R\alpha^{1/2}) \quad (158)$$

In this special case, α does not depend on k^{obs} and so Equation 158 gives the solution for the bimolecular rate constant. Taking $D_A = 1 \times 10^{-5}$ cm sec^{-1}, we find that the exponential correction to the Smoluchowski result in Equation 158 has the values 1.3 for $\tau = 10^{-8}$ sec and 2.2 for $\tau = 10^{-9}$ sec. These are typical fluorescence lifetimes for organic molecules.[14] This effect is viscosity dependent and would increase the apparent Smoluchowski radius by a factor of $\exp(R\alpha^{1/2})$.

Using the Smoluchowski reactivity, the diffusion-controlled rate constant for the general case of fluorescence quenching can be written using Equation 155 as

$$k^{obs} = 4\pi D'RC(R) \quad (159)$$

The form of C(R) is given by the reciprocal of the curly bracketed expression in Equation 155 evaluated at R. In most experimental situations, the density of the fluorophore is small and can be set equal to zero. In this case, the expression for the correction factor C(R) becomes [60,61]

$$C(R) = 2D_B[(D_B - D_A)\exp(-R\beta^{1/2}) + D'\exp(-R\alpha^{1/2})]^{-1} \tag{160}$$

with

$$\alpha = (k_u + k^{obs}\bar{\rho}_B)/D_A$$

$$\beta = (k_u + k^{obs}\bar{\rho}_B)/D' \tag{161}$$

These expressions are compared to experiment in the final section.

2. Effect of the Back Reaction

Our earlier treatment of the recombination reaction in Equation 93 ignored the effect of the back reaction, as well as the mechanism by which the species A is generated. A more complete mechanism would be

$$A_2 + h\nu \rightarrow A + A$$

$$A + A \underset{k_u}{\overset{k^{obs}}{\rightleftarrows}} A_2 \tag{162}$$

This scheme includes reactions in which both A and A_2 are reactants. Using the now-familiar techniques of the mechanistic statistical theory, we find that the density fluctuations, $\delta\rho_A$ and $\delta\rho_{A_2}$, solve two equations that are coupled dynamically. Neglecting the effect of molecular interactions and solving for the relevant radial distribution function, we find that[50]

$$g_{AA}(r) = 1 - \frac{k^{obs}k^*}{2\pi D' r(k_u + k^*)}\Bigg[\exp(-\xi_2 r) + (D_{A_2}/D_A)\left(\frac{\xi_1^2\exp(-\xi_1 r) - \xi_2^2\exp(-\xi_2 r)}{\xi_1^2 - \xi_2^2}\right)\Bigg] \tag{163}$$

where $2k^*\bar{\rho}_{A_2}$ is the rate at which A is generated from A_2, and ξ_1 and ξ_2 are defined by

$$\xi_1^2 = [(4D_{A_2}/\bar{\rho}_A) + (D_A/\bar{\rho}_{A_2})][k^{obs}\bar{\rho}_A^2 + k_u\bar{\rho}_{A_2}]/D_A D_{A_2}$$

$$\xi_2^2 = [(4/\bar{\rho}_A) + (1/\bar{\rho}_{A_2})][k^{obs}\bar{\rho}_A^2 + k_u\bar{\rho}_{A_2}]/(D_A + D_{A_2}) \tag{164}$$

The relationship between $\bar{\rho}_{A_2}$ and $\bar{\rho}_A$ at steady state is

$$\bar{\rho}_{A_2} = \bar{\rho}_A^2 k^{obs}/(k^* + k_u) \tag{165}$$

If the diffusion constant of A_2 and its density are set equal to zero and k* is small, the expression for $g_{AA}(r)$ reduces to that obtained in Equation 105. Notice, also, that when the illumination intensity vanishes, we recover the equilibrium result $g_{AA}(r) = 1$. In that case, the steady state is actually an equilibrium state and $k^{obs} = k^o$. The effect of the back reaction becomes appreciable when k^* is comparable to k_u. As with all the effects we have examined, the exact magnitude will depend on the numerical value of the kinetic constants and the diffusion coefficients. Whenever the correlation lengths are comparable to molecular dimensions, an increase in the rate constant over the Smoluchowski value by factors of the order of two to three is possible.

3. Trapping of Receptors by Clathrin-Coated Pits

An important mechanism that is used by many large molecules or molecular aggregates to enter living cells is called receptor-mediated endocytosis.[62] This mechanism involves the specific binding of the ligand molecule to a membrane-bound receptor protein. The receptors, in turn, diffuse laterally in the membrane until they encounter regions called coated pits. Coated pits are disk-shaped regions in the plasma membrane about 0.1 μm in radius which are coated with the protein clathrin. The clathrin pits serve both to trap receptors and to transport the receptors and their ligands into the interior of the cell. Transport occurs when the coated pits invaginate into the cell and pinch off a vesicle called a receptosome. The receptosomes contain the ligands which are then utilized inside the cell.

Receptor-mediated endocytosis provides an interesting example of a diffusion-controlled reaction in a membrane coupled in complex ways to other reactions. In outline, the portion of the mechanism involving receptors and coated pits is[54]

$$P + R \underset{k_u}{\overset{k^{obs}}{\rightleftarrows}} P$$

$$P \underset{\lambda'}{\overset{\lambda}{\rightleftarrows}} P^* \tag{166}$$

The bimolecular process is the binding of a receptor, R, to a coated pit, P, with k_u governing the dissociation reaction. The unimolecular reaction is the invagination process. A steady state on the membrane surface is achieved since the invaginated pits, P*, reappear and the receptors which are lost to the receptosomes are reinserted into the membrane by other mechanisms. An early hypothesis was that the invaginated pits returned to the surface at random locations. This can be modeled by assuming that the invagination back reaction occurs at a constant rate, K_p, which is independent of position. If K_r is the rate at which receptors are randomly reinserted into the membrane, then the steady-state densities of receptors, $\bar{\rho}_r$, coated pits, $\bar{\rho}_P$, and receptors in pits, $\bar{\rho}_{rp}$, satisfy:

$$\bar{\rho}_R = \bar{\rho}_{rp}(\lambda + k_u)/\bar{\rho}_p k^{obs}$$

$$\bar{\rho}_{rp} = K_r/\lambda$$

$$\bar{\rho}_P = K_p/\lambda \tag{167}$$

The diffusion constants for many membrane-bound receptors are small. For example, on human fibroblasts,[63] $D = 4.5 \times 10^{-11}$ cm^2 sec^{-1} for the low-density lipoprotein receptors. This suggests that diffusion is slow enough to make the binding reaction diffusion controlled.[64] Coated pits, on the other hand, are immobile. Thus, the problem can be thought of as the trapping of receptors by static traps, embellished by the possibility of a back reaction and the random disappearance of the traps.

Although this problem is difficult to formulate in the Smoluchowski theory,[65] it becomes a straightforward problem using the mechanistic statistical theory. The relevant radial distribution function involves receptors around the coated pits. A short calculation gives:[54]

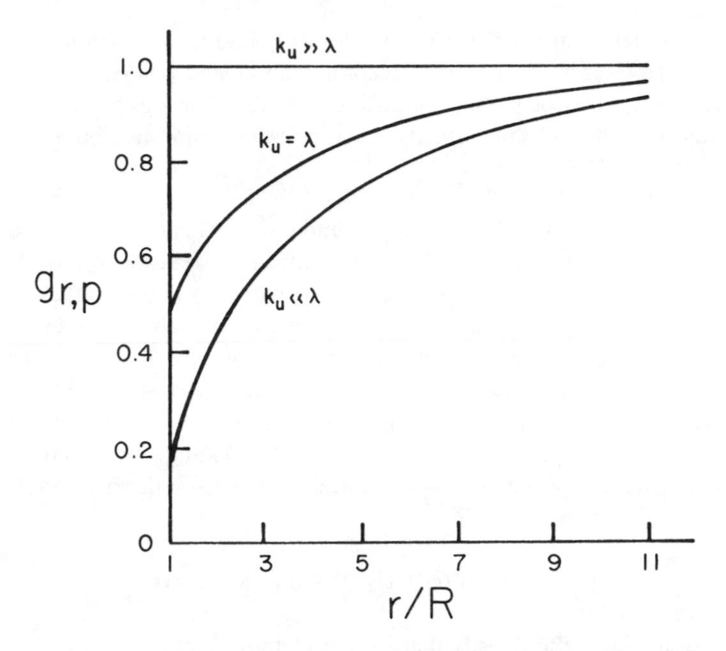

FIGURE 6. The radial distribution function of low-density lipoprotein receptors around coated pits for the model described by Equation 166 in the text. The parameter k_u is the unimolecular rate constant for dissociation of receptors from coated pits, and the lifetime of a coated pit is $\lambda^{-1} = 300$ sec. The correlation lengths for the lower curves are $\xi^{-1} = 6.5$ and 5.9 R, respectively, where R is the radius of a coated pit.

$$g_{rp}(r) = 1 - \frac{k^{obs}\bar{b}}{4\pi D} K_o(\xi r) \tag{168}$$

with

$$\bar{b} = b\lambda/(\lambda + k_u),\ b = 2\lambda/(2\lambda + k_u),$$
$$\xi = [(\bar{\rho}_p b k^{obs} + \lambda)/D]^{1/2} \tag{169}$$

If it is assumed that binding occurs at the radius of the coated pit, R, then the Smoluchowski reactivity can be used to calculate that

$$k^{obs} = \frac{4\pi D[\bar{b}K_o(\xi R)]^{-1}\, k^o}{4\pi D[\bar{b}K_o(\xi R)]^{-1} + k^o} \tag{170}$$

For low-density lipoprotein receptors, all the quantities required to solve Equation 170, e.g., D, R, $\bar{\rho}_p$, and λ either have been measured experimentally or can be inferred from measurements — except k_u, the dissociation reaction rate.

The calculated radial distribution function is shown in Figure 6 as a function of the dissociation rate constant, k_u. If k_u is much smaller than the invagination rate constant, λ, the reaction turns out to be about 84% diffusion controlled. Larger values of k_u give a smaller percentage of diffusion control since the effect of the dissociation reaction is to increase the number of receptors near the pits. Only when there is negligible dissociation during the lifetime of the pit (i.e., $k_u << \lambda$) is this effect unimportant. In this case, k^{obs} is found to be 2.3×10^{-2} μm^2 sec^{-1}. This is a lower bound for k^{obs} and if $k_u = \lambda$, then $k^{obs} = 4.6 \times 10^{-2}$ μm^2 sec^{-1}. Since both receptors and coated pits can be visualized in the electron microscope, it may be possible to differentiate these various possibilities by counting receptors around the pits and thus directly measuring the radial distribution function.

4. Rotational Diffusion

If the intrinsic reactivity is not spherically symmetric, it is necessary to take into account the rotational motion of the reactants. Using the Smoluchowski theory, we have seen that this effect can be significant even for small molecules. To describe these effects in the mechanistic statistical theory, we need only to enlarge the variables that are included in the theory. In addition to specifying the center of mass of each reactant, we must also specify their orientations. This can be done, e.g., by using the Euler angles ϕ, θ, and ψ which denote the orientation of the molecule using three independent molecular axes.[66] The statistical quantities of interest are then the average and fluctuating densities of A and B in the six-dimensional space composed of $\mathbf{X} = \mathbf{r}, \phi, \theta, \psi$. The density-density correlation function then becomes

$$\langle \delta\rho_A(\mathbf{X}, t)\, \delta\rho_B(\mathbf{X}', t)\rangle \equiv \sigma_{AB}(\mathbf{X}, \mathbf{X}', t) \tag{171}$$

Just as in position space, the density-density correlation function allows one to assess the number of pairs of A-B molecules with given spatial locations and orientations. The analog of the radial distribution functions in Equations 86 and 87, which holds when orientations are included, is[66]

$$g_{ij}(\mathbf{X} - \mathbf{X}') = 1 - \delta_{ij}\delta(\mathbf{X} - \mathbf{X}')/\bar{\rho}_i + \sigma_{ij}(\mathbf{X} - \mathbf{X}')/\bar{\rho}_i\bar{\rho}_j \tag{172}$$

This distribution function is related to the observed rate constant by the formula:

$$\hat{k}^{obs} = \int k^o(\mathbf{X})\, g_{ij}(\mathbf{X})\, d\mathbf{X} \tag{173}$$

Because the $\bar{\rho}_i$ now represent densities in position and orientation space, $\hat{k}^{obs}$ differs from the rate constant, k^{obs}, defined before by a factor of the volume of orientation space.

To calculate the density-density correlation function in position and orientation space requires the inclusion of rotational diffusion in the equations which describe the average and fluctuations. Steiger[66] has shown how to do this under rather general conditions. As with the Ŝolc-Stockmayer[40] extension of the Smoluchowski theory, the resulting differential equations are somewhat complicated. The special case of the recombination reaction between identical disk-shaped molecules in two dimensions illustrates what is involved. In this case, only a single angle, θ, is required to specify the orientation and the average density, $\bar{\rho}_A(x,y,\theta,t)$, satisfies the equation:[66]

$$\partial\bar{\rho}_A/\partial t = D\nabla^2\bar{\rho}_A + D_r\partial^2\bar{\rho}_A/\partial\theta^2 - 2\hat{k}^{obs}\bar{\rho}_A^2 + K \tag{174}$$

Because the angular volume is $\int_0^\pi \sin\theta d\theta = 2$, the relationship between $\hat{k}^{obs}$ and k^{obs} is $\hat{k}^{obs} = 2k^{obs}$. The density fluctuations around a uniform steady state satisfy the equation:

$$\partial\delta\rho_A/\partial t = D\nabla^2\delta\rho_A + D_r\partial^2\delta\rho_A/\partial\theta^2 - 2\hat{k}^{obs}\bar{\rho}_A\delta\rho_A + \tilde{f} \tag{175}$$

with

$$\langle \tilde{f}(\mathbf{X}, t)\, \tilde{f}(\mathbf{X}', t) \rangle = [4\hat{k}^{obs}\bar{\rho}_A^2 - 2D\bar{\rho}_A\nabla^2 - 2D_r\partial^2/\partial\theta^2]\, \delta(\mathbf{X} - \mathbf{X}')\, \delta(t - t') \quad (176)$$

Notice that $\hat{k}^{obs}\bar{\rho}_A = k^{obs}(2\bar{\rho}_A)$ where, according to our definition, $(2\bar{\rho}_A)$ is the bulk density of A in position space. The solution of these equations for the density-density correlation function has been carried out to yield:[66]

$$\sigma_{AA}(|\mathbf{r} - \mathbf{r}'|, |\theta - \theta'|) = \bar{\rho}_A\delta(\mathbf{r} - \mathbf{r}')\, \delta(\theta - \theta') - \bar{\rho}_A\left(\frac{\xi^2}{2(2\pi)^2}\right) K_o(\xi r)$$

$$- \bar{\rho}_A\left(\frac{\xi^2}{4\pi^2}\right) \sum_{m=1}^{\infty} \cos(m(\theta - \theta'))\, K_o(\sqrt{1 + \eta^2 m^2}\,\xi r) \quad (177)$$

where

$$\xi = (4\hat{k}^{obs}\bar{\rho}_A/D)^{1/2}$$

$$\eta = (D_r/4\hat{k}^{obs}\bar{\rho}_A)^{1/2} \quad (178)$$

Notice that ξ^{-1} is the same correlation length as in the absence of rotational diffusion, while η^{-1} is the square root of the ratio of the rotational relaxation time, $\tau_r = D_r^{-1}$, and the bimolecular lifetime of A.

A simple reactivity function in two dimensions, which still involves orientations, is

$$k^o(|\mathbf{r} - \mathbf{r}'|, |\theta - \theta'|) = k^o \frac{\delta(|\mathbf{r} - \mathbf{r}'| - R)(1 + \cos(\theta - \theta'))}{2\pi|\mathbf{r} - \mathbf{r}'|} \quad (179)$$

This is called the dipole reactivity. This reactivity maintains the Smoluchowski restriction of reaction only when $|\mathbf{r} - \mathbf{r}'| = R$, but favors reaction when the two orientation vectors of the disks are lined up, as might happen for dipoles. The reason that this reactivity is simple becomes apparent when it is used in Equation 173. It picks out only a single term from the series in Equation 177, so that one has

$$k^{obs} = k^o\left\{1 - \frac{k^{obs}}{\pi D}\,[K_o(\xi R) + K_o(\sqrt{1 + \eta^2}\,\xi R)]\right\} \quad (180)$$

Recalling that $\eta^2 = D_r/4\hat{k}^{obs}\bar{\rho}_A$, we notice that if rotational diffusion is rapid, we recover the result based on Equation 126. Indeed, in this case, $\lim_{x \to \infty} K_o(x) = 0$ and the second term in Equation 180 vanishes.

Equation 180 has been solved under conditions of diffusion control using the Stokes-Einstein expressions for the rotational and translational diffusion constants.[66] The effect of slow rotation is always to decrease the observed rate constant, although the percentage decrease disappears rapidly at low density. In fact, when the bulk density is below about 0.4% of the close-pack density, the effect of rotation is already reduced to 5%. Rotation also reduces the size of the rate constant in the analogous three-dimensional calculation. This effect, however, does not disappear at low densities where, for the dipolar reactivity, the effect is about 15%.

5. Heat Conduction

Another molecular process which might, in principle, affect the rate of a rapid chemical reaction is heat conduction. One reason for this is that all chemical reactions have activation barriers for reaction. In the case of diffusion-controlled reactions, activation energies are due mainly to the temperature dependence of the diffusion constant[7] and are the order of 15 kJ mol^{-1}. It is possible that reaction rates could be accelerated if thermal energy diffuses slowly enough to increase the rate of mass diffusion. A second reason that heat conduction might be important involves the heat of reaction. If heat diffuses sufficiently slowly, it might increase or decrease the reaction rate depending on whether the heat of reaction is positive or negative.

To include the effect of heat conduction in the present theory, we need to keep track of the local internal energy density,[47,57,68] $e(\mathbf{r},t)$. On the average, heat conduction changes according to Fourier's law, i.e.

$$\partial\bar{e}/\partial t = \kappa\nabla^2\bar{T} \tag{181}$$

where κ is the thermal conductivity. According to the mechanistic statistical theory, this means that energy fluctuations, δe, satisfy the equation:[47]

$$\partial\delta e/\partial t = \kappa\nabla^2\delta T + \tilde{f}_e \tag{182}$$

and that

$$\langle\tilde{f}_e(\mathbf{r}, t)\, \tilde{f}_e(\mathbf{r}', t')\rangle = -2\kappa k_B\nabla_{\mathbf{r}}^2\delta(\mathbf{r} - \mathbf{r}')\, \delta(t - t') \tag{183}$$

It is convenient to write Equation 182 in terms of fluctuations in the temperature and the molecule densities. Thus, we use the relationship:

$$\delta e = (\partial\bar{e}/\partial T)\, \delta T + (\partial\bar{e}/\partial\bar{\rho}_A)\, \delta\rho_A \tag{184}$$

The first term involves the heat capacity, due primarily to the solvent, and the second term involves the solute. For simplicity, we have assumed an inert solvent which contains only one kind of solute molecule, A. It is the second term that involves the heat of reaction, an effect which is small at low density. Substituting Equation 184 into Equation 182, we find that

$$\partial\delta T/\partial t = (\partial\bar{T}/\partial\bar{\rho}_A)\, \delta\bar{\rho}_A + D_T\nabla^2\delta T + \tilde{f}_T \tag{185}$$

where $D_T = \kappa/(\delta\bar{e}/\delta\bar{T})$ is called the thermal diffusivity and $\tilde{\mathbf{f}}_T = \tilde{\mathbf{f}}/(\delta\bar{e}/\delta\bar{T})$.

Equation 185 describes temperature fluctuations in an inert solvent containing A. If the A molecules are involved in a recombination reaction, their density fluctuations are coupled to the temperature through the activation energy, ϵ. Indeed, writing $K^{obs} = \hat{k}^{obs}\exp(-\epsilon/k_BT)$, a more complete linearization of the average Equation 94 for $\bar{\rho}_A$ leads to

$$\partial\delta\rho_A/\partial t = D\nabla^2\delta\rho_A - 4k^{obs}\bar{\rho}_A\delta\rho_A - \frac{4\epsilon k^{obs}\bar{\rho}_A}{k_B\bar{T}^2}\delta T + \tilde{f} \tag{186}$$

with $\tilde{\mathbf{f}}$ still as defined in Equation 96 and $\langle\tilde{\mathbf{f}}\tilde{\mathbf{f}}_T\rangle \equiv 0$. Equation 186 is dynamically coupled to Equation 185 and, thus, the thermal fluctuations will affect the density-density correlation function.

The resulting radial distribution function, which involves two correlation lengths, has been used to calculate k^{obs} for the recombination reaction.[67] Given the small size of the activation energy for diffusion-controlled reactions, the effect of the thermal diffusivity is negligible unless D_T greatly exceeds the diffusion constant. Since thermal diffusivities are usually of the order of 10^{-3} cm^2 sec^{-1}, thermal diffusion can be expected to have no appreciable effect in most solutions. The refrigerant fluoroform, CHF_3, on the other hand, has a thermal diffusivity about two orders of magnitude smaller than its self-diffusive constant at room temperature and 47 atm pressure. Even for such a solvent, the activation energy effect on the rate constant is no more than 10% even when the concentration of A is as high as 0.1 *M*. Thus, thermal diffusion appears to provide only a minor perturbation to rate constants for rapid reactions.

IV. COMPARISON WITH EXPERIMENT

A. Present Status of Verification of the Theory

We have attempted to demonstrate in this chapter that the statistical nonequilibrium thermodynamic theory of diffusion effects on rapid chemical reactions provides a useful generalization of the Smoluchowski theory. The basic ideas of the Smoluchowski theory, of course, have been corroborated by a variety of experiments.[2] It provides a qualitative explanation for the effect of changing the viscosity of a solvent and a qualitative upper bound for reaction rates in solution. The statistical nonequilibrium thermodynamic theory, on the other hand, is able to cope with a variety of effects which can be dealt with only approximately by modification of the Smoluchowski theory. In this section, we review the present status of comparison between the theory and experimental measurements of bimolecular rate constants for rapid reactions.

A good deal of experimental work has been carried out on the iodine atom recombination reaction:[69-71]

$$I + I \rightarrow I_2 \tag{187}$$

Data for this reaction taken using the rotating sector technique have been analyzed using Equation 106. This equation, as we have pointed out, leads to a formula for k^{obs} which is a factor of two too large. Subsequently, the diffusion constant of free iodine atoms was measured[70] which allowed a comparison to experiment to be made.[71] The experimental data can be reinterpreted using the Collins-Kimball formula, whose corrected form for the dimerization reaction is

$$k^{obs} = \frac{2\pi D_I R k^o}{2\pi D_I R + k^o} \tag{188}$$

Following Noyes,[71] we assume that k^o is given by the gas-phase recombination rate constant, $k^o = 5.5 \times 10^{10}$ M^{-1} sec^{-1}. The other measured values in the solvent CCl_4 are $D_I = 8.0 \times 10^{-5}$ cm^2 sec^{-1} and $k^{obs} = 8.2 \times 10^9$ M^{-1} sec^{-1}, both at 25°C. These values are compatible with Equation 188 if the encounter radius is taken to be 3.2 Å. At 38°C, the measured values are $D_I = 12 \times 10^{-5}$ cm^2 sec^{-1}, $k^{obs} = 13 \times 10^9$ M^{-1} sec^{-1}, and k^o can be taken to be 5.6×10^{10} M^{-1} sec^{-1}. Using the value R = 3.2 Å and the experimental value of D_I in Equation 188 leads to a calculated value of k^{obs} equal to 12.3×10^9 M^{-1} sec^{-1}. This is in good agreement with the measured value. Had we used instead of Equation 189 the Smoluchowski result which replaces 2π by 4π, agreement with experiment would still be found, but the encounter radius would be 1.6 Å. This value corresponds to one third of the value based on the van der Waals radius of an iodine atom.[71] This is clearly an unreasonably small value for the encounter radius.

Recent experimental work on iodine atom recombination has used laser exitation to well-defined electronic states.[72-74] These experiments, which are on the picosecond time scale and measure geminate recombination, appear to be sensitive to vibrational relaxation of the recombinant I_2 molecules.[73] The experiments involve a time scale which is too rapid for diffusional motion to be significant[74] and so do not provide useful information about diffusion effects.

Rate constant data collected for the bimolecular reaction of ferroprotoporphyrin IX (heme) with carbon monoxide[75] can be used to test the validity of the equation:

$$k^{obs} = (k_B T/\eta)\left(\frac{3R/2}{R_H + R_{CO}}\right) \tag{189}$$

This equation is equivalent to the Smoluchowski result if the Stokes-Einstein equation in Equation 51 is used to evaluate the relative diffusion constant. To obtain a broad range of viscosities, aqueous glycerol at various temperatures and pressures was used as the solvent. In these experiments, the quantity T/η was varied more than two orders of magnitude and the rate constant was obtained at low concentrations after photodissociation of the heme-CO complex. The rate constant was found to be proportional to the value of T/η with a proportionality constant very close to the Boltzmann constant. This is in agreement with Equation 189 and implies that the encounter radius is $R \approx 2(R_H + R_{CO})/3$. The wide variation of temperatures, pressures, and viscosities employed in these experiments provides an excellent confirmation of the theory of diffusion effects on binary reactions in dilute solution.

A convincing illustration of the validity of the Collins-Kimball equation in dilute solution is provided by studies of reactions of the solvated electron.[76] Free electrons that are ultimately stablized by interactions with solvent can be produced by the radiolysis of a variety of substances.[77] In water, the solvated electron has a rather typical aqueous diffusion constant, $D_e = 4.9 \times 10^{-5}$ cm^2 sec^{-1}. In certain hydrocarbon solvents, however, D_e is anomalously large. At room temperature, e.g.,[78] $D_e = 2.6$ cm^2 sec^{-1} in tetramethylsilane, $D_e = 2.9 \times 10^{-2}$ cm^2 sec^{-1} in cyclopentane, and $D_e = 3.4 \times 10^{-3}$ cm^2 sec^{-1} in benzene. By utilizing solvents like these and different temperatures, it has been possible to study the dependence of the rate constant on diffusion of the hydrated electron for several bimolecular reactions. For the reaction:

$$SF_6 + e^- \rightarrow SF_6^- \tag{190}$$

it has been possible in this way to vary the diffusion constant over a range of five orders of magnitude. The bimolecular rate constants have been analyzed[78] using the Collins-Kimball equation with an encounter radius of R = 13.8 Å and a value for $k^o = 2 \times 10^{14}$ M^{-1} sec^{-1} determined by the thermal de Broglie wavelength of the electron. Despite the fact that different solvents and temperatures were employed in this comparison, the agreement between experiment is excellent. Due to the large diffusion constants and the large value of k^o, the observed rate constants[78] are found to range between 10^{11} and 3×10^{14} M^{-1} sec^{-1}. These are the largest bimolecular rate constants known.

In the past decade,[79,80] several groups have resolved the transient change in the bimolecular rate constant[22,81] associated with the changing radial distribution function. In the limit of dilute solution, the Smoluchowski theory can be used to describe these effects. In particular, we found in Equation 12 that, for a diffusion-controlled reaction:

$$k^{obs}(t) = 4\pi D'R[1 + R/(\pi D't)^{1/2}] \tag{191}$$

In fluorescence quenching, Equation 191 leads to a nonexponential decay law with the form:[79]

$$\exp(-at - 2bt^{1/2}) \tag{192}$$

where $a = k_u + 4\pi D'R\bar{\rho}_B$ and $b = 4(\pi D')^{1/2}R^2\bar{\rho}_B$ where k_u^{-1} is the fluorescence lifetime in the absence of quencher and $\bar{\rho}_B$ is the bulk number density of quencher. Nemzek and Ware[79] were the first to verify the characteristic decay law (Equation 192) for the quenching of 1,2-benzanthracene fluorescence by CBr_4 in a variety of organic solvents. More recently, Equation 192 has been verified very accurately for the quenching of tryptophan fluorescence by iodide ions in aqueous solution.[80] The experimental data imply an encounter radius R = 3.4 Å, a relative diffusion constant $D' = 1.1 \times 10^{-5}$ cm²sec⁻¹, and a fluorescence lifetime $k_u^{-1} = 3 \times 10^{-9}$ sec. These values for D′ and k_u^{-1} are close to those obtained by independent measurements,[82] and the encounter radius agrees with that usually assumed for contact quenchers.

The Debye extension of the Smoluchowski theory provides a good description of the ionic strength dependence of rate constants for diffusion-controlled reactions in dilute solution. This is called the kinetic salt effect[2] and is described by Equation 37. When the finite size of ions is taken into account, Equation 37 can be written at 25°C as[2]

$$\log_{10}(k^{obs}/k^{o}) = 1.02z_Az_B[I^{1/2}/(1 + I^{1/2})] \tag{193}$$

where I is the ionic strength. The bimolecular rate constants for reduction of $KFe(CN)_6^{2-}$, NO_2^-, O_2, Ag^+, and H^+ by hydrated electrons have been found[83] to satisfy Equation 193 rather well for ionic strengths less than 0.03 molal. Thus, the graph of $\log_{10}(k^{obs}/k^{o})$ vs. $I^{1/2}/(1 + I^{1/2})$ for the reduction of $KFe(CN)_6^{2-}$ is a straight line with slope close to 2.04. The comparable graph for the reaction $O_2 + e_{aq}^- \rightarrow O_2^-$ has slope zero, and the graph for the reduction of H^+ has slope close to −1.02. The quality of agreement between theory and experiment for these reactions is comparable to that for the kinetic salt effect for slow reactions in solution.[2]

As we have pointed out in previous sections, for a single bimolecular reaction, the Smoluchowski theory is valid only in dilute solution and only for long-lived reactant species. Fluorescence quenching often involves fluorophores whose unimolecular lifetime is of the order of nanoseconds[84] and quenchers in concentrations exceeding 10^{-3} *M*. Thus, the uni- and bimolecular lifetime effects discussed in the previous section should be important. If the fluorophore is dilute and the quenching reaction is diffusion controlled, then Equations 159 and 160 show that

$$k^{obs} = 4\pi D'RC(R) \tag{194}$$

where

$$C(R) = \left[\frac{D_B - D_A}{2D_B}\exp(-R\beta^{1/2}) + \frac{D_A + D_B}{2D_B}\exp(-R\alpha^{1/2})\right]^{-1} \tag{195}$$

and

$$\alpha = (k_u + k^{obs}\bar{\rho}_B)/D_A$$
$$\beta = (k_u + k^{obs}\bar{\rho}_B)/D' \tag{196}$$

Since both the exponentials in Equation 193 are less than one, it is easy to see that C(R) is always greater than one. Thus, the combined unimolecular/bimolecular lifetime effects increase the observed bimolecular quenching constant. As we have pointed out previously, this effect increases with increasing quencher concentration. Thus, k^{obs} should be an increasing function of the concentration of quencher.

Effects like these have been observed in fluorescence quenching at high concentrations. Steady-state quenching experiments are usually interpreted using the Stern-Volmer equation.[14] This equation plots the ratio of the intensity of fluorescence in the absence of quencher, I_o, to the intensity in the presence of quencher, I. The usual steady-state analysis gives:

$$I_o/I = 1 + \tau_o k^{obs} \bar{\rho}_B \tag{197}$$

where $\tau_o = k_u^{-1}$ is the lifetime of the fluorophore A in the absence of quencher. Thus, if k^{obs} is constant, a Stern-Volmer plot of I_o/I vs. $\bar{\rho}_B$ should be a straight line of slope $\tau_o k^{obs}$. At high concentrations of quencher, Stern-Volmer plots are often curved upward.[85-88] This is sometimes caused by weak association between the fluorophore and quencher, which is referred to as "static" quenching.[89] In a number of cases, this reason for curvature can be ruled out. One criterion for doing so involves a comparison of lifetime measurements with intensity measurements. If τ is the measured lifetime of the fluorophore, then in the absence of complex formation, analysis of the kinetic equations shows that[90]

$$\tau_o/\tau = I_o/I \tag{198}$$

The quenching of aqueous flavin mononucleotide fluorescence by sodium thioglycollate, e.g., satisfies Equation 198 and yet the Stern-Volmer plot shows significant positive curvature abouve 0.2 *M* sodium thioglycollate.[90]

In the absence of molecular association, positive curvature appears to be a result of the bimolecular lifetime effect.[60,61] To carry out the theoretical analysis based on Equations 194 and 195, it is necessary to know D_A, D_B, τ_o, and R. The first three of these quantities are known for the quenching of quinine fluorescence by iodide ions[85] and tryptophan fluorescence by oxygen atoms in aqueous solution.[82] Values for the encounter radius, R, for each reaction can be assigned by fitting Equation 194 at $\bar{\rho}_B = 0$ to the experimental value of k^{obs}. Figure 7 shows both the experimental points and the theoretical Stern-Volmer plot based on solving Equation 194. Agreement is quite satisfactory. The dashed line represents the naive linear extrapolation of the low-density results for quinine/I^-, which is seen to differ considerably from experiment above 0.05 *M*. Figure 7 shows comparable data and calculations for oxygen quenching of 9-vinylanthracene and perylene fluorescence in dodecane solutions.[82] The diffusion constants of neither 0_2 nor the fluorophores have been measured in dodecane, so their values were estimated on the basis of other data.[91] The values of these and other parameters which were used in Equation 195 are noted in the legend of Figure 7. They are seen to produce good agreement with experiment.

We are aware of only one experiment which has been used to assess the effect of diffusion on rapid reactions in membranes. In measuring the photodimerization rate for parinaric acid in phospholipid membranes, it was found that the rate constant was insensitive to variations in the viscosity of the phospholipid.[92] Indeed, a decrease of viscosity by four orders of magnitude by decreasing the temperature actually led to a slight increase in the rate constant. This seemed anomalous since the dimerization reaction is diffusion controlled in solution. This result, however, is in qualitative agreement with the unimolecular lifetime effect discussed in previous sections. Indeed, the fluorescence lifetime of parinaric acid is 5 nsec. Thus, an excited molecule cannot move far enough in its lifetime to make diffusion effects important. This is born out by detailed calculations.[50]

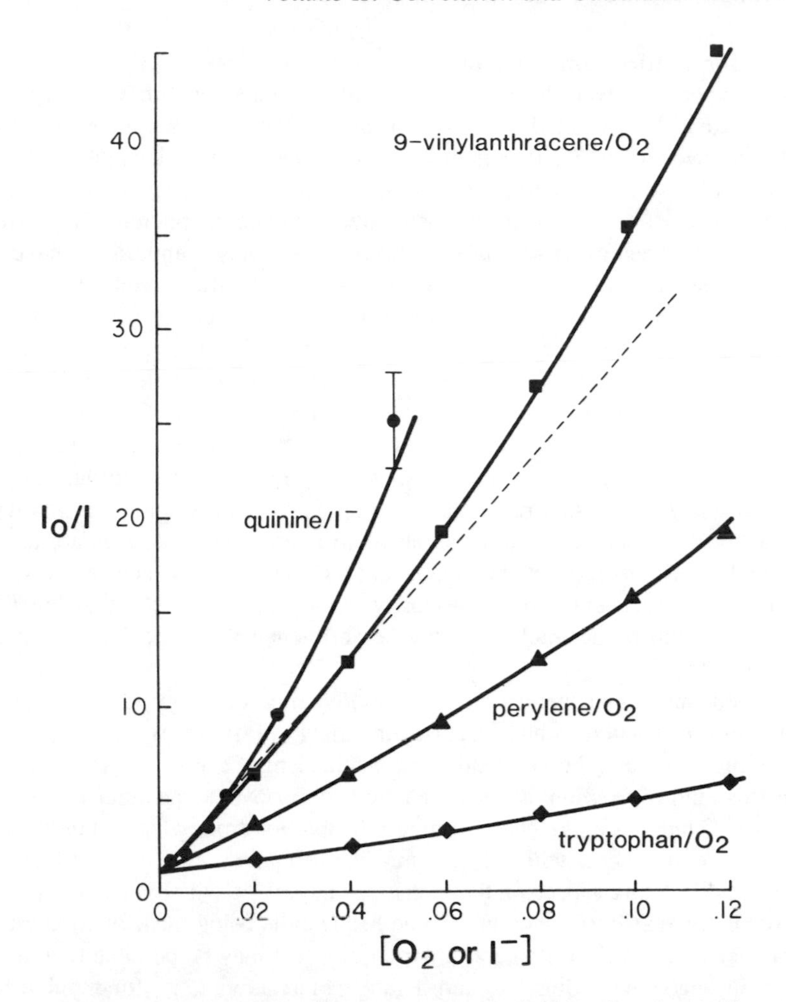

FIGURE 7. Stern-Volmer plots of the decrease in intensity of steady-state fluorescence for the quenching of four organic fluorophores by I^- or O_2. The experimental points are from References 82 and 85. The full lines are based on Equations 194 to 196 and use the following parameter values: quinine/I^- (D_Q = 4×10^{-6} cm^2 sec^{-1}, D_{I^-} = 1.2×10^{-5} cm^2 sec^{-1}, R = 5.5 Å, τ_o = 19 (nsec); 9-vinylanthracene/O_2 (D_V = 7×10^{-6} cm^2 sec^{-1}, D_{O_2} = 4×10^{-5} cm^2 sec^{-1}, R = 3.6 Å, τ_o = 11 nsec); perylene/O_2 (D_P = 9×10^{-6} cm^2 sec^{-1}, D_{O_2} = 4 $\times$ 10^{-5} cm^2 sec^{-1}, R = 3.2 Å, τ_o = 5.4 nsec); and tryptophan/O_2 (D_T = 7 $\times$ 10^{-6} cm^2 sec^{-1}, D_{O_2} = 2.6×10^{-5} cm^2 sec^{-1}, R = 2.7 Å, τ_o = 2.7 nsec). The calculated curves are a correction of those which originally appeared in Reference 60.

Agreement between the theory of diffusion effects on rapid reactions and experimental measurements is rather good for the simple reactions discussed in this section. The agreement with experiment for the dynamic component of fluorescence quenching illustrated in Figure 7 is particularly encouraging since it is difficult to account for this effect in the Smoluchowski theory without serious modifictions.[28] Nonetheless, one often encounters difficulty in making quantitative comparisons of the theory to experiment. One difficulty is the lack of reliable data for diffusion constants. Another is the unreliable nature of some methods for obtaining bimolecular rate constants. Accurate measurements of both these transport coefficients would greatly facilitate our understanding of the effect of diffusion.

B. Prospects for Future Comparisons

It is not unexpected, given the number of rapid reactions which have been investigated over the years, that the predicted deviations from the Smoluchowski theory in solution are of the order of two or three rather than ten. Nonetheless, these deviations appear to be significant and accessible to experimental measurement. The investigation of rapid reactions in membranes seems like a particularly fertile area for experimentation. Only a few quantitative studies have been made of rapid reactions in membranes and, as we have seen, the Smoluchowski theory yields results in two dimensions only after a good deal of patching-up. Even in that case, the results are correct only at zero density of reactants. Specific problems which would be interesting to investigate experimentally both in solutions and membranes include the effect of increased concentrations (the bimolecular lifetime effect); the effect of rotational diffusion; the effect of the unimolecular lifetime; the effect of molecular interactions; and long-range reactivity effects on reaction, especially energy transfer.

In order to understand these effects, accurate measurements of rate constants are required. At a glance, Table 1 shows that rate constants for many rapid reactions are known only to 10 to 50% accuracy. Some of the effects that seem interesting in solution are in this range and would be difficult to ferret out of available data. Other effects, which vary systematically with parameters like concentration or viscosity, have been easier to see. Nonetheless, in future work, it would be desirable to obtain experimental values for rate constants to an accuracy of at least 5%.

One of the advantages of the statistical thermodynamic theory of rapid reactions is that it can be applied to reactions which occur over a range of distances rather than at a single encounter radius. Although the Förster dipolar mechanism of energy transfer is an important example of this kind of reaction, all rapid reactions occur over some range of distance. This dependence on distance is undoubtedly different in solution from what it is in the gas phase. Thus, scattering experiments will not give much information about these effects. In order to take advantage of this aspect of the statistical theory, it will be necessary to develop realistic models for reactivity functions. Some headway is being made along these lines for electron transfer reactions.[93] With careful experiments, it may be possible to ascertain both a characteristic encounter radius, R, and a range of reactivity, r_o, for rapid reactions as suggested in Equation 121. Indeed, if R is of the order of 6 Å and r_o is of the order of 2 Å, such effects may be easy to see.

We have restricted most of our treatment of rapid reactions to steady state. Under conditions which lead to a steady state this is sensible since relaxation to steady state occurs on a microsecond time scale for typical solvents at room temperature. Moreover, at a steady state, the statistical distribution of fluctuations is stationary. This means that the radial distribution function and, therefore, the bimolecular rate constant are both time independent. This is not so in transient pulsed experiments. Such experiments originally relied on flash lamps and can now be done much more accurately using pulsed lasers.[80] The problem with interpreting pulsed experiments is that the average environment around reactant molecules continually changes as the experiment proceeds. For example, if local reactant concentrations exceed about 10^{-3} *M*, then a transient bimolecular lifetime effect will be important. An alternative kind of transient experiment, namely, periodic perturbation of the steady state, provides a well-defined method of obtaining the kinetic information buried in pulsed experiments. For photochemical reactions, these experiments involve periodic modulation of the intensity of the excitation beam. Data from this type of transient experiment would be much easier to interpret with present theories.

The prospects for comparison of theory and experiment in the near future seem excellent. Calculations for many complex problems are now tractable using the statistical theory, and we hope that fresh experimental data will soon become available. Appropriate experiments should provide an interesting probe of chemical reactivity in condensed matter.

ACKNOWLEDGMENTS

This work was supported by grants from the National Science Foundation (CHE-83-11583-A01) and the National Institutes of Health (PHS GM30688-03). We thank David Volman for illuminating discussions concerning the rate constant for iodine atom recombination and Susan Keizer for preparation of the illustrations.

REFERENCES

1. **Harcourt, A. V. and Esson, W.,** On the laws of connexion between the conditions of a chemical change and its amount, *Philos. Trans.*, 156, 193, 1866.
2. **Weston, R. E. and Schwarz, H. A.,** *Chemical Kinetics,* Prentice-Hall, Englewood Cliffs, N.J., 1972, chap. 6.
3. **Johnston, H. S.,** *Gas Phase Reaction Rate Theory,* The Ronald Press, New York, 1966, chap. 7
4. **Hickel, B.,** Absorption spectra and kinetics of methyl and ethyl radicals in water, *J. Phys. Chem.*, 79, 1654, 1975.
5. **Kondratiev, V. N.,** *Rate Constants of Gas Phase Reactions,* National Bureau of Standards, Washington, D.C., 1972, 165.
6. **Smoluchowski, M.,** Versuch einer Mathematischen Theorie der Koagulationskinetik Kolloider Lösungen, *Z. Phys. Chem.*, 92, 129, 1917.
7. **Hague, D. N.,** *Fast Reactions,* Wiley-Interscience, London, 1971, 14.
8. **Rosman, H. and Noyes, R. M.,** Rate constants for combination of iodine atoms in inert solvents, *J. Am. Chem. Soc.*, 80, 2410, 1958.
9. **Swallow, A. J.,** Reactions of free radicals produced from organic compounds in aqueous solution by means of radiation, *Prog. React. Kinet.*, 9, 195, 1978.
10. **Draganić, I. G. and Draganić, Z. D.,** *The Radiation Chemistry of Water,* Academic Press, New York, 1971, chaps. 3 and 4.
11. **Hague, D. N.,** *Fast Reactions,* Wiley-Interscience, London, 1971, chap. 3.
12. **Caldin, E. F.,** *Fast Reactions in Solution,* Blackwell Scientific, Oxford, 1964.
13. **Wilkins, R. G. and Eigen, M.,** The kinetics and mechanism of formation of metal complexes, in *Mechanisms of Inorganic Reactions,* Adv. Chem. No. 49, Kleinberg, J., Murmann, R. K., Fraser, R. T. M., and Bauman, J., Eds., American Chemical Society, Washington, D.C., 1965, chap. 3.
14. **Turro, N. J.,** *Modern Molecular Photochemistry,* Benjamin-Cummings, Menlo Park, 1978, chaps. 6 and 8.
15. **Balzani, V., Moggi, I., Manfrin, M. F., and Bolletta, F.,** Quenching and sensitization processes of coordination compounds, *Coord. Chem. Rev.*, 15, 321, 1975.
16. **Hague, D. N.,** *Fast Reactions,* Wiley-Interscience, London, 1971, 91.
17. **Hague, D. N.,** *Fast Reactions,* Wiley-Interscience, London, 1971, 90.
18. **Hague, D. N.,** *Fast Reactions,* Wiley-Interscience, London, 1971, 83.
19. **Keizer, J.,** Theories of diffusion controlled reactions, *Chem. Rev.*, in preparation.
20. **McQuarrie, D. A.,** *Statistical Mechanics,* Harper & Row, New York, 1976, 386.
21. **Collins, F. C. and Kimball, G. E.,** Diffusion-controlled reaction rates, *J. Colloid Sci.*, 4, 425, 1949.
22. **Noyes, R. M.,** Effects of diffusion rates on chemical kinetics, *Prog. React. Kinet,* 1, 129, 1961.
23. **Debye, P.,** Reaction rates in ionic solutions, *Trans. Electrochem. Soc.*, 82, 265, 1942.
24. **Moore, W. J.,** *Physical Chemistry,* 4th ed., Prentice-Hall, Englewood Cliffs, N.J., 1972, 107.
25. **Weston, R. E. and Schwarz, H. A.,** *Chemical Kinetics* Prentice-Hall, Englewood Cliffs, N.J., 1972, 168.
26. **Czapski, G. and Schwarz, H. A.,** The nature of the reducing radical in water hydrolysis, *J. Am. Chem. Soc.*, 66, 471, 1962.
27. **Monchick, L.,** Note on the theory of diffusion controlled reactions: application to photodissociation in solution, *J. Chem. Phys.*, 24, 381, 1956.
28. **Peak, D., Werner, T. C., Dennin, R. M., and Baird, J. K.,** Fluorescence quenching at high quencher concentrations, *J. Chem. Phys.*, 24, 381, 1956.
29. An attempt to do this in Reference 28 using an equation like Equation 56 confuses the difference between ρ and $\Delta\ \rho$.
30. **Caldin, E. F.,** *Fast Reactions in Solution,* Blackwell Scientific, Oxford, 1964, 142.

31. **Schwarz, F. P. and Moet-Ner, M.,** Fluorescence quenching of liquid alkybenzenes excited by nonionizing and ionizing ultraviolet radiation and by β radiation, *J. Phys. Chem.*, 87, 5206, 1983.
32. **Baird, J. K., McCaskill, J. S., and March, N. H.,** On the theory of the Stern-Volmer coefficient for dense fluids, *J. Chem. Phys.*, 74, 6812, 1981.
33. **Felderhof, B. U. and Deutch, J. M.,** Concentration dependence of the rate of diffusion controlled reactions, *J. Chem. Phys.*, 64, 4551, 1976.
34. **Kirkpatrick, T. R.,** Time dependent transport in a fluid with static traps, *J. Chem. Phys.*, 76, 4255, 1982.
35. **Calef, D. F. and Deutch, J. M.,** Diffusion-controlled reactions, *Ann. Rev. Phys. Chem.*, 34, 493, 1983.
36. **Naqvi, K. R.,** Diffusion controlled reactions in two dimensional fluids, *Chem. Phys. Lett.*, 28, 280, 1974.
37. **Adam, G. and Delbruück, M.,** Reduction of dimensionality in biological diffusion processes, in *Structural Chemistry and Molecular Biology*, Rich, A. and Davidson, N., Eds., W. H. Freeman, San Francisco, 1968, 198.
38. **Berg, H. C. and Purcell, E. M.,** Physics of chemoreception, *Biophys. J.*, 20, 193, 1977.
39. **Debye, P.,** *Polar Molecules*, Dover, New York, 1929, chap. 5.
40. **Ŝolc, K. and Stockmayer, W. H.,** Kinetics of diffusion-controlled reaction between chemically asymmetric molecules. I. General theory, *J. Chem. Phys.*, 54, 2981, 1971.
41. **Hill, T. L.,** Diffusion frequency factors in some simple examples of transition state theory, *Proc. Natl. Acad. Sci. U.S.A.*, 73, 679, 1976.
42. **Chou, K. and Forsén, S.,** Diffusion-controlled effects in reversible enzymatic fast reaction systems, *Biophys. Chem.*, 12, 255, 1980.
43. **Shoup, D., Lipari, G., and Szabo, A.,** Diffusion-controlled bimolecular reaction rates: the effect of rotational diffusion and orientational constraints, *Biophys. J.*, 36, 697, 1981.
44. **Hill, T. L.,** *Introduction to Statistical Thermodynamics*, Addison-Wesley, Reading, Mass., 1960, chap. 17.
45. **McQuarrie, D. A.,** *Statistical Mechanics*, Harper & Row, New York, 1976, chap. 13.
46. **Keizer, J.,** A theory of spontaneous fluctuations in macroscopic systems, *J. Chem. Phys.*, 63, 398, 1975.
47. **Keizer, J.,** Dissipation and fluctuations in nonequilibrium thermodynamics, *J. Chem. Phys.*, 64, 1679, 1976.
48. **Keizer, J.,** Fluctuations, stability, and generalized state functions at nonequilibrium steady states, *J. Chem. Phys.*, 65, 4431, 1976.
49. **Keizer, J.,** Effect of diffusion on reaction rates in solution and in membranes, *J. Phys. Chem.*, 85, 940, 1981.
50. **Keizer, J.,** Nonequilibrium statistical thermodynamics and the effect of diffusion on chemical reaction rates, *J. Phys. Chem.*, 86, 5052, 1982.
51. **Förster, T.,** *Fluoreszenz Organischer Verbindungen*, Vandenhoech & Ruprech, Göttingen, 1951.
52. **Keizer, J.,** Master equations, Langevin equations, and the effect of diffusion on concentration fluctuations, *J. Chem. Phys.*, 67, 1473, 1977.
53. **Fixman, M.,** Absorption by static traps: initial-value and steady-state problems, *J. Chem. Phys.*, 81, 3666, 1984.
54. **Keizer, J., Ramirez, J., and Peacock-Lopez, E.,** The effect of diffusion on the binding of membrane-bound receptors to coated pits, *Biophys. J.*, 47, 79, 1985.
55. **Watson, G. N.,** *A Treatise on the Theory of Bessel Functions*, 2nd ed., Cambridge University Press, London, 1962.
56. **Medina-Noyola, M. and Keizer, J.,** Spatial correlations in nonequilibrium systems: the effect of diffusion, *Physica*, 107A, 437, 1981.
57. **Keizer, J. and Medina-Noyola, M.,** Spatially nonlocal fluctuation theories: hydrodynamic fluctuations for simple fluids, *Physica*, 115A, 301, 1982.
58. **McQuarrie, D. A.,** *Statistical Mechanics*, Harper & Row, New York, 1976, 268.
59. **Frost, A. A. and Pearson, R. G.,** *Kinetics and Mechanism*, 2nd ed., John Wiley & Sons, New York, 1965.
60. **Keizer, J.,** Nonlinear fluorescence quenching and the origin of positive curvature in Stern-Volmer plots, *J. Am. Chem. Soc.*, 105, 1494, 1983.
61. **Keizer, J.,** Theory of rapid bimolecular reactions in solution and membranes, *Acc. Chem. Res.*, in press.
62. **Paston, I. and Willingham, M. C.,** Receptor mediated endocytosis: coated pits, receptosomes, and the Golgi, *Trends Biochem. Sci.*, 8, 250, 1983.
63. **Barak, L. S. and Webb, W. W.,** Diffusion of low density lipoprotein-receptor complex on human fibroblasts, *J. Cell. Biol.*, 95, 846, 1982.
64. **Goldstein, B., Wofsy, C., and Bell, G.,** Interactions of low density lipoprotein receptors with coated pits on human fibroblasts: estimate of the forward rate constant and comparison with the diffusion limit, *Proc. Natl. Acad. Sci. U.S.A.*, 78, 5695, 1981.
65. **Wofsy, C. and Goldstein, B.,** Coated pits and low density lipoprotein recycling, in *Cell Surface Phenomena*, Perelson, A., De Lisi, C., and Wiegel, F., Eds., Marcel Dekker, New York, 1984, 405.

66. **Steiger, U. R. and Keizer, J.,** Origin and effects of orientational correlations in chemical reactions, *J. Chem. Phys.,* 77, 777, 1982.
67. **Peacock-Lopez, E. and Keizer, J.,** in preparation.
68. **Keizer, J.,** A theory of spontaneous fluctuations in viscous fluids far from equilibrium, *Phys. Fluids,* 21, 198, 1978.
69. **Booth, D. and Noyes, R. M.,** The effect of viscosity on the quantum yield for iodine dissociation, *J. Am. Chem. Soc.,* 82, 1872, 1960.
70. **Levison, S. A. and Noyes, R. M.,** Diffusion coefficient of iodine atoms in carbon tetrachloride by photochemical space intermittency, *J. Am. Chem. Soc.,* 86, 4525, 1964.
71. **Noyes, R. M.,** Validity of equations derived for diffusion-controlled reactions, *J. Am. Chem. Soc.,* 86, 4529, 1964.
72. **Dutoit, J.-C., Zellweger, J.-M., and van den Bergh, H.,** The photolytic cage effect of iodine in gases and liquids, *J. Chem. Phys.,* 78, 1825, 1983.
73. **Nesbitt, D. J. and Hynes, J. T.,** Slow vibrational relaxation in picosecond iodine recombination in liquids, *J. Chem. Phys.,* 77, 2130, 1982.
74. **Bado, P., Dubuy, C., Magdl, D., Wilson, K. R., and Malley, M. M.,** Molecular dynamics of chemical reactions in solution: experimental picosecond transient spectra for I_2 photo dissociation, *J. Chem. Phys.,* 80, 5531, 1984.
75. **Caldin, E. F. and Hasinoff, B. B.,** Diffusion controlled kinetics in the reaction of ferroprotoporphyrin IX with carbon monoxide, *J. Chem. Soc. Faraday Trans. 1,* 515, 1975.
76. **Hughes, G. and Lobb, C. R.,** Reactions of solvated electrons, in *Comprehensive Chemical Kinetics,* Vol. 18, Bamford, C. H. and Tipper, C. F. H., Eds., Elsevier, Amsterdam, 1976, chap. 7.
77. **Buxton, G. V.,** Basic radiation chemistry of liquid water, in *The Study of Fast Processes and Transient Species by Electron-Pulse Radiolysis,* Baxendale, J. H. and Busi, F., Eds., D. Reidel Publishing, 1982, 241.
78. **Warman, J. M.,** The dynamics of electrons and ions in non-polar liquids, in *The Study of Fast Processes and Transient Species by Electron Pulse Radiolysis,* Baxendale, J. H. and Busi, F., Eds., D. Reidel Publishing, 1982, 433.
79. **Nemzek, T.L. and Ware, W. R.,** Kinetics of diffusion-controlled reactions: transient effects in fluorescence quenching, *J. Chem. Phys.,* 62, 477, 1975.
80. **van Resandt, R. W. W.,** Picosecond transient effects in fluorescence quenching of tryptophan, *Chem. Phys. Lett.,* 95, 205, 1983.
81. **Schwarz, H. A.,** Some applications of time-dependent rate constant theory to radiation chemistry, *J. Chem. Phys.,* 55, 3647, 1971.
82. **Lakowicz, J. R. and Weber, G.,** Quenching of protein fluorescence by oxygen, *Biochemistry,* 12, 4171, 1973.
83. **Hague, D. N.,** *Fast Reactions,* Wiley-Interscience, London, 1971, 97.
84. **Baird, J. K. and Escott, S. P.,** On departures from the Stern-Volmer law for fluorescence quenching in liquids, *J. Chem. Phys.,* 74, 6993, 1981.
85. **Jette, E. and West, W.,** Studies on fluorescence and photosensitization in aqueous solution. II., *Proc. R. Soc. London,* A121, 299, 1928.
86. **Frank. J. M. and Wawilow, S. J.,** Über die Wirkungssphäre der Auslöschungsvoränge in den Fluoreszierenden Flüssigkeiten, *Z. Phys.,* 69, 100, 1931.
87. **Williamson, B. and La Mer, V. K.,** The kinetics of activated-diffusion controlled reactions in solution, *J. Am. Chem. Soc.,* 70, 717, 1948.
88. **Lakowicz, J. R. and Weber, G.,** Quenching of fluorescence by oxygen. A probe for structural fluctuations in macromolecules, *Biochemistry,* 12, 4161, 1973.
89. **Bowen, E. J. and Metcalf, W. S.,** The quenching of anthracene fluorescence, *Proc. R. Soc. London,* A206, 437, 1951.
90. **Weber, G.,** Intramolecular complexes of flavins, in *Flavins and Flavoproteins,* Slater, E. C., Ed., Elsevier, Amsterdam, 1966, 15.
91. **Ware, W.,** Oxygen quenching of fluorescence in solution: an experimental study of the diffusion process, *J. Am. Chem. Soc.,* 66, 455, 1962.
92. **Morgan, C. G., Hudson, B., and Wolber, P. K.,** Photochemical dimerization of parinaric acid in lipid bilayers, *Proc. Natl. Acad. Sci. U.S.A.,* 77, 26, 1980.
93. **Tembe, B. L., Friedman, H. L., and Newton, M. D.,** The theory of the Fe^{2+}-Fe^{3} electron exchange in water, *J. Chem. Phys.,* 76, 1490, 1982.

Chapter 8

PREDICTION OF RATE CONSTANTS FOR RADICAL REACTIONS USING CORRELATIONAL TOOLS

Jeffrey Steven Gaffney and Kristie Bull

TABLE OF CONTENTS

I. INTRODUCTION

Researchers are becoming increasingly aware of the important roles that small radicals play in organic oxidation processes, flame chemistry, discharge systems, polymer formation, and biochemical processes, as well as the chemistry of the atmosphere.[1-18] In order to determine the importance of specific mechanistic pathways for the formation and removal of small organic radicals, detailed kinetic data as well as product distribution are necessary.[1,5,7] Due to the wide variety of organic classes of compounds (alkenes, alkanes, alkynes, aromatics, halogen- , nitrogen- , oxygen- , and sulfur-containing hydrocarbons, etc.) and the astronomical number of individual molecules and their corresponding small organic radicals, it would be inconceivable to carry out detailed kinetic and product studies on all the possible reactions with all the possible species.[1]

This, of course, is not unique for organic radical reactions, but has been a serious problem in organic chemistry in general. In order to simplify his work, the organic chemist has developed the functional approach to classifying organic molecule reactivities. This approach has allowed him to generalize and to predict qualitatively the types of products expected from certain types of reactions. The need to predict the relative distributions of products and the reaction rates more quantitatively, and to determine the important mechanistic pathways has led to the development of physical organic chemistry.[5,7] The name itself implies the combination of two approaches to understanding chemistry: the quantitative and theoretical approach of physical chemistry with the more empirical and qualitative studies of traditional organic chemistry.

The most successful approaches in simplifying the studies of organic reactions have involved the concept of the "reaction series".[5,7] Reactions involving only minor changes from one to the next, such that the mechanisms of the reactions are so similar that they can be considered a single reaction pathway, constitute such a series. This approach has led to the study and identification of a number of structure-reactivity correlations (SRC)[1,5-18] SRCs are empirical relations between the physical properties of reactants and products (e.g., reactivities with another reagent, ionization potentials, bond strengths, and electronegativity) and the observed reaction rate constants. By extrapolating this relationship to other members of the reaction series, unknown reaction rate constants may be predicted by the researcher or modeler in need of a rate constant estimate. The most effective theoretical treatment aimed at explaining these observed correlations involves transition state theory. An equilibrium between the reactants and the products is assumed, with the transition state existing at the peak of the energy barrier along the reaction coordinate. Thermodynamic equations which treat the reaction rate in terms of the free energy barrier (ΔG) are formulated and the Arrhenius equation, which directly relates the measured reaction rates to ΔG, can then be used. This theoretical development has led to the term linear free energy (LFE) correlation. A number of LFE correlations have been derived and demonstrated over the years, with the Hammett equation for prediction of the ionization of substituted benzoic acids probably the most well known.

This simple treatment, however, does not take into account the possible interference of steric factors, resonance, and other effects which will lead to nonlinear behavior. With varying degrees of success, these problems have been resolved by deriving the steric and resonance factors needed for correcting the reaction rates predicted by linear methods.

Most of the work that has led to the identification of SRCs has dealt with solution reactions involving nonradical chemistry. This has occurred mainly due to the dominance of ionic chemistry in the kinetic and mechanistic data available, and the difficulty in measuring the rates and products of radical reactions. Techniques recently developed for studying radicals in both gas and solution phases, however, have prompted the reexamination of various radical reaction series to determine if SRCs can be developed for radical reactions. These advances include improvements in a variety of spectroscopic tools (such as electron spin resonance, and application of lasers to kinetic spectroscopy), as well as use of rapid data acquisition systems which have enabled rapid reactions to be studied (for some examples, see Reference 19 to 36). Thus, accurate data and product analyses for a number of radical and related reaction systems can and are being obtained, which is allowing correlational studies to be pursued. This chapter briefly overviews both the theoretical and empirical approaches to the identification of SRCs for radical reaction series. Specific considerations and associated difficulties in radical rate predictions are examined, and the most recent SRCs that have been identified for organic radical formation and reaction rates are discussed.

II. STRUCTURE-REACTIVITY CORRELATIONS

Rates of chemical reactions have been dealt with on a theoretical basis by two essential theories: the collision theory and the transition state theory. The collision theory is a simple mechanical model for explaining rates of reaction; it is derived from gas kinetic theory and assumes that molecules are spherical masses which undergo elastic collisions. This is reviewed in most kinetics textbooks (e.g., Gilliom[5]). For bimolecular reactions collision theory predicts that the bimolecular rate constant for reaction will have the form:

$$\ln k = \ln Z - E^*/RT \qquad (1)$$

where Z is the mean number of bimolecular collisions, E* is the minimum energy necessary for the reacting molecules, R is the gas constant, and T is the temperature in Kelvin. The form of this equation can be compared directly with the Arrhenius equation:

$$\ln k_{obs} = \ln A - E_{\alpha}/RT \qquad (2)$$

It can be shown that Z has a $T^{-1/2}$ dependence, whereas A in Equation 2 is temperature independent. Rates predicted from collision theory are almost always too large when compared to experimental data even when the temperature dependence of Z is taken into account. To correct for this discrepancy, a probability factor correction (P) has been derived.[5] Reported values for P range from close to 1 to as low as 10^{-6}. These discrepancies arise from various factors which are difficult to incorporate into collisional theory; these include steric factors, inelastic collisions resulting in vibrational and rotational energy conversions, resonance effects, and so on. Because of the difficulties involved in calculating P, transition state theory has been more successful in leading to predictive correlations.

Transition state theory treats the reaction in terms of a thermodynamic equilibrium. Statistical equilibrium is assumed to hold between the reactants and the "activated complex" existing at the height of the energy barrier. The rate of reaction is simply proportional to the number of activated molecule complexes and is therefore just the frequency with which the "transition state complex" becomes products. Because of the equilibrium assumption, thermodynamic functions are employed which have been derived from studies of established equilibrium systems. Thus, the equilibrium between reactants and activated complexes forming products can be expressed in the form of a bimolecular reaction:[5]

$$\ln k = \ln(Kk_bT/h) - \Delta G\dagger/RT \qquad (3)$$

where K is the transmission coefficient (usually 1), k_b is Boltzmann's constant, h is Planck's constant, and $\Delta G\dagger$ is the free energy of activation for the transition state complex. Equation 3 can be further expanded using the relationship between free energy and the enthalpy and entropy of reactions:

$$\ln k = [\ln(Kk_bT/h) + \Delta S\dagger/R] - \Delta H\dagger/RT \quad (4)$$

Comparing this with Equation 2, the observed form of the Arrhenius equation incorporates the steric or entropy factor directly into the frequency factor. Because the free energy of reaction incorporates the enthalpy and entropy factors for the reaction, it is the free energy, not the enthalpy of reaction, which determines the reaction rate. This is an important result since the entropy of reaction rather than the enthalpy may often be the dominant factor in determining the change in free energy.

This treatment, however, oversimplifies the potential energy surface of a chemical reaction. For complex organic radical systems in which a number of reaction pathways exist and the actual potential energy surface can be quite complex, quantum mechanical calculations are the only means to an exact solution. Because these types of calculations are difficult at best for large molecules and because it is not clear how a specific detailed solution for one system will relate to another system, transition state theory has been applied to organic reaction systems in a phenomenological manner.[6] One approach has been to develop simple models for the energy profile along the reaction coordinate to determine the transition state energy barrier height and location along the reaction pathway as a function of the free energy, ΔG,[6,37-42] thus serving to develop a more unified formulation for a number of free energy correlations.

A. Linear Free Energy Relationships

Probably the best known of the free energy correlations is the LFE relationships. These are derived for reaction series in which similar reaction mechanisms are known or thought to dominate. It had been known for some time that plotting the natural logarithm of the rate constant or equilibrium constant for one closely related reaction series vs. another often yielded a straight line (see Hammett[7]). These linear relationships are expressed as

$$\ln k_1 = m(\ln k_2) + b \quad (5)$$

where m is the slope, b is the intercept, and k_1 and k_2 are the appropriate rate constants for the reaction series being compared. For the comparison of radical A with radical B, reaction series are usually examined in which both radical reactants are likely to react via the same mechanism (i.e., addition, insertion, or abstraction).

From transition state theory, it can be shown that linear correlations of this type occur when the activation free energy or change in energy for one reaction system is linearly related to the free energy of activation or change in energy to the second reaction system. The relationship between the equilibrium constant and the difference between the standard free energies of the products and reactants can be written as

$$\ln K = -\Delta G^o/RT \quad (6)$$

which is simply Equation 3 in a slightly different form. For cases in which one set of reaction rates is found to be linearly related to the equilibrium constants of another reaction series, direct substitution of Equations 3 and 6 into Equation 5 leads to the following general form of LFE relations:

$$\Delta G = m\Delta G^{o} + bRT - RT\ln(Kk_bT/h) \tag{7}$$

Equation 7 is a special case of the general equation:

$$\Delta G = n\Delta G + M(n) \tag{8}$$

derived previously.[6]

For a number of LFE correlations, many researchers have used a specific compound as a standard or reference reaction with a well-known rate or equilibrium constant. Using this reaction as a normalization factor, then, the LFE relation between this reaction and a similar reaction is given by

$$\ln k - \ln k_{std} = s(\Delta G/RT) \tag{9}$$

where s is the effect of the substituent change or solvent or medium change. A similar equation can be written for another reaction series, (k′). If the reaction series are linearly correlated, the differences in free energy changes must also be linear so that

$$\ln(k/k_{std}) = R\ \ln(k'/k'_{std}) \tag{10}$$

The constant R in Equation 10 is a measure of the comparable reactivity of the systems being studied and is often referred to as the reaction constant. For the standard reaction system, the value of R is one, by definition. As shown in Equation 9, the value of ln (k'/k'_{std}) is dependent upon the change in the substituent or the medium. This factor is often referred to as a substituent or media constant. Of course, an analogous expression for equilibrium constants may also be written.

This general form has been used for many investigations of solution reactions. Familiar equations having the same general form are found in the Hammett equation (so-called extended or modified Hammett equations), the Okamoto-Brown equation, the Taft equation, the Swain-Scott equation, the Edwards equation, the Winstein-Grunwald equation, etc.[5,7] These equations contain slight modifications in order to evaluate field, inductive, resonance, steric, and solvent effects for a wide variety of electrophilic and nucleophilic substitution reactions in solution, as well as for acid/base equilibria of organics. In fact, whole compendia are available for these types of substituent constant evaluations in chemistry and biology.[8]

Studies of reactions involving cation, anion, or polar transition states have led to σ^+ and σ^- substituent constants. These constants are a measure of the ability of an added functional substitutent to stabilize these types of intermediates. The comparable influences on the free radical transition states ($\sigma\cdot$ or σ-H) have not been as well characterized and are not as well understood.

For the case of small organic radical reactions, few LFE correlations have been derived. A number of efforts have been made to define special substituent constants for radical reactions,[8-18] particularly in the field of polymer chemistry. Although some of these approaches have been successful for specific reaction systems, no single extended Hammett approach has found general use in radical reaction correlational analysis. In general, σ or σ^+ substituent constants from the Hammett equation or the Okamoto-Brown equation have been used to correlate radical reactions in solution with moderate success.[8] More recently, there have baeen a number of successful studies using these approaches, and specific examples are given for solution and gas phase chemistry in Section IV. The problems with correlation in solution probably stem from the high reactivity of radicals in solution and the problems with solvent-cage effects, which can lead to experimental difficulties in determining reactivities. Recently, gas-phase radical data have become available which has led to a number of LFE correlations which are discussed in Section IV. Some of the difficulties in obtaining correlations for radical reactions are discussed in Section III.

B. Other Correlational Tools

A number of other correlations have been identified which predict reactivities for a series of organics reacting with a particular reactant. Many of these correlations are based upon some physical characteristic of the reaction series being studied. For example, gas-phase reactions of the electrophilic reactants $O(^3P)$ and OH with olefins have been shown to correlate with ionization potentials.[1,43-47] This correlation has since been extended by showing that the vertical ionization potential correlates with OH reactivity for 168 organic molecules.[47]

For the case of abstraction reactions, the bond energy (or bond order) of the bond being broken during the radical attack has been used to correlate rate constants and activation energies of reaction.[1,44,48-55] In addition, other thermochemical parameters such as heats of formation have been used to correlate reactivities.[56,57]

These observations and other correlations (with acidity, spectral intensities, bond order or bond strengths, etc.) are all related to the free energy correlations in the sense that these physical constants reflect the free energy of the reactant molecules. For electrophilic reactants, one would expect correlations with ionization potential and other physical measures of the reaction series' ability to give up an electron to the reactant of interest. For example, one would predict a negative correlation for the electron capture cross section of a series of reactants' electrophilicity in the opposite way that the ionization potential does. All of these other correlations, then, can be seen to reflect the same idea as the free energy approach, since they are comparing a physical property of the reactants to their reactivity just as one is comparing the free energy of reactants to the transition state.

These approaches are very useful since they tend to give the researcher another handle on the predicted rate constants. Because it is often easier to obtain an accurate measure of a spectral absorption coefficient or ionization potential than an accurate reaction rate for a radical in solution or in the gas phase (which are necessary for the free energy correlation evaluations), these correlations assist in the evaluation of both LFE, the available rate constant data, and techniques and methodologies used to obtain these data. In a number of instances, the reverse can also be true; i.e., one can use an established correlation to predict bond energies, redox potentials, ionization potentials, etc., where these are difficult to determine directly or where data are lacking.[1,51,58,59]

III. DIFFICULTIES IN RADICAL REACTION RATE PREDICTIONS

A number of problems arise when attempting to predict radical reaction rate constants, particularly when dealing with larger organic radicals in solution. Many radical reactions are very rapid, some approaching collisional frequencies. Reactions with solvent molecules can become important, and so-called solvent cage effects can lead to appreciable differences in the predicted reaction rates and product distributions. These types of effects are known to be important in a number of solution radical reaction systems and lead to difficulties in the identification of predictive correlations since, even though the reaction mechanism is well established, these second-order (medium) effects can cause significant departures from the predicted rates and products. It has been difficult to study the reactions of small organic radicals in solution for the reasons mentioned above, as well as limitations in the analytical instrumentation needed to monitor key radical intermediates. As mentioned earlier, a number of advances in electron spin resonance spectroscopy and in optical detection of radical species using lasers (resonance fluorescence) are beginning to yield better data sets for correlational analysis.[19-36]

In particular, analytical improvements and the use of high-speed data acquisition have led to significant improvements in the sensitivities of detection of many small radicals, especially in the gas phase. This has been of great use when attempting to determine the actual reaction mechanisms and their associated inherent substituent coefficients when using predictive Hammett-type correlations. The obvious advantages of gas-phase data are the absence of solvent and subsequent interactions from the reaction media. Thus, cage effects and solvent side reactions can be eliminated as well as any polar or ionic effects associated with solvent usage.

One of the most powerful tools which has been developed is resonance fluorescence (or other variations of excited state spectroscopy) to determine the absolute rates of gas and solution radical reactions. These methods are attractive due to time scale and sensitivity factors. Absolute rates are generally determined using pseudo first-order kinetic situations, assuming that the observed excited state rate of loss is directly proportional to the reaction rate. In fact, what is measured is the total fluorescence quenching rate, which includes physical and chemical quenching of the excited state being monitored. One must be careful to remember this when using existing rate data to establish predictive correlations or when comparing a predicted rate of reaction for a particular mechanistic pathway (e.g., addition, insertion, abstraction, or intramolecular rearrangement) with rate data obtained using these methods.

It is always useful, if possible, to intercompare relative rate determinations with the absolute methods to determine if physical quenching or some secondary reaction which leads to excited state species is important. In some situations, established correlations can be used to help validate the absolute methods. Some examples of this are discussed in the following sections dealing with gas-phase reactions.

As indicated previously, many solution reactions are expected to have pH dependencies, solvent interactions, temperature effects, etc. Many aqueous radical reaction systems can also react with ionic species such as H^+ leading to charged transition states (both positive and negative), which have led to interpretive problems in separating out the relative abilities of different substituents to stabilize radical vs. charged radical transition states. This has led to a variety of extended or modified Hammett equations as outlined earlier in this chapter.

Due to steric effects, resonance effects, charged transition states and polarity considerations, solvent effects, and associated problems in intercomparisons of model reactions systems, there have been more than 20 types of Hammett substituent constants described in the literature. As pointed out previously by other investigators,[11] this is obviously too many if we are to develop a unified approach to classify reaction systems and are going to establish useful substituent tables for reliable empirical predictions.

Attempts have been made to simplify and unify the observed Hammett substituent correlations.[8,11] For the case of aromatic reactions (which are easily the best studied), the overall substituent effects can be divided into three types; field, resonance, and steric. These types can in turn be separated into specific cases which provide a measure of their contribution to the overall effect upon the reaction rate. For example, the three substituent effects σ^+, σ^-, and $\sigma^{\cdot}$ provide a measure of the overall direct resonance interactions in aromatic reaction systems involving positive, negative, and radical transition states, respectively. Thus, the evaluation of substituent coefficients necessarily reflects the reaction system being studied and its reaction mechanism.

The evaluation of many of the substituent constants then involves the subtraction (or addition) of one evaluation of a reaction system from another evaluation of the appropriate reaction system. Therefore, the evaluated constants are dependent on the careful choice of the model reaction systems which are used to ''fish out'' the inherent substituent constants of interest. All of these constraints make it important that the model systems chosen for study are well understood in terms of the actual reaction mechanism. With the experimental difficulties involved in the investigation of radical reactions, particularly in solution, the evaluation of the organic radical sigma values has been difficult at best, even for the well-studied aromatic systems.

Reactions of radicals with organic reaction systems other than aromatic or mixed aromatic systems have not been as well studied. Although a number of correlational tools for the reactions of simple radicals with alkanes and alkenes have been identified (these will be discussed in the following section, particularly with regard to gas-phase reactions), there is a sparsity of data for the reaction of small organic radicals with organics other than aromatic derivatives. More data will need to be acquired before the reactions of these important systems can be adequately characterized and appropriate empirical relations realized for predictive purposes.

IV. EXAMPLES OF ESTABLISHED RADICAL PREDICTIVE TOOLS

In this section, an overview is presented of some specific attempts to establish radical reaction rate correlations. Radical reactions correlations have been reported for reactions in the gas phase,[1,35,36,43-74] in radical polymerizations,[9,10,16,75-81] and solution chemistry.[82-125] Examples have been chosen from gas-phase and solution studies which hopefully will serve to highlight the types of correlational tools which have been identified. The intent is not to comprehensively review all the radical chemistry which has been studied, but instead to summarize the types of correlational tools which may be useful to researchers in these areas.

A. Gas-Phase Reactions

Free radical reactions are known to occur quite readily in gas and solution phases. As outlined earlier in this text, the basic electronic theory that the organic chemist has relied upon to predict reaction rates and invoke mechanisms has been based largely upon experimental results obtained from solution chemistry. These results have been dominated by studies of ionic reactions in solution, due to the difficulties in studying radical species. The solvent's potential for playing major roles in affecting the apparent kinetics of these reactions has caused some difficulties in interpretation of the date (see Section III). As pointed out in the review article by Tedder and Walton,[60] gas-phase studies of organic radical reactions have two major advantages: no solvent effects, and activation energies for reactions can be more readily determined since wider temperature ranges are easily studied. The advantage of the gas phase, then, is its relative simplicity compared to the solution phase.

Until recently, very few systematic studies of gas-phase radical reactions with organics have been reported.[18,60] Indeed, gas-phase kineticists and researchers involved in establishing substituent effects in solution chemistry have just begun to communicate. Reviewed here are some of the successes in establishing correlation tools for prediction of gas-phase radical reactions with organics.

Three basic initial reaction mechanisms are of interest for organic/free radical systems: addition, abstraction, and insertion. The first two, addition and abstraction, are the best known.[1,60] Given in Table 1 is a list of some of the small organic radical species for which kinetic data exist for addition and abstraction reactions in the gas phase. Listed also are

Table 1
SOME EXAMPLES OF ATOMS AND RADICALS FOR WHICH GAS-PHASE KINETIC DATA FOR REACTIONS WITH ORGANICS HAVE BEEN DETERMINED[1,60,63,64,111]

Atomic Species				
Cl	H	$S(^3P)$	$O(^P)$	$O(^1D)$
$O(^1S)$	$Se(^3P)$	Br		
Radicals				
OH	$C_2(a^3II_u)$	C_3	CF_3	CF_2Br
CH_2F	CCl_3	CH_3	CH_3CH_2	$(CH_3)_2CH$
$(CH_3)_3C$	HO_2	CH_3O	CH_3O_2	ClO
C_2F_5	NO_3			
Molecules With Radical Character				
$O_2(^1\Delta_g)$	O_3	NO_2		

some atomic and molecular species which have radical character. These species are of interest here as they are likely to act similarly to small organic radicals with regard to their mechanistic behavior.

As indicated by previous reviewers,[3,31,33,60,125] addition reactions for small organic radicals are complicated in many cases by ejection of the added radical or subsequent loss of an alternate radical from the initial adduct. In many instances, secondary reactions, and chain reactions leading to polymer formations, have made these systems difficult to study and interpret. This has especially been the case for studies which have investigated alkyl radical reactions and other radical systems where the rates of reaction are relatively slow (i.e., secondary reactions can, in many cases, readily compete with the reactions of interest). Concurrent abstraction reactions can also occur which make it difficult to isolate the various pathways. These aspects of small organic radical reactions are reviewed in detail in the other chapters of this book. Despite the problems of disproportionation and multiple reaction pathways, a number of useful correlations have been identified.[8,60]

Work through 1978 has been reviewed previously,[60] and will be summarized briefly here. The gas-phase addition reactions of simple alkyl radicals to simple olefins have found methyl substituent effects (on the olefin or on the radical) to be small (less than a factor of ten). The more substituted radicals or olefins are slower to react, indicating the possible importance of polar and/or steric efects. Thus, unlike many of the other electrophilic atomic and radical species,[1,60] alkyl radical systems appear to be more nucleophilic in nature. Increasing the polarity of the radicals or the olefins by addition of halogen atoms has been found to increase the relative substituent effects. Here the relative reactivities have been observed to vary up to two orders of magnitude.

In Table 2, the reaction rate parameters for a number of atomic and radical species are given for the addition reaction with ethylene. The preexponential factors (ln A), activation energies, and room temperature rates constants (300 K) are given for comparison. As indicated previously,[60] with the exception of OH and $C_2(a^3II\mu)$ radicals, all of the atomic species react faster than the larger organic radicals. It is apparent from Table 2 that there are large variabilities in the reactivities of both radical and atomic species. Room temperature rates can vary by over eight orders of magnitude. This is due to both entropy (ln A) and

Table 2
SOME KINETIC PARAMETERS FOR ATOM AND RADICAL REACTIONS WITH ETHENE AT 300 K

	ln A	E (kcal/mol)	k^{300K}(ℓ $mol^{-1}sec^{-1}$)	Ref.
		Atoms		
Cl	24.4	0	4×10^{10}	60
H	25.3	2.8	9×10^{8}	60
$S(^3P)$	23.3	1.5	1×10^{9}	60
$O(^3P)$	21.9	1.1	5×10^{8}	126
$Se(^3P)$	23.2	2.8	1×10^{8}	60
Br	21.9	2.9	2×10^{7}	60
		Radicals		
OH	21.0	−0.7	4×10^{9}	126
$C_2(a^3\Pi_u)$	24	0	7×10^{10}	64
CF_3	18.4	2.9	8×10^{5}	60
CF_2Br	18.4	3.1	5×10^{5}	60
CH_2F	17.5	4.3	3×10^{4}	60
CCl_3	18.0	6.3	2×10^{3}	60
CH_3	19.6	7.7	8×10^{2}	60
CH_3CH_2	19.1	7.5	7×10^{2}	60
$(CH_3)_2CH$	18.0	6.9	6×10^{2}	60
$(CH_3)_3C$	17.3	7.1	2×10^{2}	60

free energy effects (activation energies). For many of the organic radical gas-phase systems that have been studied, the differences in reactivity are consistent with polar effects playing the dominant role.[60] For the case of organic radical addition reactions to olefins in the gas phase, available data indicate that steric effects may also be important.[1,60] Studies of the product distributions from radical addition to substituted olefins are also strongly indicative of the importance of steric effects in the gas phase, as the orientation of attack is clearly dependent upon the size of the attacking radical species.[60] However, as pointed out by Tedder and Walton,[60] the polar and steric effects are not strictly decoupled, as the size of the organic radical is strongly dependent upon the electron density at the trivalent carbon atom.

A number of useful correlations have been examined to attempt to separate these terms, as well as resonance effects. For example, the orientation ratio of radical attack has been correlated with the "radical diameter" for reaction with vinyl fluoride as an argument for the importance of steric effects in addition reactions.[60] Similiar arguments have been made using the correlation of the logarithm of the orientation ratio of perfluoroalkyl radical reactions with the pK_a of the corresponding perfluoroalkanes.[60] An extension of the Hammett approach from polymer radical reaction correlations have been used to derive an expression for the prediction of the orientation ratio, O_r, for the reaction of a radical with an unsymmetrical olefin.[60] The equation:

$$\log O_r = \log k'/k = \sigma(\alpha' - \alpha) + (\beta' - \beta) \qquad (11)$$

has been found to correlate radical reactions with three fluoroethenes by plotting log O_r against σ to obtain the factors for $(\alpha' - \alpha)$ and $(\beta' - \beta)$ which are characteristic values for the respective ends of the substituted ethene.[60] The experimental data from the gas phase are in reasonable agreement with data derived from polymerization studies for olefins.[60] Thus, correlations with σ and σ^+ are to be expected for gas-phase reactions where polar effects are likely to dominate steric and resonance effects.

The gas-phase reactions of OH radicals with organic reactions have received a great deal of attention, principally because of their importance in atmospheric and combustion chemistry.[1] These systems, although not organic radicals, are indicative of the types of correlations that can be obtained. In addition, they allow the prediction of the formation rates of organic radicals which is of interest to researchers involved in the study of organic radicals. The availability and extent of data for OH reactions with organic molecules have allowed gas-phase correlations to be examined in more detail than most other radical species.

Other nonradical reactions can also be of use to the investigator when attempting to determine reaction mechanisms (see Table 1). In particular, atomic species (H, O, Cl, etc.) and molecular species which have some radical nature can be of use in separating out reaction pathways.

This has been demonstrated recently for the reaction of OH with unsaturated aldehydes.[71,72] The possibility exists for the abstraction reaction (aldehydic proton attack) to compete and dominate the addition reaction for these systems. LFE correlations with ground state oxygen atoms were used to predict the OH addition rate and found to be in disagreement (lower by a factor of two to three).[1,43,44] Measurements of the rates of ozone reaction with these compounds and other olefins were found to correlate well with OH, except for the unsaturated aldehydes acrolein and crotonaldehyde. Comparison of the expected addition rates from $O(^3P)$, O_3, and ionization potential correlations with OH addition to olefins was found to be in excellent agreement.[72] The apparent discrepancies between addition and observed rates are due to aldehydic abstraction by OH. The use of these established addition rate correlations allows the relative amounts of addition and abstraction to be estimated directly (approximately 30% addition reaction). This example again indicates the strength of examining multiple correlations when interpeting kinetic data. Thus, by correlating reaction series for selective reagents (i.e., ozone) with less selective agents (i.e., OH), one can begin to sort out reaction pathway branching (e.g., between addition and abstraction).

Note also that many small organic radical systems are likely to be electrophilic reagents (with the major exception being the alkyl radical systems which are relatively nucleophilic),[60] thus correlations with other electrophilic rectants OH, $O(^3P)$, ozone, nitrate radical, etc. are likely. Thus, established correlational tools for these inorganic radical, atomic, and molecular reaction systems are likely to be excellent candidates for improving and establishing reaction correlations for small organic radicals (as the data become available). Direct comparison of the relative reactivities for specific reaction series can lead to direct comparison of the electrophilicities of radicals and atomic species as well.[1] which is of use in interpreting changes in mechanisms.

The use of these tools can also aid the kineticist in evaluating kinetic data, particularly for experimental determinations that may be subject to possible error due to secondary reactions and other systematic errors. For example, the reaction of $O(^3P)$ with methanol was found to be in question using the LFE abstraction correlation with hydroxy radical.[1] Remeasurement of this rate constant found the LFE prediction to be quite accurate.[1,61] Other predicted abstraction rate constants for aldehyde/OH reactions have been experimentally confirmed as well.[1,62] A recent determination of the OH/methyl nitrate reaction rate has indicated that the available $O(^3P)$/methyl nitrate rate data are in error; i.e., the LFE prediction is in substantial disagreement with the available literature data.[127] In this work, it is suggested that this rate constant should be remeasured in the near future.

As mentioned in Section II.B olefin ionization potential correlations with OH, $O(^3P)$, and other electrophilic species have been used to predict reaction rates.[1,43-47] This initially was identified for the case of substituted and cyclic olefins.[1,43-46] However, this correlation has been extended for OH to include a wide variety of organics,[47] including substituted aromatics and aliphatics. For highly reactive electrophilic radicals (e.g., OH), this type of correlation with electronic characteristics (in this case, ionization potential) is to be expected in the

Table 3
SOME EXAMPLES OF CARBON-HYDROGEN BOND DISSOCIATION ENERGIES (BDE) WHICH HAVE BEEN ESTIMATED FROM OH ABSTRACTION REACTION CORRELATIONS WITH BOND ENERGIES

C–H bond	BDE (kcal/mol)	Estimated error (kcal/mol)	Ref.
CH_3CHF–H	96.3	1.5	51
CH_2FCHF–H	98.8	1.0	51
CF_3CHF–H	103.5	1.0	51
CHF_2CF_2–H	103	1.5	51
cyclo-C_3H_5–H	101	2	50
O_2NOCH_2–H	101	2	127

Note: BDE are given for 298 K.

absence of steric or other complicating effects. For the reactions of organic radicals, the occurrence or lack of correlation with ionization potential may be of use in corroboration of LFE or other predictive methods.

For hydrogen abstraction reactions, bond dissociation energies have been used to correlate electrophilic radical and atom reactivites.[1,44,48-55] This approach is one of the oldest methods for estimating radical reaction abstraction rates. The method is often referred to as the bond energy/bond order (BEBO) approach.[1,48,67] A number of approaches have used statistical analysis to empirically derive the relative rate contributions from primary, secondary, and tertiary hydrogens.[48,53,54,67]

More quantitative approaches have attempted to use transition state theory to extrapolate reaction rates[128] (for examples of this approach for the case of OH/C-H abstraction reactions, see References 48 and 67). These methods have been reasonably successful for systems in which competing reactions (i.e., insertion or addition) are not important. A general expression that allows rate estimates over the temperature range 200 to 400 K has been developed by Heicklen[48] for gas-phase OH hydrogen abstraction. Indeed, this and other simpler empirically derived relationships between radical and atom abstraction rates and bond order have allowed the carbon-hydrogen bond strengths to be estimated for a number of compounds that are difficult to determine directly, or where the bond strengths have been in question. Some examples of hydrogen bond strengths estimated in this manner are presented in Table 3. It should also be pointed out that the correlations with ionization potential (IP) can also be used to predict IP and not only rates of reaction.[127]

As implicated above, simple LFE correlations for addition and abstraction reactions of ground state oxygen atoms and hydroxyl radical have been identified and used to predict reaction rates within a factor of two.[1] Figure 1 shows the simple LFE correlation obtained for OH and $O(^3P)$ addition to a number of substituted alkenes and alkylated aromatics with ethene as the reference reaction.[1] For the case of abstraction reactions, the correlation is given graphically in Figure 2, with methane as the reference.[1,70] It is clear from these graphs that quite reasonable correlations can be obtained for these gas-phase reaction systems.

The correlation of OH addition reactions to aromatics has been extended to include a wide variety of substituted aromatic compounds.[66] This has allowed the comparison of solution-derived substituent effects to be compared directly with the gas-phase data. As expected, polar effects are apparently the dominant factors in the gas phase. Thus, reasonably good correlations are found for the observed reaction rates with σ^+ (see Figure 3). For the available data, a comparison of $O(^3P)$ with OH for aromatic reactions is also in agreement with the

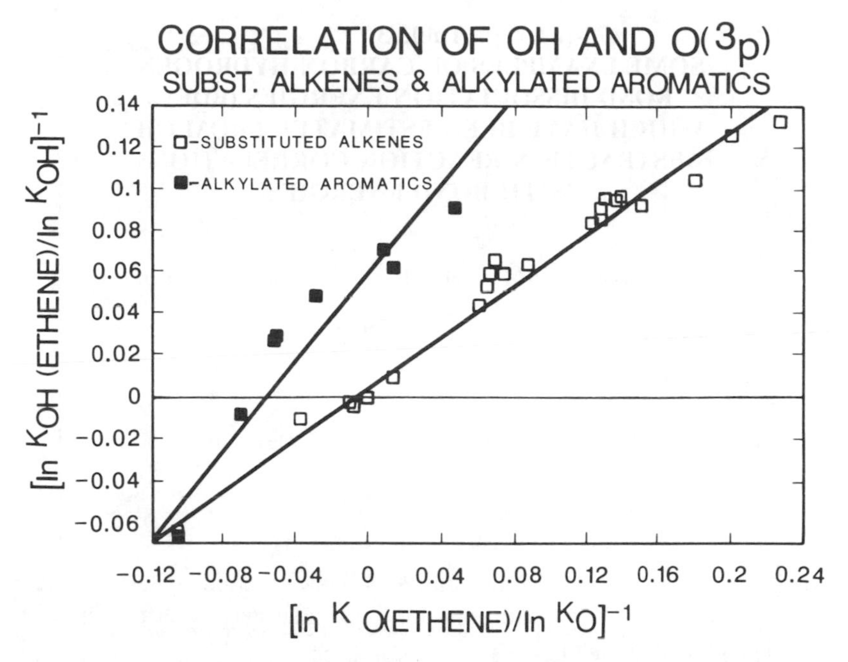

FIGURE 1. LFE plot of OH and O(^{3}P) kinetic data for addition reactions to alkenes and alkylated aromatics. Ethene is used as the reference reaction.[1,66]

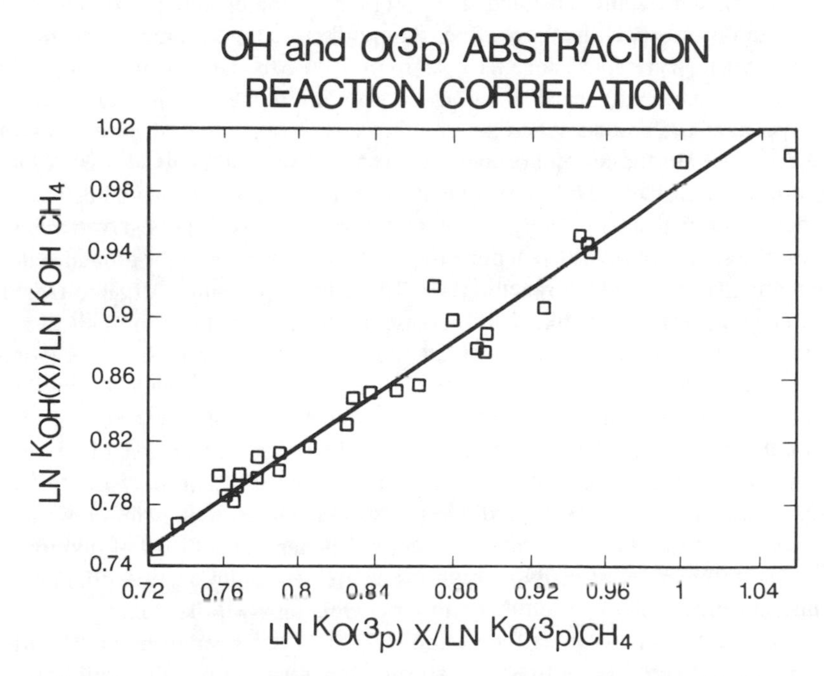

FIGURE 2. LFE plot for the abstraction reaction correlation between OH and O(^{3}P) with a variety of substituted alkanes. Methane is used as the reference reaction.[1,70]

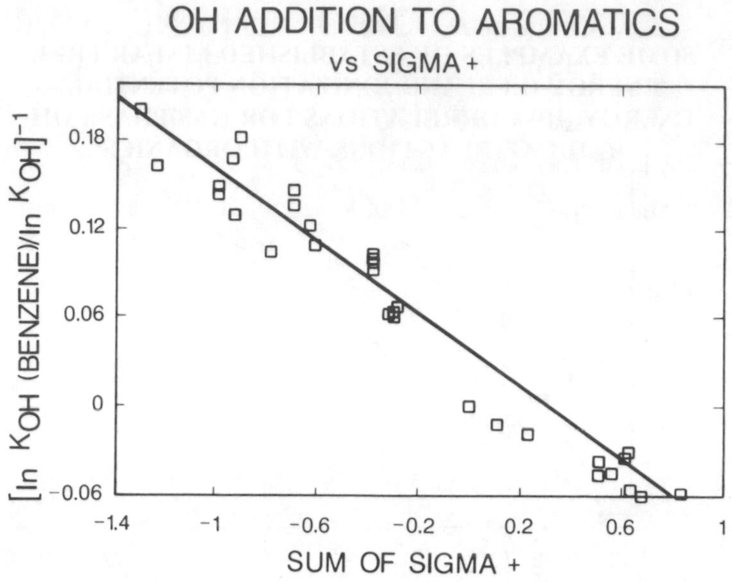

FIGURE 3. Correlation of gas-phase OH addition reaction to aromatics with σ^+ derived from solution studies. Data are normalized to benzene as the reference reaction.[8,66]

σ^+ correlation. The differences between the observed and predicted are likely due to solvent and other solution effects which may affect the σ^+ factors, or to some abstraction reaction with the aromatics (particularly for OH). As more data become available, ozone reactions may be useful in determining the abstraction contributions to the overall radical reactivities.[73]

An extension of the ground state oxygen atom correlations with alcohol reactivities has found good agreement for correlation with the Taft polar substituents (σ^*).[65] Although the data are limited, Taft plots for OH and $O(^3P)$ abstraction reactions have been observed to be linear for primary alkanes, aldehydes, and primary alcohols. The observed values for ρ^* indicate that the sensitivity of the reactions to polar effects is greater for ground state oxygen atoms than for hydroxyl radical reactions.[1,65] Thus, for electrophilic organic radicals, it is expected that polar effects are likely to increase as the selectivity of the radical increases. These data also suggest that solution-derived substituent effects can be of use in gas-phase correlations. When available, gas-phase correlations of this sort can act as good indicators of the validity of these solution-derived inherent substituent values.

Some of the established linear correlations that have been identified for the gas-phase reactions of OH with organic molecules are summarized in Table 4. These relationships have been found to be accurate in prediction of gas-phase reactivities to within a factor of three. In most cases, the estimated rate constants are better than a factor of two for the cases of OH-LFE addition and abstraction correlations.[1] As increasing data become available for other radical species, it is expected that similar correlations will be found.

Abstraction reaction correlations for Cl and HO_2 have also been examined[70] with respect to their correlation with OH and oxygen atoms. The correlations observed, while not as good as the examples given here, deserve to be considered in the correlational arsenal. This is particularly true for the reactions of HO_2 since the available data are scarce, and the experimental systems used to determine reaction rates are inherently subject to errors (principally due to its chemical and physical properties).

Use of IP and LFE correlations has been an aid in the interpretation of the mechanisms by which $C_2(a^3\Pi_u)$ and C_3 radicals react with simple alkanes and alkenes.[63,64] For the case of C_3 radical reaction with alkenes, alkynes, and allenes, unlike OH and ground state oxygen

Table 4
SOME EXAMPLES OF ESTABLISHED LINEAR FREE ENERGY (LFE) AND IONIZATION POTENTIAL ENERGY (IP) CORRELATIONS FOR GAS-PHASE OH RADICAL REACTIONS WITH ORGANICS

Correlation type	X	Y	m	b	Ref.
Olefins (IP)	IP	log k_{OH}	−0.6	−4.7	1
Aromatics (IP)	IP	log k_{OH}	1.5	−2	47
Aliphatics (IP)	IP	log k_{OH}	0.8	3	47
Addition (LFE)	log k_O	log k_{OH}	0.6	−4	1
Abstraction (LFE)	log k_{OH}	log k_O	1.4	2.3	1
Aromatics (LFE)	$\Sigma\sigma^+$	log $k(add)_{OH}$	−1.4	−11.6	66

atoms, the electrophilic radical C_3 reactivities were not found to correlate well with IPs. This observation was used to suggest that additional pathways (such as insertion and intramolecular rearrangements leading to fragmentation) are arising in the species reactions.[63] Correlations of the reactions of $C_2(a^3\Pi_\mu)$ radicals with hydrogen and saturated hydrocarbons have been observed to correlate linearly with OH and $O(^3P)$ reaction rates consistent with an abstraction mechanism.[64] For the case of reaction with ethene, the reactivity was observed to lie outside the abstraction correlational line, and has been used to imply a mechanism change. These interpretative results based upon direct comparisons with the well-studied OH and $O(^3P)$ reaction series and their LFE and IPs are indicative of the usefulness of these correlational tools as mechanistic indicators for carbon-containing radicals.

More subtle effects, such as the effects of ring strain and/or steric effects, have been studied for the reaction of OH with a series of cyclic olefins.[74] No apparent effects were observed for the OH reaction and by analogy, little are expected for $O(^3P)$. However, the more selective reagent, ozone, was found to correlate with ring strain.[68] This indicates that less reactive electrophilic organic radicals may also show some selectivity for reaction with cyclic olefins. Separation of polar and ring strain effects can likely be accomplished by comparison of their reactivities with OH and ozone. It is clear that detailed product studies, which can yield the position of initial radical attack are also needed to yield information regarding steric and other orientational effects.

From the gas-phase examples presented here, it is clear that predictive tools are beginning to be identified for gas-phase radical reactions with organic molecules. These tools are expected to expand their use for prediction of small organic radical reactivites as more and better data become available. The correlational methods should also continue to aid the investigator in identification of reaction mechanisms, as well as in rate data evaluations.

B. Solution Reactions

As indicated, earlier solution studies of radical reactions have often been difficult to interpret due to solvent interactions, radical rearrangements and disproportionations, polymerization reactions, and difficulties in producing the radicals of interest cleanly.[3,125] Up until recently, almost all of the Hammett-type correlations for radical reactions had been carried out in solution, and were developed from relative rate data based principally upon product distributions.[3,125] These studies typically took reactions to greater than 10% completion due to analytical difficulties, so that secondary reactions, if competitive, would often lead to errors. Solvent effects can also be a potential problem in interpreting the kinetic results.[3,106,125]

Advances in analytical instrumentation have improved this situation. This is particularly true for the development of electron spin resonance (ESR) techniques that allow the researcher

Table 5
EXAMPLES OF SOLUTION HAMMETT-TYPE CORRELATIONS FOR ATOM AND RADICAL-HYDROGEN ATOM ABSTRACTION REACTIONS

Atom or radical	Reaction systems	Solvent	$\rho^+(\rho)$ values	T (°C)	Ref.
Br	Substituted toluenes	Reactants	−1.78	19	118
	Substituted toluenes	Reactants	−1.39	80	118
	Substituted cumenes	CCl_4	−0.29(−0.38)	70	123
Cl	Substituted toluenes	CCl_4	−0.66(−0.76)	40	102
CH_3	Substituted toluenes	CCl_4	−0.14	100	117
t-Butyl	Substituted toluenes	Reactants	(0.99)	30	113
CCl_3	Substituted cumenes	CCl_4	−0.67(−0.89)	70	123
	Substituted cumenes	ClC_6H_5	−1.46	50	119
CH_2COOH	Substituted toluenes	CH_3COOH	−0.63	130	102
C_6H_5	Hydrogen donors	Reactants	−0.3(−0.4)	60	114
	Hydrogen donors	Reactants	(−0.1)	60	115
$pCH_3C_6H_5$	Hydrogen donors	Reactants	−0.1(−0.1)	60	114
$pBr\text{-}C_6H_5$	Hydrogen donors	Reactants	−0.25(−0.32)	60	114
$pNO_2\text{-}C_6H_5$	Hydrogen donors	Reactants	−0.44(−0.59)	60	114
RO_2	Substituted cumenes	ClC_6H_5	−0.41	60	102
	Substituted styrenes	Benzene	−0.3(−0.4)	60	102
t-ButylO	Substituted toluenes	Reactants	−0.6(−0.75)	39.6	122
	Substituted toluenes	Reactants	(−0.83)	40	121
	Substituted aromatics	Reactants	(−0.5)	110	116
t-ButylO$_2$	Substituted toluenes	Reactants	−0.56(−0.78)	30	112
$HN^+(CH_2)_5$	Substituted toluenes	H_2SO_4/CH_3COOH	−1.36	20	120
di-*t*-ButylNO	p-C_6H_5S rads	Hexanes	0.75	23	104

the capability to directly monitor the radicals. A large number of radical systems have been and continue to be studied using ESR, with and without the use of spin-trap reagents.[31-33] These methods are capable of examining structure, as well as reactivity, and constantly are being improved to enable faster reactions and possible radical intermediates to be monitored during reaction.[31-33] Improvements in optical laser-based radical systems, other analytical methods, and improved methods for the clean generation of radicals are discussed in this text. All of the recent data indicate that our capabilities to study radical reactions in solution are improving rapidly.

Despite the difficulties described above and in Section III, a number of useful correlations have been identified, especially for the hydrogen abstraction reactions. Some examples of the Hammett-type correlations that have been reported in the literature are given in Table 5. A typical Hammett-type plot (vs. σ^+) for the correlation of *t*-butylperoxy radical abstraction reaction with substituted toluenes is given in Figure 4.[88] Although there have been reports of correlation of radical reactions with σ^-(see References 100 and 105) and σ^* (see Reference 99) most of the reported solution-phase radical reactions have been found to correlate well with the Hammett σ substituent factors or the Okamoto-Brown σ^+ values. For many of the radical reaction systems, the differences between the correlations using σ or σ^+ are small (see Table 5), and difficult to choose between, especially if one considers measurement errors.

In a number of cases, subtle differences in the correlations between σ^+ and σ have been used to argue the amount of polar nature in the transition state and the existence of a number of more or less polar transition state structures. In most cases, the differentiation for these abstraction reactions has relied on the use of nitro or methoxy functional groups to amplify the polar effect differences between the two correlational constants (for examples, see References 82, 112, 120, and 122).

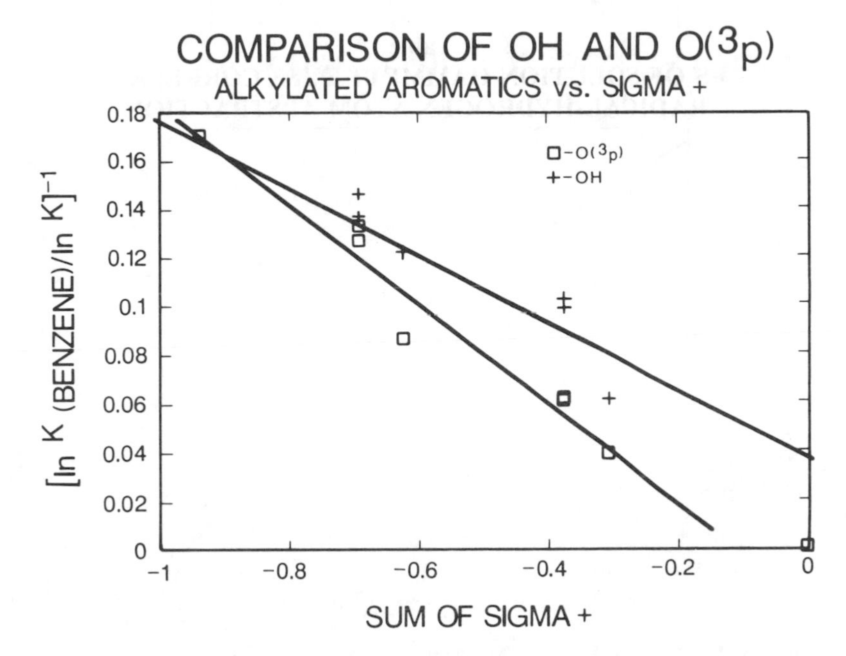

FIGURE 4. Comparison of hydroxyl and ground state oxygen atom correlation with solution-derived σ^+ values.[1,8,66]

In many instances, the correlations with σ or σ^+ have been found to deviate substantially from linearity. The deviations from linearity have been attributed to radical stabilization by the functional substituents studied. These researchers have proposed a $\sigma^{\cdot}$ substituent scale to correct for these deviations.[92,93,95-98,109,110] As pointed out previously, [109] the agreement between the proposed $\sigma^{\cdot}$ scales is poor. This is likely due to the fact that the substituent factors are derived using an extended Hammett approach after the assumption of some polar contribution scale.[109] This apparently has led to as great as a factor of ten discrepancy between the proposed scales.[109] With the advent of ESR spectroscopy, Wayner and Arnold[92] and Dust and Arnold[109] have proposed a $\sigma_a^{\cdot}$ scale based upon the changes in the hyperfine splitting constants of benzylic σ hydrogens caused by substituents. This approach attempts to measure the increases in stability directly. The ESR work indicates that 4-F, 4-CH_3, 4-t butyl, 4-CN, and 3-CN functional groups should be studied in substituted aromatic/radical studies to determine radical stabilization vs. polar factors. It will be interesting to see how useful this direct approach will be in future solution work.

For most of the radical-abstraction reactions studied in solution, the following generalization can be made; the more reactive the radical, the better the correlation with σ^+. The ρ values determined for most radical reactions in solution have been found to be negative, although a few positive values have been reported (see Table 5).[104,113] The general trend is for the more stable radicals to have less negative ρ values which tend to approach zero for the lowest radical reactivities. These values are typically smaller than comparable ionic reaction mechanisms, as expected.

For the case of addition reactions, the relative reactivities of CCl_3 radicals with substituted styrenes and stilbenes,[30] with 3-phenylprop-1-enes and 4-phenylbut-1-enes,[90] and with alkenylsilanes,[91] have all led to reasonable correlations with polar factors. These studies have indicated that the CCl_3 is electrophilic in its reactions with these types of compounds from the negative ρ values determined. The ρ values observed for this species reaction have been found to be smaller than comparable ionic reaction mechanisms, as expected. A similar correlation with σ^* for addition of CCl_3 radicals to substituted alkenes has measured ρ^* values ranging from -0.024 to -0.15, dependent upon the number of methylene groups

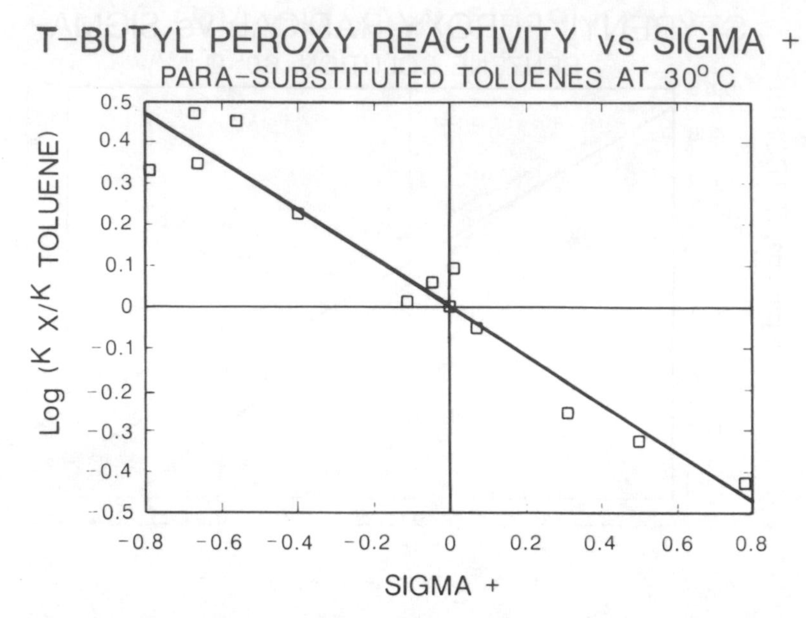

FIGURE 5. A typical Hammett-type plot for a radical reaction in solution. Correlation of *t*-butylperoxy radical hydrogen abstraction from para-substituted toluenes with σ^+ at 30°C.[5,8,112]

between the substituent and the double bond.[99] It is interesting to compare the gas phase $O(^3P)$ and OH addition-derived ρ^+ for alkylated and substituted aromatics (rates normalized to ethene; see Figures 3 and 5). For these gas-phase addition reactions, the ρ values range from -0.12 to -0.16, consistent with the small negative ρ values determined for solution reactions. These studies, again, indicate the smaller effects on radical reactions by polar substituents when compared to ionic reactions in solution.

For organoperoxy radicals, it is interesting to compare the results from solution-[88,102,106,112] and gas-phase studies.[20-23] Similar observations have been made in both phases. It is observed that the reactions of RO_2 species with various substrates have about the same reactivity irrespective of the nature of the R group. For example, the *t*-butylperoxy radical has a ρ^+ value of -0.56 for hydrogen abstraction reactions with para-substituted toluenes (see Figure 4.[112] Figure 6 is a plot of the reaction of styrenylperoxy radical reaction with substituted styrenes.[88] Although the data are limited, the apparent ρ^+ is approximately -0.3, indicating the similarity of reactivity between peroxy radicals. However, termination rate constants for organic peroxy radicals, particularly with themselves, are apparently quite dependent upon the nature of the substituent, indicating the importance of steric effects in both phases for these type processes. This behavior appears to merit more investigation since the peroxy radical reactions are very important in combustion processes, atmospheric chemistry, and auto-oxidation of organics in solution. The basis for these observations are limited data sets which need to be extended to determine the overall validity of the conclusions.

Similar to the reactions of OH and $O(^3P)$ with olefins, correlations with ionization potential for reactions with substituted phenylallenes have been observed.[94] This observation and Hammett correlations using σ^+ (ρ^+ of -0.36) have been used to argue the importance of biradical intermediates in thermal cycloaddition of thiobenzophenone to the phenylallenes. This is another example of how these correlations have been used to invoke or determine mechanistic pathways. The correlation of structure and reactivity with this physical characteristic and others (e.g., electron capture response),[103] as pointed out earlier, is related to the electronic nature of the reactants and therefore the free energy of the reaction system.

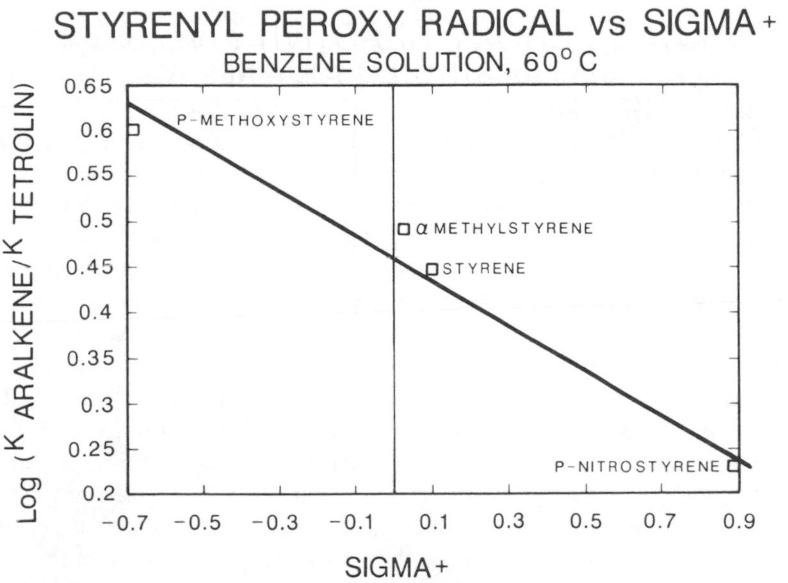

STYRENYL PEROXY RADICAL vs SIGMA+
BENZENE SOLUTION, 60° C
Log (K ARALKENE/K TETROLIN)
0.65 0.6 0.55 0.5 0.45 0.4 0.35 0.3 0.25 0.2
P-METHOXYSTYRENE
α METHYLSTYRENE
STYRENE
P-NITROSTYRENE
−0.7 −0.5 −0.3 −0.1 0.1 0.3 0.5 0.7 0.9
SIGMA+

FIGURE 6. Correlation σ^+ with reaction of styrenylperoxy radical reaction with substituted styrenes.[5,8,88]

It is expected that as improvements in experimental methods continue, more examples of these types of predictive correlations will be determined for solution reactions. From the available data, it appears that correlations with σ^+ or σ are likely to yield the best predictions. This is consistent with the available gas-phase data, which continue to support polar effects as being the dominant factors, with other effects playing secondary roles.

V. CONCLUSIONS AND SUGGESTIONS FOR FUTURE WORK

It should be apparent from the previous remarks that a number of identified correlational tools for radical and atom reactions in gas phase and solution systems have been identified and are available to the interested researcher. In particular, the physical organic chemist has developed an arsenal of LFE correlational tools which can be used to identify mechanisms, mechanism changes, estimates of abstraction and addition branching ratios, etc. In the continuing need for rate constant evaluations (not just predictions), LFE and other correlational tools can also be useful in identification of outliers which need to be reinvestigated.[127]

It is important to realize that no one correlational tool is perfect. The derived correlations are based upon experimental data which, in some instances, are in error. Thus, the researcher using these tools should not consider them a panacea. As in the use of any method, the researcher should use adequate caution and common sense in application. For this reason, it is important that we continue to improve our fundamental understanding of radical reactions, and use multiple correlations when possible to improve the reactivity predictions.

With our increased abilities to study these systems in both gas and solution phases, it is clear that these improvements will be forthcoming for a wide variety of atom and radical reactions. It should be noted that both kinetic and product studies should be carried out to unambiguously determine reaction mechanisms in both phases. The correlation of gas-phase data, as it continues to become available, with solution data will be very useful in the further evaluation of solution-derived substituent factors. Comparative analysis of rate and product studies for radical and atomic reactions with organics will, hopefully, lead to simplified Hammett-type and other correlational approaches for these rather complicated systems.

For the case of gas-phase reactions (as well as solution), it is clear that substituent factors for abstraction reactions with substituted alkanes, and alkenes as well as aromatic reaction systems are needed. In this work, gas-phase abstraction reactions have been normalized to methane, while gas-phase addition reactions have been normalized to ethene for the case of olefins and benzene for aromatic reactions. It is suggested that these reaction systems be used as standard systems for these dominant types of radical reactions for internal comparison, in order to standardize these types of reactions.

As indicated in the earlier part of this chapter, kinetic parameters can be expressed in terms of changes in the free energies. These relationship are in turn functions of enthalpy, entropy, and temperature. Depending upon the temperature dependence of the reaction system entropy, it is important to compare systems below the isokinetic temperature where the rates are controlled mainly by the changes in enthalpy.[5] If one compares reaction systems above the isokinetic temperature where entropy effects are controlling, errors in interpretation can occur. Since many LFE relationships are likely to be temperature dependent (see Br in Table 5), it is important for the kineticist to study these relationships as a function of temperature to determine if the enthalpy and entropy are temperature dependent, or independent as well.[5] Otherwise, the possibility for misleading interpretation exists for correlations determined at only one temperature, or in comparison of correlations at different temperatures. It is hoped that kineticists in this field will recognize the importance of determining the temperature dependences of the radical reactions in both phases.

With the above caveats, the correlational tools discussed in this chapter are becoming increasingly valuable to the modeler in need of a predicted rate constant or product distribution, to the mechanistic chemist, and to the kineticist in data evaluation. It is important to continue the iterative use of these methods to improve existing tools and to identify new and better tools for the future.

ACKNOWLEDGMENTS

The author wishes to thank Drs. John Hall and Ines Triay for helpful discussions, as well as for proofreading and editing this manuscript. The graphic art capabilities of Mr. Garth Tietjen are also appreciated. This work was initiated at Brookhaven National Laboratory and completed at Los Alamos National Laboratory. The work was performed under the auspices of the U. S. Department of Energy.

REFERENCES

1. **Gaffney, J. S. and Levine, S.Z.,** Predicting gas phase organic molecule reaction rates using linear free energy correlations. I. O(^{3}P) and OH addition and abstraction reactions, *Int. J. Chem. Kinet.*, II, 1197, 1979 (and references therein).
2. **Bradley, J. N.** *Flame and Combustion Phenomena,* Chapman & Hall, London, 1969.
3. **Pryor, W. A.** *Free Radicals,* McGraw-Hill, New York, 1966.
4. **Calvert, J. G., Lazrus, A., Kok, G. L., Heikes, B. G., Walega, J. G., Lind, J. and Cantrell, C. A.** Chemical mechanisms of acid generation in the troposphere, *Nature (London),* 317, 27, 1985.
5. **Gilliom, R. D.** *Introduction to Physical Organic Chemistry,* Addison-Wesley, Reading, Mass., 1970 (and references therein).
6. **Agmon, N.,** From energy profiles to structure-reactivity correlations, *Int. J. Chem. Kinet.*, 13, 333, 1981 (and references therein).
7. **Hammett, L. P.** *Physical Organic Chemistry,* McGraw-Hill, New York, 1940.
8. **Hansch, C. and Leo, A. J.** *Substituent Constants for Correlation Analysis in Chemistry and Biology,* John Wiley & Sons, New York, 1979.
9. **Alfrey, T. and Price, C. C.,** Relative reactivities in vinyl copolymerization *J. Polym. Sci.*, 2, 101, 1947.
10. **Cammarata, A. and Yau, S. J.,** A comment on monomer reactivity indices for copolymerization, *J. Polym. Sci.*, 8, (Part A-1), 1303, 1970.

11. **Fisher, T. and Meierhoefer, A. W.,** On the nature of the phenylazo group, *Tetrahedron,* 31, 2019, 1975.
12. **Yamamoto, T. and Otsu, T.,** Effects of substituents in radical reactions: extension of the Hammett equation, *Chem. Ind. (London),* 787, 1967.
13. **Hansch, C. and Kerley, R.,** Interrelation between the extended Hammett equations for correlation of homolytic and heterolytic reactions, *Chem. Ind. (London),* 294, 1969.
14. **Hansch, C.,** The use of homolytic, steric, and hydrophobic constants in a structure-activity study of 1.3-benzodioxole synergists, *J. Med. Chem.,* 11, 920, 1968.
15. **Cammarata, A. and Yau, S. J.** Predictability of correlations between in vitro tetracycline potencies and substituent indices, *J. Med. Chem.,* 13, 93, 1970.
16. **Otsu, T., Ito, T., Fujii, Y., and Imoto, M.,** The reactivities of nuclear-substituted phenyl methacrylates in radical copolymerization with styrene, *Bull. Chem. Soc. Jpn.,* 41, 204, 1968.
17. **Streitwieser, A., Jr. and Perrin, C.,** Acidity of hydrocarbons. XIV. Polarographic reduction of substituted benzyl chlorides and polycyclic arylmethyl chlorides, *J. Am. Chem. Soc.,* 86, 4938, 1964.
18. **Jenkins, A. D.,** The reactivity of polymer radicals, *Adv. Free-Radical Chem.* 2, 139, 1967.
19. **Basco, N., James, D. G. L., and James, F. C.,** A quantitative study of alkyl radical reactions by kinetic spectroscopy. II. Combination of the methyl radical with the oxygen molecule, *Int. J. Chem. Kinet.,* 4, 129, 1972.
20. **Nangia, P. S. and Benson, S. W.,** The kinetics of the interaction of peroxy radicals. II. Primary and secondary alkyl peroxy, *Int. J. Chem. Kinet.,* 12, 43, 1980.
21. **Adachi, H. and Basco, N.,** The reaction of ethylperoxy radicals with NO_2, *Chem. Phys. Lett.,* 67, 324, 1979.
22. **Nangia, P. S. and Benson, S. W.,** The kinetics of the interaction of peroxy radicals. I. The tertiary peroxy radicals, *Int. J. Chem. Kinet.,* 12, 29, 1980.
23. **Adachi, H. and Basco, N.,** The reaction of CH_3O_2 radicals with NO_2, *Int. J. Chem. Kinet.,* 12, 1, 1980.
24. **Eaton, G. R. and Eaton, S. S.,** Relaxation times for the organic radical signal in the EPR spectra of oil shale, shale oil, and spent shale, *J. Magn. Reson.,* 61, 81, 1985.
25. **Kemp, T. J.,** Organic radicals in solids, *Electron Spin Reson.,* 6, 208, 1981.
26. **Kemp, T. J.,** Organic radicals in solids, *Electron Spin Reson.,* 7, 252, 1982.
27. **Freyholdt, T. und Mockel, H. J.,** Liquid-Chromatographische Trennung freier Organischer Radikale, *Fresenius Z. Anal. Chem.,* 305, 405, 1981.
28. **Kurreck, H., Kirste, B., and Lubitz, W.** ENDOR spectroscopy - a promising technique for investigating the structure of organic radicals, *Angew. Chem. Int. Ed. English,* 23, 173, 1984.
29. **Elliot, A. J. and Simsons, A. S.,** Reactions of NO_2 and nitrite ion with organic radicals, *Can. J. Chem.,* 62, 1831, 1984.
30. **Cadogan, J. I. G. and Sadler, I. H.,** Quantitative aspects of radical addition. IV. Rate constant ratios for addition of trichloromethyl and thiyl radicals to olefins, *J. Chem. Soc. (B),* 1191, 1966.
31. **Gilbert, B. C.,** Organic radicals; structure, *Electron Spin Reson.,* 7, 174, 1982 (and references therein).
32. **Sealy, R. C.,** Organic radicals: kinetics and mechanisms of their reactions, *Electron Spin Reson.,* 6, 177, 1981 (and references therein).
33. **Ayscough, P. B.,** Organic radicals: kinetics and mechanisms of their reactions, *Electron Spin Reson.,* 7, 216, 1982 (and references therein).
34. **Nagaoka, T., Okazaki, S., Itoh, T., and Fuginaga, T.,** CEESR method for rapid and precise tracing of organic radicals, *J. Electroanal. Chem.,* 127, 289, 1981.
35. **Bouma, W. J., Macleod, J. K., Nobes, R. H., and Radom, L.,** Unusual gas-phase isomers of simple organic radical cations, *Int. J. Mass Spectrom. Ion Phys.,* 46, 235, 1983.
36. **Maier, J. P.,** Structure and decay of gaseous organic radical cations studied by their radiative decay, exemplified by the 1.3-pentadiyne cation., *Angew. Chem. Int. Ed. English,* 20, 638, 1981.
37. **Agmon, N.,** Generating reaction coordinates by the Pauling relation, *Chem. Phys. Lett.,* 45, 343, 1977.
38. **Agmon, N. and Levine, R. D.,** Empirical triatomic potential energy surfaces defined over othogonal bond order coordinates, *J. Chem. Phys.,* 71, 3034, 1979.
39. **Agmon, N.,** Quantitative Hammond postulate, *J. Chem. Soc. Faraday Trans.* 2, 74, 388, 1978.
40. **Levine, R. D.,** Free energy of activation. Definition, properties, and dependent variables with special reference to "linear" free energy relations, *J. Phys. Chem.,* 83, 159, 1979.
41. **Agmon, N. and Levine, R. D.,** Energy, entropy and the reaction coordinate: thermodynamic-like relations in chemical kinetics, *Chem. Phys. Lett.,* 52, 197, 1977.
42. **Kurz, J. L.,** The relationship of barrier shape to "linear" free energy slopes and curvatures, *Chem. Phys. Lett.,* 57, 243, 1978.
43. **Gaffney, J. S. and Levine, S. Z.,** Reply to comments on predicting gas phase organic molecule reaction rates using linear free energy correlations. I. $O(^3P)$ and OH addition and abstraction reactions *Int. J. Chem. Kinet.,* 12, 767, 1980.

44. **Atkinson, R.**, Comments on the paper "predicting gas phase organic molecule reaction rates using linear free energy correlations. I. O.(^{3}P) and OH addition and abstraction reactions" *Int. J. Chem. Kine.*, 12, 765, 1980.
45. **Atkinson, R. and Pitts, J. N., Jr.**, Absolute rate constants for the reaction of O(^{3}P) atoms with allene, 1.3-butadiene, and vinyl methyl ether over the temperature range 297 to 439 K, *J. Chem. Phys.*, 67, 2492, 1977.
46. **Gaffney, J. S.**, Kinetics, products and mechanism of the gas phase reactions of O(^{3}P) atoms with unsaturated organic compounds. Ph.D. dissertation in chemistry, University of California, Riverside, June 1975, *Diss. Abst. Int. B*, 37(6), 2854-B, 1976.
47. **Gusten, H., Klasinc, L., and Maric, D.**, Prediction of the abiotic degradability of organic compounds in the troposphere, *J. Atmos. Chem.*, 2, 83, 1984.
48. **Heicklen, J.**, The correlation of rate coefficients for H-atom abstraction by OH radicals with C-H bond dissociation enthalpies. *Int. J. Chem. Kine.*, 13, 651, 1981.
49. **Stephens, R. D.**, Quenching of OH(A2$^+$) by alkanes and chlorofluorocarbons at room temperature, *J. Phys. Chem.*, 89, 2630, 1985.
50. **Jolly, G. S., Paraskevopoulos, G., and Singleton, D. L.**, Rates of OH radical reactions. XII. The reactions of OH with c-C_3H_6, c-C_5H_{10} and c-C_7H_{14}. Correlation of hydroxyl rate constants with bond dissociation energies, *Int. J. Chem. Kinet.*, 17, 1, 1985.
51. **Martin, J. P. and Paraskevopoulos, G.**, A kinetic study of the reactions of OH radicals with fluoroethanes. Estimates of C-H bond strengths in fluoroalkanes, *Can. J. Chem.*, 61, 861, 1983.
52. **Truhlar, D.**, Test of bond order methods for predicting the position of the minimum-energy path for hydrogen atom transfer reactions, *J. Am. Chem. Soc.*, 94, 7584, 1972.
53. **Atkinson, R., Aschmann, S. M., Carter, W. P. L., Winer, A. M., and Pitts, J. N., Jr.**, Kinetics of the reactions of OH radicals with n-alkanes at 299 ± 2 K, *Int. J. Chem. Kinet.*, 14, 781, 1982.
54. **Greiner, N. R.**, Hydroxyl radical kinetics by kinetic spectroscopy. VI. Reactions with alkanes in the range 300 to 500 K. *J. Chem. Phys.*, 53, 1070, 1970.
55. **de P. Nicholas, A. M. and Arnold, D. R.**, Thermochemical parameters for organic radicals and radical ions. III. The relationship between bond dissociation enthalpy and radical stability in alkyl systems, *Can. J. Chem.*, 62, 1850, 1984.
56. **de P. Nicholas, A. M., Boyd, R. J., and Arnold, D. R.**, Thermochemical parameters for organic radicals and radical ions. II. The protonation of hydrocarbon radicals in the gas phase. *Can. J. Chem.*, 60, 3011, 1982.
57. **Holmes, J. L. and Lossing, F. P.**, Heats of formation of organic radicals from appearance energies, *Int. J. Mass Spectrom. Ion Phy.*, 58, 113, 1984.
58. **de P. Nicholas, A. M., and Arnold, D. R.**, Thermochemical parameters for organic radicals and radical ions. I. The estimation of the pKa of radical cations based on the thermochemical calculations, *Can. J. Chem.*, 60, 2165, 1982.
59. **Kimura, M., Yamabe, S., and Minato, T.**, Kinetic studies of the oxidation reactions of o-, m-, and p-benzendiols with tris (1,10-phenathroline) iron (III). An estimation of the redox potentials of the organic radicals by application of the Marcus theory, *Bull. Chem. Soc. Jpn.*, 54, 1699, 1981.
60. **Tedder, J. M. and Walton, J. C.**, Directive effects in gas-phase radical addition reactions, *Adv. Phys. Org. Chem.*, 16, 51, 1978 (and references therein).
61. **Failes, R. L., Singleton, D. L., Paraskevopoulos, G., and Irwin, R. S.**, Rate constants for the reaction of ground-state oxygen atoms with methanol from 297 to 544 K, *Int. J. Chem. Kinet.*, 14, 371, 1982.
62. **Audley, G. J., Baulch, D. L., and Campbell, I. M.**, Gas-phase reactions of hydroxyl radicals with aldehydes in flowing $H_2O_2+NO_2+CO$ mixtures. *J. Chem. Soc. Faraday Trans. 1* 77, 2541, 1981.
63. **Nelson, H. H., Pasternack, L., Eyler, J. R., and McDonald, J. R.**, Reactions of C_3 with alkenes, alkynes, and allenes, *Chem. Phys.*, 60, 231, 1981.
64. **Pasternack, L., Pitts, W. M., and McDonald, J. R.**, Temperature dependence of reactions and intersystem crossing of $C_2(a^3II_\mu)$ with hydrogen and small hydrocarbons from 330 to 600 K, *Chem. Phys.*, 57, 19, 1981.
65. **Roscoe, J. M.**, The reactions of O(^{3}P) with the butanols, *Can. J. Chem.*, 61, 2716, 1983.
66. **Atkinson, R., Aschmann, S. M., Winer, A. M., and Pitts, J. N., Jr.**, Atmospheric gas phase loss processes for chlorobenzene, benzotrifluoride, and 4-chlorobenzotrifluoride, and generalization of predictive techniques for atmospheric lifetimes of aromatic compounds *Arch. Environ. Contam. Toxicol.*, 14, 417, 1985.
67. **Cohen, N.**, The use of transition state theory to extrapolate rate coefficients for reaction of OH with alkanes, *Int. J. Chem. Kinet.*, 14, 1339, 1982.
68. **Atkinson, R., Aschmann, S. M., Carter, W. P. L., and Pitts, J. N., Jr.**, Effect of ring strain on gas-phase rate constants. I. Ozone reactions with cycloalkenes, *Int. J. Chem. Kinet.*, 15, 721, 1983.
69. **Sanhueza, E. and Lissi, E.**, On the prediction of reaction rates by simple correlations, *Int. J. Chem. Kinet.*, 13, 317, 1981.

70. **Sanhueza, E. and Lissi, E.**, Unmeasured hydrogen abstraction reactions and rates of potential atmospheric importance; their estimation from related data, *Acta Cient. Venez.*, 6, 445, 1978.
71. **Atkinson, R., Aschmann, S. M., Winer, A. M., and Pitts, J. N., Jr.**, Rate constants for the gas-phase reactions of O_3 with a series of carbonyls at 296 K. *Int. J. Chem. Kinet.*, 13, 1133, 1981.
72. **Gaffney, J. S. and Levine, S. Z.**, Comments on the linear free energy correlation between O_3 and OH addition reactions reported in "rate constants for the gas-phase reactions of O_3 with a series of carbonyls at 296 K", *Int. J. Chem. Kinet.*, 14, 1281, 1982.
73. **Toby, S., Van de Burgt, and Toby, F. S.**, Kinetics and chemiluminescence of ozone-aromatic reactions in the gas phase, *J. Phys. Chem.*, 89, 1982, 1985.
74. **Atkinson, R., Aschmann, S. M., and Carter, W. P. L.**, Effects of ring strain on gas-phase rate constants. II. OH radical reactions with cycloalkenes, *Int. J. Chem. Kinet.*, 15, 1161, 1983.
75. **Otsu, T., Inoue, M., Yamada, B., and Mori, T.**, Structure and reactivity of vinyl monomers: radical reactivities of n-substituted acrylamides and methacrylamides, *Polym. Lett. Ed.*, 13, 505, 1975.
76. **Oishi, T., Maruyama, S., Momoi, M., and Fujimoto, M.**, Syntheses of N-(4-substituted phenyl) isocitraconimides and reactivites in their radical copolymerizations with styrene, *Makromol. Chem.*, 185, 479, 1984.
77. **Patnaik, L. N., Rout, M. K., Rout, S. P., and Rout, A.**, Rate parameters for transfer reactions of some polymer radicals; in terms of Swain and Lupton's F and R and the unique positional weighting factors of Williams and Norrington, *Eur. Polym. J.*, 15, 509, 1979.
78. **Rout, S. P., Rout, A., Singh, B. C., and Rout, M. K.**, Substituent parameters for radical reactions: cerium (IV)-substituted toluenes-acrylonitrile system, *Colloid Polym. Sci.*, 258, 949, 1980.
79. **Fujihara, H., Shindo, T., Yoshihara, M., and Maeshima, T.**, Solvent effects on the radical copolymerizabilities of styrene with *p*-substituted *N,N*-diethylcinnamamides and of p-substituted styrenes with methyl vinyl sulfoxide, *J. Macromol. Sci. Chem.*, A14(7), 1029, 1980.
80. **Asakura, J.**, Solvent effect on the radical copolymerizability of styrene with p-substituted methyl cinnamates, *J. Macromol. Sci. Chem.*, A19(2), 311, 1983.
81. **Schriver, J. and German, A. L.**, Structure-reactivity relations of conjugated and unconjugated monomers: acrylates and methyl vinyl ketone on copolymerization with styrene compared with vinyl esters in copolymerization with ethylene: *J. Polym. Sci. Poly. Chem. Ed.*, 21, 341, 1983.
82. **Zavitsas, A. A. and Pinto, J. A.**, The meaning of the "polar effect" in hydrogen abstractions by free radicals. Reactions of the tert-butoxy radical, *J. Am. Chem. Soc.*, 94, 7390, 1972.
83. **Bartlett, P. D. and Ruchardt, C.**, Peresters. IV. Substituent effects upon the concerted disposition of t-butyl phenylperacetates, *J. Am. Chem. Soc.*, 82, 1756, 1960.
84. **Ito, R., Migita, T., Morikawa, N., and Simamura, O.**, Influence of substituent groups on the arylation of substituted benzenes by aryl radicals derived from *p*-substituted *N*- nitrosoacetanilides, *Tetrahedron*, 21, 955, 1965.
85. **Afanas'ev, I. B., Prigoda, S. V., Mal'tseva, T. Y., and Samokhvalov, G. I.**, Electron transfer reactions between the superoxide ion and quinones, *Int. J. Chem. Kinet.*, 6, 643, 1974.
86. **de P. Nicholas, A. M. and Arnold, D. R.**, Thermochemical parameters for organic radicals and radical ions. IV. The relationship between bond dissociation enthalpy and radical stability in benzenoid systems, *Can. J. Chem.*, 62, 1860, 1984.
87. **Miller, J., Paul, D. B., Wong, L. Y., and Kelso, A. G.**, Phenylation of azobenzene, *J. Chem. Soc. (B)*, 62, 1970.
88. **Russell, G. A. and Williamson, R. C., Jr.**, Directive effects in aliphatic substitutions. XXV. Reactivity of aralkanes, aralkenes and benzylic ethers toward peroxy radicals, *J. Am. Chem. Soc.*, 86, 2364, 1964.
89. **Bandlish, B. K., Garner, A. W., Hodges, M. L., and Timberlake, J. W.**, Substituent effects in radical reactions. III. Thermolysis of substituted phenylazomethanes, 3.5-diphenyl-1-pyrazolines, and azopropanes, *J. Am. Chem. Soc.*, 97, 5856, 1975.
90. **Martin, M. M. and Gleicher, G. J.**, The addition of trichloromethyl radical to ω-phenylalkenes, *J. Am. Chem. Soc.*, 86, 238, 1964.
91. **Sakurai, H., Hayashi, S., and Hosomi, A.**, Relative reactivities in the addition of free trichloromethyl radicals to substituted styrenes. An attempt to separate polar and resonance effects, *Bull. Chem. Soc. Jpn.*, 44, 1945, 1971.
92. **Wayner, D. D. M. and Arnold, D. R.**, Substituent effects on benzyl radical hyperfine coupling constants. II. The effect of sulphur substituents, *Can. J. Chem.*, 62, 1164, 1984.
93. **Leigh, W. J., Arnold, D. R., Humphreys, R. W. R., and Wong, P. C.**, Merostabilization in radical ions, triplets, and biradicals. IV. Substituent effects on the half-wave reduction potentials and n, II* triplet energies of aromatic ketones, *Can. J. Chem.*, 58, 2537, 1980.
94. **Kamphuis, J., Grootenhuis, P. D. J., and Bos, H. J. T.**, Thermal cyclo addition of thiobenzophenone to substituted phenylallenes. Hammett correlation of rates and ionization energies, *Tetrahedron Letters*, 24, 1101, 1983.

95. **Dincturk, S., Jackson, R. A., and Townson, M.,** An improved σ scale. The thermal decomposition of substituted dibenzylmercury compounds in alkane solutions, *J. Chem. Soc. Chem. Commun.*, 172, 1979.
96. **Rout, S. P., Mallick, N., Rout, A., and Rout, M. K.,** Substituent parameters for radical reactions; cerium(IV)-substituted acetophenones-acrylonitrile system, *J. Polym. Sci. Polym. Chem. Ed.*, 17, 3859, 1979.
97. **Fisher, T. H. and Meierhoefer, A. W.,** Substituent effects in free-radical reactions. A study of 4-substituted 3-cyanobenzyl free radicals *J. Org. Chem.*, 43, 225, 1978.
98. **Creary, X.,** Rearrangement of 2-aryl-3,3-dimethylmethylenecyclopropanes. Substituent effects on a non-polar radical-like transition state. *J. Org. Chem.*, 45, 280, 1980.
99. **Martin, M. M. and Gleicher, G. J.,** A. Hammett-Taft study of the addition of the trichloromethyl radical to substituted alkenes. *J. Am. Chem. Soc.*, 86, 242, 1964.
100. **Tanner, D. D., Plambeck, J. A., Reed, D. W., and Mojelsky, T. W.,** Polar radicals. XV. Interpretation of substituent effects on the mechanism of electrolytic reduction of the carbon-halogen bond on series of substituted benzyl halides, *J. Org. Chem.*, 45, 5177, 1980.
101. **Cristol, S. J. and Bindel, T. H.,** Photochemical transformations. XXVI. Sensitized and unsensitized photoreactions of some benzyl chlorides in tert-butyl alcohol, *J. Org. Chem.* 45, 951, 1980.
102. **Russel, G. A. and Williamson, R. C., Jr.,** Nature of the polar effect in reactions in atoms and radicals. II. Reactions of chlorine atoms and peroxy radicals. *J. Am. Chem. Soc.*, 86, 2357, 1964 (and references therein).
103. **Kojima, T., Satouchi, M., and Tanaka, Y.,** Correlation between electron capture response and chemical structure for benzyl chlorides, *Bull. Chem. Soc. Jpn.*, 52, 277, 1979.
104. **Nakamura, M., Ito, O., and Matsuda, M.,** Substituent effect on the rate constants for the reactions between benzenethiyl radicals and stable free radicals estimated by flash photolysis, *J. Am. Chem. Soc.*, 102, 698, 1980.
105. **Blackburn, E. and Tanner, D. D.,** Polar radicals. XIV. On the mechanism of trialkylstannane reductions. Positive ρ values for the tri-n-butylstannane reduction of benzyl halides. A correlation with σ−, *J. Am. Chem. Soc.*, 102, 692, 1980.
106. **Hendry, D. G. and Russel, G. A.,** Solvent effects in the reactions of free radicals and atoms. IX. Effect of solvent polarity on the reactions of peroxy radicals, *J. Am. Chem. Soc.*, 86, 2368, 1964.
107. **Abeywickrema, R. S. amd Della, E. W.,** Substituent effects on electroreduction of the carbon-halogen bond, *Aust. J. Chem.*, 34, 2331, 1981.
108. **Finke, R. G., Schiraldi, D. A., and Hirose, Y.,** $(C_5Me_5)_2UCl$*THF oxidative-additive reactions. II. A kinematic and mechanistic study, *J. Am. Chem. Soc.* 103, 1875, 1981.
109. **Dust, J. M., and Arnold, D. R.,** Substituent effects on benzyl radical ESR hyperfine coupling constants. The σα scale based upon spin delocalization, *J. Am. Chem. Soc.*, 105, 1221, 1983 (and references therein).
110. **Dincturk, S. and Jackson, R. A.,** Free radical reactions in solution. VII. Substituent effects on free radical reactions; comparison of the sigma radical scale with other measures of radical stabilization, *J. Chem. Soc. Perkin Trans. 2*, 1127, 1981.
111. **Heiba, E. I., Dessau, R. M., and Koehl, W. J., Jr.,** Oxidation by metal salts. III. The reaction of manganic acetate with aromatic hydrocarbons and the reactivity of the carboxymethyl radical, *J. Am. Chem. Soc.*, 91, 138, 1969.
112. **Howard, J. A. and Chenier, J. H. B.,** Absolute rate constants for the reaction of tert-butylperoxy radicals with some meta- and para-substituted toluenes, *J. Am. Chem. Soc.*, 95, 3054, 1973.
113. **Pryor, W. A., Davis, W. H., Jr., and Stanley, J. P.,** Polar effects in radical reactions. II. A positive ρ for the reaction of tert-butyl radicals with substituted toluenes, *J. Am. Chem. Soc.*, 95, 4754, 1973.
114. **Pryor, W. A., Echols, Jr., and Smith, K.,** Rates of the reactions of substituted phenyl radicals with hydrogen donors, *J. Am. Chem. Soc.*, 88, 1189, 1966.
115. **Bridger, R. F. and Russell, G. A.,** Directive effects in the attack of phenyl radicals on carbon-hydrogen bonds, *J. Am. Chem. Soc.*, 85, 3754, 1963.
116. **Johnston, K. M. and Williams, G. H.,** Homolytic reactions of aromatic side chains. II. Relative rates of α-hydrogen abstraction by t-butoxy-radicals, *J. Chem. Soc.*, 1446, 1960.
117. **Pryor, W. A., Tonellato, U., Fuller, D. L., and Jumonville, S.,** Relative rate constants for hydrogen abstraction by methyl radicals substituted toluenes, *J. Org. Chem.*, 34, 2018, 1969.
118. **Pearson, R. E. and Martin, J. C.,** The mechanism of benzylic bromination with n-bromosuccinimide. *J. Am. Chem. Soc.*, 85, 354, 1963.
119. **Huyser, E. S.,** Relative reactivities of substituted toluenes toward trichloromethyl radicals, *J. Am. Chem. Soc.*, 82, 394, 1960.
120. **Neale, R. S. and Gross, E.,** The chemistry of nitrogen radicals. VII. The abstraction of hydrogen from substituted toluenes by the piperidinium radical, *J. Am. Chem. Soc.*, 89, 6579, 1967.
121. **Walling, C. and Jacknow, B. B.,** Positive halogen compounds. II. Radical chlorination of substituted hydrocarbons with t-butyl hypochlorite, *J. Am. Chem. Soc.*, 82, 6113, 1960.

122. **Gilliom, R. D. and Ward, B. F., Jr.,** Homolytic abstraction of benzylic hydrogen, *J. Am. Chem. Soc.*, 87, 3944, 1965.
123. **Gleicher, G. J.,** Free-radical hydrogen atom abstractions from substituted cumenes, *J. Org. Chem.*, 33, 332, 1968.
124. **Schuh, H.-H. and Fischer, H.,** The kinetics of the bimolecular gel-reaction of t-butyl radicals in solution. I. Termination rates, *Helv. Chim. Acta*, 61, 2130, 1978.
125. **Walling, C.,** *Free Radicals in Solution*, John Wiley & Sons, New York, 1957.
126. **Hampson, R. F., Jr. and Garvin, D., Eds.,** Reaction Rate and Photochemical Data for Atmospheric Chemistry - 1977, Spec. Publ. No. 513. U. S. Department of Commerce, National Bureau of Standards 1978.
127. **Gaffney, J. S., Fajer, R., Senum, G. I., and Lee, J. H.,** Measurement of the reactivity of OH with methyl nitrate: implications for prediction of alkyl nitrate-OH reaction rates, *Int. J. Chem. Kine.* in press.
128. **Benson, S. W.,** *Thermochemical Kinetics*, 2nd ed. John Wiley & Sons, New York, 1976.

Chapter 9

ACTIVATION ENERGIES FOR METATHESIS REACTIONS OF RADICALS

Tibor Bérces and Ferenc Márta

TABLE OF CONTENTS

I. INTRODUCTION

Much attention has been paid in the last 15 to 20 years to the experimental and theoretical study of elementary free radical reactions and especially to the investigation of metathesis reactions. Expansion and progress in this area are only partly due to the spectacular improvement in experimental techniques (general use of discharge flow and flash photolysis techniques, etc.), and it is particularly stimulated by the ever and more increasing need for kinetic data in practical and technological applications (e.g., industrial processes, pyrolysis and combustion, and atmospheric chemistry and pollution).

Metathesis reactions represent the oldest and largest group of bimolecular chemical reactions. In these processes, an atom is transferred from a stable molecule to an attacking species.

$$A + BC \rightarrow AB + C \qquad (1)$$

Both A and C may be an atom, a free radical, or a molecule. Here we discuss only metathesis reactions in which univalent atoms are abstracted from substrates by small organic radicals. Reactions of atoms and diatomic radicals are treated only to the extent required by a clear and comprehensive presentation of theoretical and experimental aspects of the determination of activation energies for metathesis reactions of organic free radicals. Discussion of the transfer of a multivalent atom or a group, and attack by a stable molecule, is entirely omitted.

The number of quantitative kinetic data on metathesis reactions has increased very rapidly during the past two decades, and a survey of the literature is very difficult today. Fortunately, a number of excellent compilations and a few evaluations have been published and most of these appeared in the 1980s. Of the earlier papers dealing with gas-phase data, the tables of Kondratiev[1] and compilations by Kerr, Trotman-Dickenson, and co-workers[2-5] are mainly noteworthy. The latter also appeared in one volume edited by Kerr and Moss.[6] An evaluation dealt with methyl radical reactions.[7] More recent compilations contain elementary reaction rate data which are of particular importance in special fields. Thus, reactions involved in paraffin pyrolyses[8] and high temperature oxidations,[9] and evaluations of kinetic parameters for metathesis reactions involved in atmospheric chemistry and pollution[10-12] are among the subjects considered. The numerous compilations are supplemented by the two essays written by Kerr.[13,14]

Compilations dealing with kinetic data of metathesis reactions in the liquid phase are scarce. Among the earlier ones, the compilations of Denisov[15] and Hendry et al.[16] should be mentioned, while the tables which appeared in a recent volume[17] of the Landolt-Börnstein series deserve mentioning. Finally, reference is made to the comprehensive review by Ingold.[18] It turns out from these compilations and reviews that reliable kinetic results in liquids are available often at only a single temperature — frequently, room temperature. For such cases, activation energies can be derived merely by using estimated or assumed A factors.

The present essay does not deal with the compilation and evaluation of experimental rate coefficients and Arrhenius parameters for metathesis reactions of organic free radicals. Such kinetic data may be found in the literature quoted above, and are the subject also of the reviews found in those parts of this monography, which are devoted to reactions of particular organic radicals. The aim of the present work is to consider the essential factors that determine the activation energies of atom transfer reactions, and to survey the methods whether theoretical or empirical which are available today for the calculation of the Arrhenius activation energies of metathesis reactions of organic free radicals. Since the overwhelming majority of experimental and theoretical information on activation energies of atom transfer reactions derives from gas kinetics investigations, naturally we place more emphasis in our work on literature sources describing gas-phase studies.

The present review relies first of all on papers published during the last one and half decades. Literature is covered up to the end of 1984 and, in a few cases, reference is made to articles that appeared in 1985.

II. ARRHENIUS ACTIVATION ENERGY

A. Concept of Activation Energy

One of the oldest and most fundamental laws in chemical kinetics is the Arrhenius law:

$$k(T) = A \exp(-E_A/RT) \tag{2}$$

which represents the temperature dependence on the rate coefficient by a simple equation with only two parameters. These are the preexponential factor, A, and the experimental or Arrhenius activation energy, E_A.

The Arrhenius activation energy was introduced originally (and is used today) as an empirical parameter defined by the equation:

$$E_A = -R \frac{d \ln k(T)}{d(1/T)} \tag{3}$$

Many activation energy concepts are used in the literature with meaning similar to that of the Arrhenius activation energy.[19] Their relation to E_A is demonstrated in Figure 1 for an $A + BC \rightarrow AB + C$ type reaction. The height of the potential barrier, V_b, is the difference in potential energy of the saddle point on the potential energy surface and the potential energy of the reactant state A + BC. All molecules possess at least zero-point vibrational energy. The difference between the zero-point level of the activated complex and that of the reactants is designated by $\Delta E_o^\ddagger$ and is related to V by

$$\Delta E_o^\ddagger = V_b + E_{zpe}(ABC^\ddagger) - E_{zpe}(A) - E_{zpe}(BC) \tag{4}$$

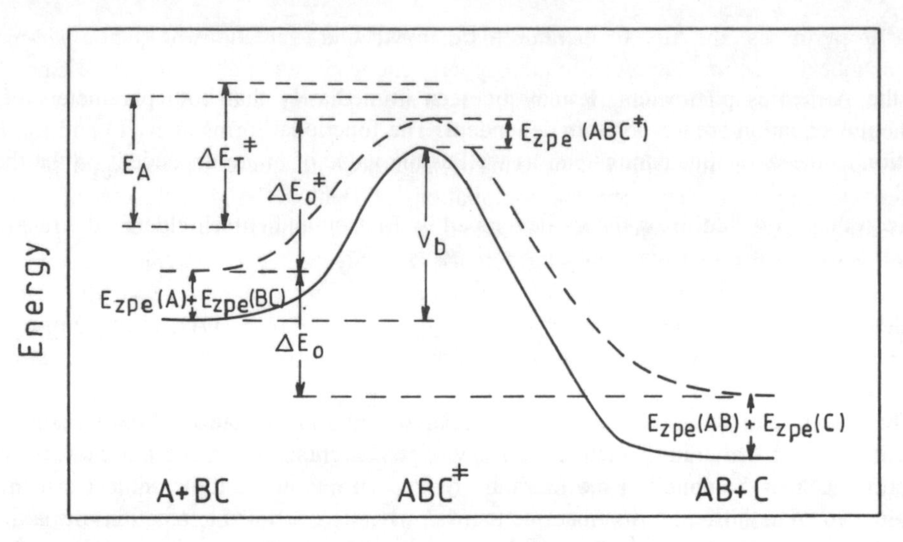

FIGURE 1. Activation energy and barrier height. (For definition of symbols, see text.)

At temperatures above 0 K, molecules have thermal energies which may be obtained from $\int_o^T C_v dT$, provided that the Boltzmann distribution is assumed for the molecules. Thus, the internal energy of activation is obtained by

$$\Delta E_T^{\ddagger} = \Delta E_o^{\ddagger} + \int_o^T C_v(ABC^{\ddagger})\, dT - \int_o^T C_v(A)\, dt - \int_o^T C_v(BC)\, dT \qquad (5)$$

The activation enthalpy for a bimolecular metathesis reaction is

$$\Delta H_T^{\ddagger} = \Delta E_T^{\ddagger} - RT \qquad (6)$$

It is evident from the definitions that the quantities $\Delta E_o^{\ddagger}$, $\Delta E_T^{\ddagger}$ and $\Delta E_T^{\ddagger}$ are closely related to the concept of activation complex and the transition state theory.

The physical meaning of the Arrhenius activation energy and its relationship to the above-introduced energy quantities are best revealed by means of a kinetic theory. Most theoretical approaches yield a rate equation of the type:

$$k(T) = B(T) \exp(-E_B/RT) \qquad (7)$$

where B(T) is a temperature-dependent preexponential factor and E_B is an energy term which is related to the Arrhenius activation energy. The preexponential factor B(T) usually can be given as a product of a temperature-independent factor and a power of the temperature:

$$B(T) = B'T^m \qquad (8)$$

Using the definition of the experimental activation energy, with the assumption of temperature-independent E_B, one obtains from Equations 7 and 8:

$$A = B'(eT)^m \qquad (9)$$

and

$$E_A = E_B + mRT \tag{10}$$

for the Arrhenius parameters. It may be seen immediately that both parameters of the Arrhenius equation are temperature dependent. The functional forms of B (T) and E_B, their relation to the Arrhenius parameters, as well as the value of m are dependent on the theory used.

According to the collision theory developed by Eliason and Hirschfelder,[20] the rate coefficient for a bimolecular reaction can be given as

$$k = \int_o^\infty S(c)\, c\, f(c)\, dc \tag{11}$$

where c and f (c) designate the relative speed of colliding reactants and its distribution function, respectively, and S (c) is the energy-dependent reaction cross section or excitation function. Under the conditions where the energy distribution of the reactants can be approximated by equilibrium distribution, the thermal rate coefficient for a bimolecular metathesis reaction may be given by

$$k(T) = \left(\frac{1}{\pi\mu}\right)^{1/2} \left(\frac{2}{k_B T}\right)^{3/2} \int_o^\infty S(\epsilon)\, \epsilon \exp(-\epsilon/k_B T)\, d\epsilon \tag{12}$$

where ϵ is the relative kinetic energy of the interacting reactants, $S(\epsilon)$ designates the thermal average reaction cross section,* μ stands for the reduced mass of the reactants, and k_B is the Boltzmann constant. Combining the rate expression (Equation 12) with the definition of the Arrhenius activation energy given by Equation 3, one obtains:

$$E_A = \left[\frac{\int_o^\infty S(\epsilon)\, \epsilon^2 \exp(-\epsilon/k_B T)\, d\epsilon}{\int_o^\infty S(\epsilon)\, \epsilon \exp(-\epsilon/k_B T)\, d\epsilon} - \frac{3}{2} k_B T\right] N_A \tag{13}$$

This equation was first derived by Tolman.[21] It reveals the physical meaning behind the concept of the Arrhenius activation energy. The first term on the right-hand side is the average energy of the reactive collisions, while the second term is the average energy of all collisions:

$$E_A = \langle E^* \rangle - \langle E \rangle \tag{14}$$

Interpretation of the Arrhenius activation energy as a difference of the energy of collisions resulting in chemical reaction and the energy of all collisions immediately shows the activation energy to be temperature dependent. This temperature dependence is discussed in detail in the next section.

The rate coefficient expression (Equation 12) and the activation energy expression (Equation 13) are general equations of the collision theory which involve no restrictions with

* $S(\epsilon)$ is an average of the state-selected cross section $S(\epsilon, i, j)$ taken according to the equilibrium distribution of the internal quantum states of the reactants. Since internal state distribution is a function of the temperature, $S(\epsilon)$ is also temperature dependent. For the sake of simplicity, $S(\epsilon)$ is regarded temperature independent in the discussion that follows.

respect to the excitation function. A widely used functional form of S (ϵ) manifests itself in the so-called "line-of-centers" hard-spheres collision model:

$$S(\epsilon) = 0 \qquad \text{if} \quad \epsilon \leq \epsilon_c$$

$$S(\epsilon) = \sigma \frac{\epsilon - \epsilon_c}{\epsilon} \qquad \text{if} \quad \epsilon > \epsilon_c \tag{15}$$

where σ is the hard-spheres collision cross section and ϵ_c is the threshold energy, i.e., the minimum relative kinetic energy of the reactants required in order to react.[22] With this reaction cross section model, the rate coefficient expression obtained from Equation 12 is

$$k(T) = \left(\frac{8k_BT}{\pi\mu}\right)^{1/2} \sigma \exp(-\epsilon_c/k_BT) \tag{16}$$

and, accordingly,

$$E_A = E_c + \frac{1}{2} RT \tag{17}$$

This equation is to be compared with Equation 10.

Transition state theory has been used frequently for the evaluation and interpretation of kinetic data of metathesis reactions. According to the thermodynamic formulation of the fundamental transition state theory rate equation:

$$k(T) = \frac{k_BT}{h} (R'T) \exp(\Delta S^{\ddagger}/R) \exp(-\Delta H^{\ddagger}/RT) \tag{18}$$

where $\Delta S^{\ddagger}$ and $\Delta H^{\ddagger}$ are the activation entropy and enthalpy of the reaction, respectively, and the term (R′ T) arises from the conversion of the equilibrium constant K_c to K_p ($R' = 0.082\ dm^3\ mol^{-1}\ atm\ K^{-1}$). Hence, one derives

$$E_A = \Delta H^{\ddagger} + 2RT \tag{19}$$

provided that $\Delta S^{\ddagger}$ and $\Delta H^{\ddagger}$ are assumed to be temperature independent. However, such an assumption is not really justified.[23] Simple thermodynamics gives:

$$\frac{d\,\Delta H^{\ddagger}}{dT} = \Delta C_p^{\ddagger} \tag{20}$$

where $\Delta C_p^{\ddagger}$ is the heat capacity of activation. A treatment which takes into account the temperature dependence of $\Delta H^{\ddagger}$ has been described by Heneghan et al.[24] The simple considerations presented here, as well as the more detailed treatments given in Section II.B show clearly that transition state theory predicts temperature dependence for the Arrhenius activation energy.

Finally, it can be stated that no matter which theory is considered, the Arrhenius activation energy (as well as the A factor) appears to be temperature dependent, giving rise to deviations from the simple Arrhenius law.

B. Non-Arrhenius Behavior and Temperature Dependence of the Experimental Activation Energy

Introduction of the concept of the experimental activation energy and consideration of its physical meaning (see preceding section) revealed that E_A defined by Equation 3 may be temperature dependent. In this section, we attempt to answer the question how common non-Arrhenius behavior is among metathesis reactions, and consider possible causes of the observed deviations from the simple Arrhenius equation.

Reliable knowledge of the temperature dependence of the reaction rate is especially important for modeling studies where required rate data are often estimated by extrapolation of measurements made at considerably different temperatures.

Non-Arrhenius behavior in various types of reactions has been discussed by Hulett[25] and Perlmutter-Hayman,[26] while a number of more recent reviews deal with the temperature dependence of the rates of bimolecular elementary reactions.[27-30]

Curiously enough, the first well-documented example of non-Arrhenius behavior[31] was the inversion of cane sugar, which itself played an essential role in the formulation and establishment of the Arrhenius equation.[32] Hitherto, convincing evidence has accumulated for a number of elementary reactions in both the gas and condensed phases which proves that there are significant deviations from the simple Arrhenius temperature dependence. This is mainly the consequence of the progress made in the development of new experimental techniques which rendered possible considerable improvement in the accuracy and precision of rate measurements and the extension of the accessible temperature range. Today, several reaction families are known to exhibit non-Arrhenius behavior.[26] The more important types are (1) composite reactions, e.g., processes consisting of concurrent reaction paths, chemical changes preceded by preequilibrium, and reactions proceeding via unstable intermediates; (2) unimolecular isomerizations, dissociations, and bimolecular combinations where chemical reaction is influenced by the rate of energy transfer; (3) solvolysis reactions; and (4) bimolecular exchange reactions.

Bimolecular exchange reactions showing non-Arrhenius behavior, which are our concern here, may be divided into three categories according to the functional form of temperature dependence of the rate coefficient.[9]

Reactions with temperature dependence of the type $k(T) = aT^b \exp(-c/T)$ belong to the first category. Most metathesis reactions show this type of non-Arrhenius behavior. Two well-known members of the category are[27]

$$OH + H_2 \rightarrow H_2O + H$$

$$CH_3 + C_2H_6 \rightarrow CH_4 + C_2H_5$$

Temperature dependence of the form $k(T) = \exp(a + bT)$ represents the second category; the best-known examples are[28]

$$OH + CO \rightarrow CO_2 + H$$

Table 1
GAS-PHASE METATHESIS REACTIONS CHARACTERIZED BY NON-ARRHENIUS BEHAVIOR

No.	Reaction	Temperature range (K)	E_a/R at T_{low} (K^{-1})	E_A/R at T_{high} (K^{-1})	Ref.
1	$D + H_2 \rightarrow HD + H$	250—2400	3.4 (300 K)	7.0 (2400 K)	34
2	$H + CH_4 \rightarrow H_2 + CH_3$	400—1600	5.4 (400 K)	9.2 (1600 K)	35
3	$H + C_2H_6 \rightarrow H_2 + C_2H_5$	360—1400	3.9 (360 K)	7.5 (1400 K)	35
4	$Cl + CH_4 \rightarrow HCl + CH_3$	200—685	1.2 (250 K)	1.6 (400 K)	37
5	$OH + H_2 \rightarrow H_2O + H$	250—2000	1.9 (200 K)	3.8 (1050 K)	41
6	$OH + CH_4 \rightarrow H_2O + CH_3$	298—1300	1.9 (300 K)	3.9 (1300 K)	42
7	$OH + C_2H_6 \rightarrow H_2O + C_2H_5$	248—472	0.9 (300 K)	1.3 (500 K)	44
8	$CH_3 + H_2 \rightarrow CH_4 + H$	372—1370	5.6 (400 K)	7.5 (1350 K)	45
9	$CH_3 + C_2H_6 \rightarrow CH_4 + C_2H_5$	390—1350	5.2 (480 K)	10.8 (980 K)	47
10	$CH_3 + neo\text{–}C_5H_{12} \rightarrow CH_4 + neo\text{–}C_5H_{11}$	404—953	5.2 (460 K)	8.1 (870 K)	53

$$OH + OH \rightarrow H_2O + O$$

Finally, there are reactions with zero or small negative temperature dependence, e.g.,[29]

$$HO_2 + NO \rightarrow OH + NO_2$$

$$O + OH \rightarrow O_2 + H$$

Our following discussion is restricted to metathesis reactions generally obeying the first type of temperature dependence.

1. Experimental Evidence for Non-Arrhenius Behavior in Metathesis Reactions

A compilation of some gas-phase metathesis reactions for which strong evidence of Arrhenius plot curvature exists is presented in Table 1.

Nonlinear Arrhenius plots were found experimentally[33] for reaction 1, however, more detailed information on the temperature dependence of the activation energy is available from theoretical studies. In their quasiclassical trajectory calculations and improved canonical variational theory studies, Blais et al.[34] found an increase of the activation energy from 28.5 to 58.5 kJ/mol in the temperature range 300 to 2400 K.

Clark and Dove found[35] that the curved Arrhenius plots calculated by the bond-energy-bond-order (BEBO) version of the transition state theory (see Section III.C) very well described all available data on the temperature dependence of rate coefficients of reaction 2 and 3.

Temperature dependence of reaction 4 was investigated between 200 and 685 K using a variety of experimental methods including the direct high-precision techniques of flash photolysis-resonance fluorescence,[36,37] discharge flow-resonance fluorescence,[38,39] and very low-pressure reactor (VLPR)[40] technique. In these studies, curvature in the Arrhenius plot has been established.

Again, a number of investigations prove the non-Arrhenius character of reaction 5. Flash photolysis-resonance fluorescence measurements,[41] carried out in a very wide temperature range of 250 to 1050 K, suggested that the temperature dependence of the rate coefficient is best described by the three-parameter function $k(T) = aT^b\exp(-c/T)$. The equation $k(T) = 1.66 \times 10^{-16} T^{1.6} \exp(-1660/T)$ $cm^3molecule^{-1} sec^{-1}$ suggested by Zellner,[28] appears to provide the best fit to all published kinetic data in the temperature range 250 to 2000 K (see also Figure 14 in Reference 29). Similarly to Reaction 5, the existence of curvature in the Arrhenius plot of OH + CH_4 reaction is also clearly demonstrated,[28,42,43] and recent experimental observations[44] indicate that a three-parameter function is required to fit the temperature dependence of the specific rate of reaction OH + C_2H_6, also.

Kobrinsky and Pacey determined[45] the rate coefficient for metathesis reaction CH_3 + H_2 in pyrolysis experiments in a flow system at 926 and 829 K. These authors found the rate coefficient at 829 K to be 3 to 4 times as high as the value predicted from Arrhenius straight-line extrapolation based on low-temperature results in the 370 to 600 K range. The high-temperature Arrhenius parameters were also significantly higher than those determined at low temperatures. The temperature dependence of the rate coefficient in the temperature range 372 to 1370 K was best described by the three-parameter function $k(T) = 1.32 \times 10^{-18} T^2 \exp(-4810/T)$ cm^3 molecule^{-1} sec^{-1}. Shock tube study of the CH_3 + H_2 reaction at 1340 K by Clark and Dove[46] supplied a rate coefficient considerably higher than the value expected by a straight-line Arrhenius extrapolation of the low-temperature measurements. All these results proved that the CH_3 + H_2 metathesis reaction shows significant deviations from the Arrhenius equation.

The hydrogen atom transfer reaction $CH_3 + C_2H_6 \rightarrow CH_4 + C_2H_5$ has been studied widely. Most of the studies were carried out in the temperature range 300 to 600 K and provided Arrhenius parameters in the range E_A = 30 to 50 kJ/mol and log (A/dm^3 mol^{-1} sec^{-1}) = 8 to 9. In contrast to this, the high-temperature alkane pyrolysis studies around 1000 K[47] and shock tube studies around 1350 K[48] suggest higher kinetic parameters of about[47] E_A = 90 ± 20 kJ/mol and log (A/dm^3 mol^{-1} sec^{-1}) = 11.7 ± 1.0. All available data, which cover a temperature range of nearly 1000 K, fall on a curve, indicating a non-Arrhenius behavior. Clark and Dove[35] came to a similar conclusion, and their experimentally observed curved Arrhenius plot agreed with predictions derived from the BEBO method. The calculated BEBO curve was described by the equation $k(T) = 6.40 \times 10^{-23} T^{4.0} \exp(-4170/T)$ cm^3 molecule^{-1} sec^{-1}.

The metathesis reaction between methyl radical and neopentane has been studied in detail at both low and high temperatures; thus, an extensive set of data is available today from 404 up to 953 K. The low-temperature results[49-52] in the temperature range 400 to 600 K yield activation energies of 42 to 48 kJ/mol and log(A/dm^3 mol^{-1} sec^{-1}) of 8.0 to 9.3, with some indication of increase in the Arrhenius parameters with increasing temperature. On the other hand, E_A = 67 kJ/mol and log(A/dm^3 mol^{-1} sec^{-1}) = 10.5 was obtained by Pacey[53] in neopentane pyrolysis experiments around 900 K. Linear extrapolation of the low-temperature data yields a rate coefficient at 900 K which is almost an order of magnitude lower than the directly measured value. Considering all measurements available in the temperature range of 404 to 953 K, one finds strong curvature in the Arrhenius plot which can be taken as convincing evidence of non-Arrhenius behavior, in spite of the fact that measurements from several laboratories had to be utilized in order to span the 500 K range.

Curved Arrhenius plots have been recognized in a number of other metathesis reactions between alkyl radicals and H_2 or alkanes as, e.g., for the CH_3 + C_3H_8 reaction[54,55] for reactions CH_3 + n − C_4H_{10},[56] CH_3 + i − C_4H_{10},[57,58] and C_2H_5 + n − C_4H_{10}.[56] Deviation from the Arrhenius equation in some of these studies is more convincing, and in others, a final conclusion can be drawn after the completion of further investigations at higher temperatures. Nevertheless, non-Arrhenius behavior is considered a common symptom. Pacey and Purnell[47,56] stated that the temperature dependence of the Arrhenius parameters is probably a general behavior of the alkyl radical-alkane metathetical reactions, and suggested that this may be a characteristic feature of many metathetical reactions.

There are several well-established examples known in liquid-phase kinetics for the deviation from the Arrhenius equation. We would only refer to the investigation of Larson and Gilliom,[59] who studied the metathesis reaction of *tert*-butoxy radicals with toluene in chlorobenzene solution. The Arrhenius plot of the k_H/k_D isotope effect in the range 212 to 353 K showed significant curvature; dramatic nonlinearity was found below 273 K. These authors explained their observations by the quantum mechanical tunnel effect (see below).

There is abundant evidence available for non-Arrhenius behavior among metathesis reactions of atoms and diatomic free radicals, e.g., OH, however, so far only a few really well-established examples are known for curved Arrhenius plots in the reactions of alkyl and other organic free radicals. This difference is understood if the usual experimental techniques used in these investigations are considered. In the investigation of atom and OH radical reactions, sensitive absolute methods are commonly applied which often can span a wide temperature range. Today, however, absolute methods are seldom used in studies of organic radical reactions, and an Arrhenius plot extending over a wide temperature range has to be constructed from measurements carried out in different laboratories often with different techniques affording only moderately consistent results.

2. Essential Factors Responsible for Non-Arrhenius Behavior

There are several possible explanations for the Arrhenius graph curvature of bimolecular metathesis reactions. We have seen in Section II.A that, from a theoretical viewpoint, Arrhenius activation energy is expected to depend on the temperature. Apart from the intrinsic factors which predetermine a non-Arrhenius behavior for practically all atom transfer reactions, occasionally a number of other causes, which could be designated exterior factors, may be responsible for deviations from the simple Arrhenius temperature dependence of the rate coefficient. Below we discuss briefly five major causes: (1) change in mechanism with increasing temperature; (2) nonequilibrium effects; (3) intrinsic factors related to the type of excitation function; (4) intrinsic factors related to the heat capacity of activation; and (5) quantum mechanical tunneling.

Naturally, Arrhenius plot curvature is expected when the mechanism of the reaction changes with the temperature:

1. In the simplest case, this occurs if two or more parallel reactions, characterized by different Arrhenius parameters, contribute. Thus, e.g., Hay[60] explained curvatures in the Arrhenius plots by postulating the simultaneous existence of planar and pyramidal forms of alkyl radicals with different reactivities whose relative contributions changed strongly with the temperature.
2. In a broader sense, one may reckon as parallel routes the elementary reactions in which different vibrational-rotational quantum states of the reactants are involved. The relative importance and, therefore, the contribution to the reaction of various internal states of different reactivity may change with temperature thereby causing curvature in the Arrhenius plot.
3. Finally, non-Arrhenius behavior, is expected if more than one transition state with significantly different characteristics contributes to the reaction.

It is a precondition of Arrhenius-type temperature dependence for rate coefficients that the energy distributions of the reactants correspond to thermal equilibrium. Clark et al.[30] investigated conditions which may lead to nonequilibrium internal energy distributions of the reactants and studied the extent of perturbation. They found indications that substantial nonequilibrium effects are uncommon among atom transfer reactions. Significant deviations from the Arrhenius temperature dependence can be expected to occur only when the degree of nonequilibrium behavior is itself temperature dependent. Well-established evidence of nonequilibrium effects in metathesis reactions have not been reported so far.

Some of the causes which result in non-Arrhenius behavior belong to the intrinsic characteristics of elementary metathetical reactions. As these effects are intimately associated with the dynamics of the particular elementary reaction, such questions are best discussed in terms of chemical kinetic theories of the elementary reactions, i.e., in terms of collision theory, transition state theory, etc. This is done in the next three sections.

a. Explanation of Non-Arrhenius Behavior in Terms of Collision Theory

In the discussion of the concept of Arrhenius activation energy, we have seen that the collision theory rate coefficient expression for Maxwell-Boltzmann energy distribution of the reactants is given by Equation 12. Making use of this equation, we may easily demonstrate the temperature dependents of the Arrhenius activation energy or the extent of non-Arrhenius behavior, provided the reaction cross section function, $S(\epsilon)$, is known. However, only in exceptional cases is $S(\epsilon)$ available from independent sources. This is why sometimes the opposite happens, i.e., the $S(\epsilon)$ function is derived by inverse Laplace transformation[61,62] using the observed temperature dependence of the rate coefficient. Even if the real reaction cross section is not known, one can assume various functional forms for $S(\epsilon)$ and inquire how the extent of non-Arrhenius behavior depends on the excitation function. Exactly this was done by Menzinger and Wolfgang[63] and by LeRoy[64] who investigated a variety of functional forms for $S(\epsilon)$. Among the excitation functions studied, we shall discuss briefly those two types which may be of major importance for atom transfer reactions between neutral species (metathesis reactions) characterized by a definite threshold energy.

The class I reaction cross section model:

$$
\begin{aligned}
S(\epsilon) &= 0 && \text{if } \epsilon < \epsilon_c \\
S(\epsilon) &= C(\epsilon - \epsilon_c)^n \exp - m(\epsilon - \epsilon_c) && \text{if } \epsilon \geqslant \epsilon_c
\end{aligned}
\tag{21}
$$

where C is a proportionality factor, n and $m \geqslant 0$ are empirical parameters, and ϵ_c designates the threshold energy. This general model was dealt with by LeRoy,[64] the special case of $m = 0$ was studied by Menzinger and Wolfgang,[63] and the simple limiting case of $n = m = 0$ with $C = \sigma$ (where σ is the hard-sphere collision cross section) yields the hard-sphere reaction model. The latter assumes that the energy requirement of the reaction has to be provided as relative kinetic energy. Substituting the function Equation 21 into the fundamental Equation 12 and performing the integration we obtain the bimolecular rate coefficient expression, which gives — by means of the experimental activation energy definition (Equation 3) — the expression:

$$
E_A = E_c + \left(n + \frac{1}{2}\right) RT - \frac{m(n + 2)\,(RT)^2}{1 + mRT} - \frac{E_c}{(n + 1) + (1 + mRT)\,E_c/RT} \tag{22}
$$

The class II reaction cross section model:

$$
\begin{aligned}
S(\epsilon) &= 0 \\
S(\epsilon) &= C\,\frac{(\epsilon - \epsilon_c)^n}{\epsilon} \exp - m(\epsilon - \epsilon_c)
\end{aligned}
\tag{23}
$$

where $n, m \geqslant 0$, and other designations are the same as in the class I model. (For the discussion of this general form of the excitation function, see Reference.[64]) A study of the special cases $m = 0$, $n = 1$ as well as the $m = 0$, $n = {}^1/_2$ is described in the paper of Menzinger and Wolfgang.[63] The former (namely, when $m = 0$, $n = 1$, and $C = \sigma$) represents the very well-known line-of-centers hard-sphere reaction model (see Equation 15 and attached discussion in Section II.A), while the latter (when m = 0 and n = $^1/_2$) leads to a rate coefficient expression of the same form as the Arrhenius equation. Again, after substitution of the excitation function in the fundamental collision theory equation of k(T), integration gives:

$$E_A = E_c + \left(n - \frac{1}{2}\right) RT - \frac{m(n + 1)(RT)^2}{1 + mRT} \tag{24}$$

Provided that the excitation functions of atom transfer reactions resemble the functional forms of Equations 21 or 23, it is immediately evident from Equations 22 and 24 that temperature-dependent activation energy and non-Arrhenius behavior should be considered as the natural conditions for metathesis reactions. The equations can be further simplified if some limiting cases are considered.

If the temperature is high enough or if m is sufficiently large, the threshold energy will be higher than the Arrhenius activation energy ($E_A < E_c$) for both class I and II reactions, and the activation energy decreases with increasing temperature. In this limit:

$$E_A \simeq E_c - \frac{3}{2} RT \tag{25}$$

and E_A becomes negative when $\frac{E_c}{RT} < \frac{3}{2}$. The negative activation energy indicates that the average relative translational energy of the reactants exceeds the threshold energy.

On the other hand, if the temperature is low enough, $\frac{E_c}{RT} \gg \frac{3}{2}$ and $n > \frac{1}{2}$, the Arrhenius activation energy exceeds the threshold energy ($E_A > E_c$) for both class I and II reactions, and the activation energy increases with increasing temperature:

$$E_A \simeq E_c + \left(n - \frac{1}{2}\right) RT \tag{26}$$

The temperature effect is greater for cases characterized with higher n values. Temperature-independent activation energy and Arrhenius-type behavior can only be expected for class II reaction cross sections in the special case when n = $^1/_2$ and m = 0.

Typical variations of the Arrhenius activation energy with reaction temperature for different functional forms of the reaction cross section may be found in the literature.[40,63] All such explanations based on the collision theory attribute the non-Arrhenius behavior to the different temperature dependence of the average energy of reactive collisions and that of all collisions (see Tolman's interpretation of the concept of Arrhenius activation energy in Section II.A).

b. Explanation of Non-Arrhenius Behavior in Terms of Transition State Theory

The transition state theory (TST) provides a clear and simple description of the rates of bimolecular atom transfer reactions; therefore, TST is used widely in understanding and interpreting the temperature dependence of the rate coefficient for these elementary reactions.[26,27,29,35,65] The fundamental TST rate coefficient expression for a bimolecular reaction of the type:

$$A + BC \rightarrow [ABC^{\ddagger}] \rightarrow AB + C$$

where A may designate an atom or a polyatomic free radical and BC may be a diatomic or a more complex molecule is given* by

$$k(T) = \frac{k_B T}{h} \frac{Q^{\ddagger}_{ABC}}{Q_A\, Q_{BC}} \exp(-\Delta\epsilon^{\ddagger}_o/k_B T) \tag{27}$$

In Equation 27, Qs are the unit volume partition functions (with $Q^{\ddagger}_{ABC}$ not including the contribution of the degree of freedom which leads to decomposition of the complex) and $\Delta\epsilon^{\ddagger}_o$ denotes the difference in zero-point energies of the complex and the reactants. The partition functions can be factored into contributions from translation, external rotation, and vibration.** Each of these is temperature dependent. The temperature dependence per translational degree of freedom is $T^{1/2}$ and we shall assume the same for external rotational degrees of freedom (i.e., we treat these classically, which is usually a reasonably good approximation at room temperature and above). The approximations then yield:

$$k(T) = \frac{k_B T}{h} C(T^{1/2})^{(\Delta r^{\ddagger} - 3)} \frac{Q^{\ddagger}_{vib,ABC}}{Q_{vib,A}\, Q_{vib,BC}} \exp(-\Delta\epsilon^{\ddagger}_o/k_B T) \tag{28}$$

In the derivation of Equation 28 it has been taken into account that on forming the activated complex from the reactants, three translational degrees of freedom are lost. The change in the number of rotational degrees of freedom has been designated $\Delta r^{\ddagger} = r^{\ddagger}_{ABC} - r_A - r_{BC}$. One internal mode is considered the reaction coordinate, and thus the change in the number of vibrations, $\Delta v^{\ddagger} = v^{\ddagger}_{ABC} - v_A - v_{BC}$, is given as $\Delta v^{\ddagger} = 2 - \Delta r^{\ddagger}$.

Further simplification of Equation 28 is not easy since this requires detailed and precise knowledge of the vibrational frequencies of the reactants and of the complex which is often not available. This is a serious shortcoming, all the more so since the extent of curvature of the Arrhenius plot is known[30,65] to depend very strongly on the activated complex frequencies. Under such circumstances, one has to be satisfied with general conclusions on the expected tendencies and with approximate predictions.

Using the harmonic oscillator approximation† for the vibrational mode of ν_i frequency:

$$Q_{vib} = [1 - \exp(-u_i)]^{-1} \tag{29}$$

where $u_i = h\nu_i/k_B T$, we can obtain the derivative of the vibrational partition function ratio required in the calculation of the activation energy:

$$\frac{d}{d(1/T)} \ln\left\{\frac{Q^{\ddagger}_{vib,ABC}}{Q_{vib,A}\, Q_{vib,BC}}\right\} = T \sum_i s_i \frac{u_i}{\exp(u_i) - 1} \tag{30}$$

Here summation is taken over all vibrational modes of the complex and reactants, however, $s_i = 1$ and $s_i = -1$ should be used for the complex and reactants, respectively. Finally, by operating the activation energy definition (Equation 3) on Equation 28 and taking into account Equation 30, we have:

* In this treatment, we neglect tunneling. Quantum mechanical tunneling as a potential cause of non-Arrhenius behavior is discussed in the succeeding section.

** In this treatment, for the sake of simplicity, we have assumed that neither the complex nor the reactants have internal rotations. If interal rotations exist, a somewhat modified analysis is required, but the major conclusion regarding the temperature dependence remains unchanged.

† Anharmonicity may become significant in the transition state, and this may be an additional cause of non-Arrhenius behavior.[66]

Table 2
THE $\Delta r^\ddagger$ VALUES FOR SOME TYPICAL METATHESIS REACTIONS

Reactant A	Reactant BC	ABC* complex	$\Delta r^\ddagger$
Atom	Diatomic	Linear	0
Diatomic	Diatomic	Bent	−1
Bent polyatomic	Diatomic	Bent	−2
Bent polyatomic	Bent polyatomic	Bent	−3

$$E_A = \Delta E_o^\ddagger + RT\left[\frac{\Delta r^\ddagger - 1}{2} + \sum_i s_i \frac{u_i}{\exp(u_i) - 1}\right] \tag{31}$$

Comparison with Equation 10 indicates that the parameters of the frequently used rate coefficient functions (Equations 7 and 8) are in terms of the TST:

$$E_B = \Delta E_o^\ddagger \tag{32a}$$

and

$$m = \frac{\Delta r^\ddagger - 1}{2} + \sum_i s_i \frac{u_i}{\exp(u_i) - 1} \tag{32b}$$

Since a given reaction $\Delta E_o^\ddagger$ has a given constant value, it follows from Equation 31 that the temperature dependence of the Arrhenius activation energy is a function of three types of parameters: (1) the change in the number of rotational degrees of freedom on forming the activated complex; (2) the vibrational frequencies of the transition complex corresponding to the new vibrational modes; and (3) the reaction temperature.

The value of $\Delta r^\ddagger$ depends on the structure of the reactants and the activated complex. Some typical cases are represented in Table 2. These data show that $\Delta r^\ddagger$ may have very different values. For metathesis reactions of organic free radicals with di- and polyatomic molecules, $\Delta r^\ddagger$ is -2 and -3, respectively. Consequently, for these classes of reactions, the respective values of the term $RT(\Delta r^\ddagger - 1)/2$ on the right-hand side of Equation 31 at room temperature can be predicted to be -3.7 and -4.9 kJ/mol.

The third term on the right-hand side of Equation 31 representing the vibrational contribution, is more difficult to evaluate since the temperature dependence of the vibrational partition function per one vibrational degree of freedom changes between T^0 and T^1 depending on the extent of vibrational excitation. Thus, we discuss here only two limiting cases.

1. At low temperature, all vibrational partition functions $Q_{vib} \rightarrow 1$ and all u_i values become large. Thus, the last term on the right-hand side of Equation 31 is negligible, and

$$E_A \simeq \Delta E_o^\ddagger - \frac{1 - \Delta r^\ddagger}{2} RT \tag{33}$$

Hence, we obtain:

$$E_A \simeq \Delta E_o^\ddagger - \frac{3}{2} RT \tag{33a}$$

and

$$E_A \simeq \Delta E_o^\ddagger - 2RT \tag{33b}$$

for metathesis reactions of polyatomic organic free radicals with di- and polyatomic molecules, respectively (assuming bent transition state). Equations 33a and 33b show that the Arrhenius activation energy is always less than $\Delta E_o^\ddagger$; however, since T is small, the difference is usually not significant. On the other hand, E_A decreases with increasing temperature and at sufficiently high temperature and, in the case of small potential barrier, E_A may become negative.

2. At high temperature, vibrational partition functions and u_i values become small and $u_i/[\exp(u_i) - 1] \simeq 1$. Thus, the value of the third term in Equation 31 depends on the change in the number of vibration modes on formation of the transition complex, i.e., $\Delta v^\ddagger = 2 - \Delta r^\ddagger$. Consequently,

$$E_A \simeq \Delta E_o^\ddagger + \frac{3 - \Delta r^\ddagger}{2} RT \tag{34}$$

Hence, we obtain:

$$E_A \simeq \Delta E_o^\ddagger + \frac{5}{2} RT \tag{34a}$$

and

$$E_A \simeq \Delta E_o^\ddagger + 3RT \tag{34b}$$

for reactions of organic free radicals with di- and polyatomic molecules, respectively, via bent transition complexes. At high enough temperature, the Arrhenius activation energy is always higher than $\Delta E_o^\ddagger$, and E_A increases with increasing temperature.

At intermediate temperatures, there is a gradual switch-over from low- to high-temperature behavior. The temperature range of switch-over depends on the frequencies of the activated complex. It has been demonstrated[30] via TST calculations that if complex frequencies are lower than vibrational frequencies of the reactants, then the increase in E_A starts at lower temperatures (usually around or below room temperature) and is more gradual. It has also been established[30] that a more pronounced non-Arrhenius behavior is predicted for reactions characterized by more negative $\Delta r^\ddagger$ values. This is the case for metathesis reactions of organic free radicals with polyatomic molecules.

TST often was found to account quantitatively for non-Arrhenius behavior of metathesis reactions. In such TST studies, empirical or semiempirical potential energy surfaces usually were employed; e.g., London-Eyring-Polanyi-Sato (LEPS) surfaces,[67] BEBO functions of Johnston,[65] or bond-strength-bond-length (BSBL) potential energy functions of Bérces and Dombi[68] were used frequently. Typical examples of TST calculations with experimental data have been given, e.g., in References 28 to 30, 35, 46, 69, and 70 and are presented for reactions $CH_3 + H_2 \rightarrow CH_4 + H$ and $CH_3 + C_2H_6 \rightarrow CH_4 + C_2H_5$ in Figures 2 and 3, respectively.

c. Quantum Mechanical Tunneling as a Cause of Arrhenius Plot Curvature

According to classical mechanical theories of rate processes, chemical reactions occur only if the energy of the reacting system is greater than the height of the potential barrier. In contrast to this, quantum mechanics predicts a finite probability for passing through the

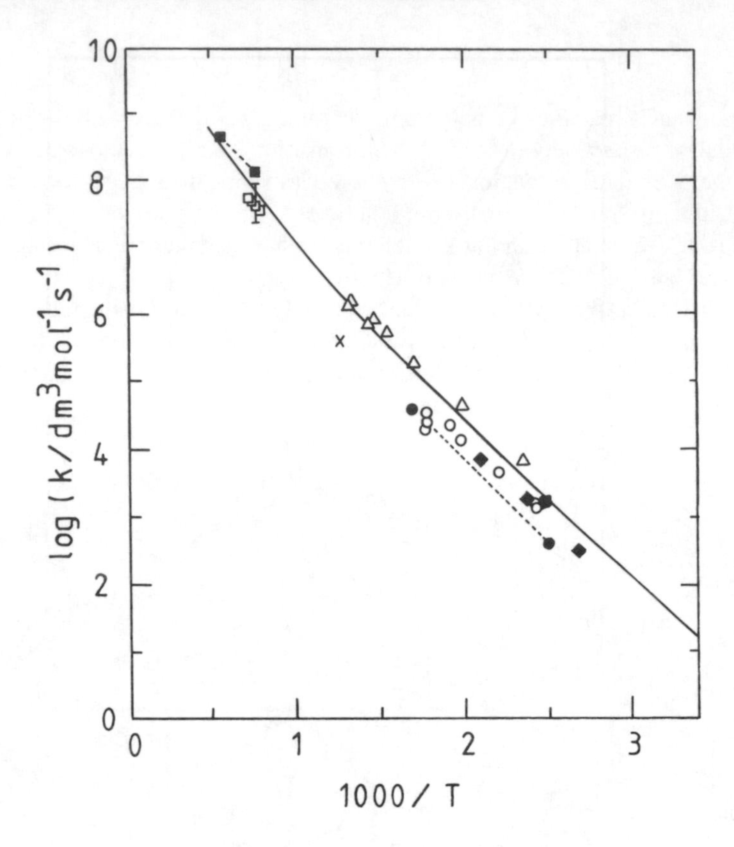

FIGURE 2. Arrhenius plot for reaction $CH_3 + H_2 \rightarrow CH_4 + H$. Solid curve is the result of a BEBO TST calculation by Clark and Dove.[46] Source of experimental data is given in Reference 46.

barrier, whatever the energy of the system is. The ratio of classical and quantum mechanical rates is represented by the tunneling factor:

$$\Gamma^* = k_{QM}/k_{CM} \tag{35}$$

where k_{QM} and k_{CM} are the rate coefficients including and neglecting tunneling, respectively.

In the simplest approximation, assuming one-dimensional motion across a potential barrier, Γ^* can be given by Wigner's formula (see Reference 65):

$$\Gamma^* = 1 + |u^*|/24 \tag{36}$$

or by Bell's high-temperature correction:[71]

$$\Gamma^* = \frac{|u^*|/2}{\sin(|u^*|/2)} \tag{37}$$

where $|u^*| = h\,|\nu^*|/k_BT$ and ν^* designates an imaginary frequency associated with the motion along the reaction coordinate. As an alternative possibility, one might prefer the use of Eckart's tunneling factor.[72] Each of these treatments predicts Γ^* to be temperature dependent and postulates an increase in importance with decreasing temperature. As a result, significant tunneling contribution to the rate causes non-Arrhenius behavior. However, it is shown that Arrhenius-plot curvature is not a simple function of the extent of tunneling; it

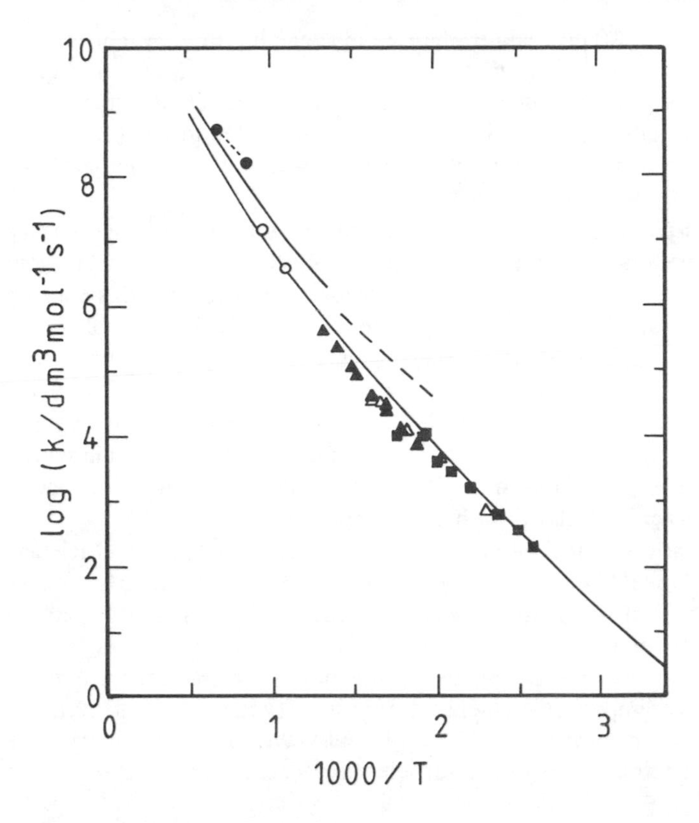

FIGURE 3. Arrhenius plot for reaction $CH_3 + C_2H_6 \rightarrow CH_4 + C_2H_5$. Solid curve is the result of a BEBO TST calculation by Clark and Dove.[35] Source of experimental data is given in Reference 35.

is more important in cases of high potential barriers.[73,74] Tunneling increases with the decrease of the mass of the atom transferred; therefore, it contributes particularly to the rate of light atom transfer (e.g., the rate of metathesis reactions involving hydrogen or proton transfer).

Although theoretical treatments predict nonlinearity caused by tunneling in the Arrhenius plots of metathesis reactions, it is very difficult to obtain conclusive experimental evidence and verification for the effect. To obtain unambiguous evidence, rate measurements have to be made over a wide temperature range extending to adequately low temperatures. Furthermore, potential causes of nonlinearity other than tunneling have to be excluded. It has been argued[75,76] that appreciable tunneling in an atom transfer reaction will be indicated by the kinetic phenomena listed below: (1) high values of the kinetic isotope effect, k_H/k_D; (2) great difference in the activation energies for H- and D-atom transfer, $E_D - E_H$; (3) low A factor of H-atom transfer relative to D-atom transfer, so that $A_H/A_D < 0.5$; (4) nonlinear Arrhenius plot which is more pronounced for H- than for D atom transfer reactions. In this diagnostic list of phenomena, non-Arrhenius behavior is only one time and in itself is not sufficient evidence for the occurrence of quantum mechanical tunneling. In spite of the difficulties, contribution by tunneling to the rate of a number of H-atom transfer reactions at low temperatures appears well established today. Since tunneling correction is more important at low temperature, most of the conclusive evidence for quantum mechanical tunneling is available for liquid- and solid-phase reactions. In the case of gas reactions taking place at higher temperatures, often the contribution from other factors to the Arrhenius-plot curvature cannot be excluded entirely. Some examples of quantum mechanical tunneling are given.

Much attention has been paid to the role of tunneling in the exchange reaction between atomic and molecular hydrogen.[33,77] Experimental data indicate some tunneling, but the results are not sufficient to supply unambiguous evidence. On the other hand, experimental and theoretical studies of the $OH + H_2$ reaction from 250 to 1050 K seem to indicate that tunneling is a major contributor to the overall rate, and tunneling contribution increases dramatically with decrease in temperature.[41]

In most gas-phase studies of the tunnel effect, experimental rate coefficients obtained over a wide temperature range are compared with theoretical rate data calculated with and without tunneling correction, and the conclusion is derived from this comparison. Usually, transition state calculations are made, often using the BEBO formalism. In a number of cases, the kinetic isotope effect, k_H/k_D, is considered instead of the rate coefficient. Using this type of treatment, Johnston[65,78] studied several reactions, e.g., $CH_3 + H_2$, $CF_3 + CH_4$, $CD_3 + C_2H_6$, and $CH_3 + CH_3COCH_3$. This author found that tunneling gives quantitative account of the temperature dependence of the deuterium isotope effect in several atom transfer reactions; however, tunneling in itself is not sufficient to provide satisfactory explanation for the non-Arrhenius behavior at higher temperatures.

The treatment of comparing experimental and calculated rate coefficients was also used by Clark and Dove[35] in a study of $H + CH_4$, $H + C_2H_6$, and $CH_3 + C_2H_6$ reactions; in an investigation[53] of gas-phase metathesis reaction $CH_3 + neo\text{-}C_5H_{12}$; and in several other theoretical considerations.

To sum up the results of gas-phase investigations, one may conclude that at temperatures from about room temperature to higher temperatures (where most gas reactions were studied), tunneling may contribute to Arrhenius-plot curvature; however, in most cases, quantum mechanical tunneling alonc is not sufficient to account for the observed deviations from Arrhenius law.

An investigation[59] of hydrogen (deuterium) abstraction from toluene by *tert*-butoxy radical in chlorobenzene has provided an excellent example of metathesis reactions in solution where curvature in the Arrhenius plot is caused (or at least mainly caused) by tunneling. The intermolecular isotope effect for side-chain hydrogen (deuterium) abstraction was studied between 212 and 353 K by the technique of competitive chlorination with *tert*-butyl hypochlorite. Considerable curvature was found in the Arrhenius plot of k_H/k_D, taken over the temperature range of 141 K, and leading to $E_D - E_H$ differences higher than and A_H/A_D values lower than those predicted by classical theory. All observations can be explained only by assuming that quantum mechanical tunneling is a major contributor to the reaction rate.

Williams and co-workers[79-82] studied metathesis reactions in solid phase at low temperatures extending to liquid nitrogen temperature:

$$\dot{C}H_3 + CH_3CN \rightarrow CH_4 + \dot{C}H_2CN \tag{38}$$

$$\dot{C}H_3 + CH_3NC \rightarrow CH_4 + \dot{C}H_2NC \tag{39}$$

$$\dot{C}H_3 + CH_3OH \rightarrow CH_4 + \dot{C}H_2OH \tag{40}$$

These investigators reported[79,80] an activation energy of 5.9 kJ/mol between 77 and 87 K for reaction Equation 38, and found[81] similar temperature dependence and strongly curved Arrhenius plot for reaction Equation 39. These low-temperature results may be compared with the much higher gas-phase activation energy $E_A = 42$ kJ/mol of reaction Equation 38. Furthermore, an activation energy of 3.8 kJ/mol was obtained[82] between 67 and 77 K, for hydrogen abstraction reaction Equation 40, which is again much lower than the gas-phase

Table 3
VIBRATIONAL ENHANCEMENT IN ATOM + DIATOMIC MOLECULE METATHESIS REACTIONS

Reaction	ΔH°_o (kJ/mol)	E_{vib} (kJ/mol)	E_A (kJ/mol)	$\frac{k(v)}{k(v=0)}$	α	Ref.
$H + Cl_2\ (v = 1) \rightarrow HCl + Cl$	−188		4.9	≤1.8		89
$F + HCl\ (v = 1) \rightarrow HF + Cl$	−136	34	~4	3.1	0.08	89
$H + HCl\ (v = 1) \rightarrow H_2 + Cl$	−4.6	34	14.6	58	0.30	90
$H + H_2\ (v = 1) \rightarrow H_2 + H$	0	50	31	1×10^3	0.33	91
$O + HCl\ (v = 1) \rightarrow OH + Cl$	3.4	34	19	4×10^3	0.60	90
$Cl + H_2\ (v = 1) \rightarrow HCl + H$	4.2	50	5.9	4×10^2	0.29	92
$O + H_2\ (v = 1) \rightarrow OH + H$	7.5	50	42	5×10^3	0.42	93
$Br + HCl\ (v = 2) \rightarrow HBr + Cl$	65.0	68	65	$\sim 6 \times 10^{10}$		94

value of 34.3 kJ/mol. The difference is due to the very different reaction temperatures rather than the phases as indicated by the similar behavior in crystalline and glassy matrix. The considerable temperature dependence of the activation energy was explained[80,82] by tunneling through the potential barrier. It appears that hydrogen transfer can proceed only by tunneling at very low temperatures.[83]

C. Effect of Vibrational Excitation on Reactivity

The rate coefficients and Arrhenius parameters determined in typical kinetic experiments and found in various kinetic data compilations are thermal average kinetic parameters, i.e., rate data which are characteristic for reactants in thermal equilibrium or with nearly equilibrium energy distribution. It is, however, known that chemical reactivity depends on the amount of energy possessed by the reactants, as well as on the distribution of this energy among the different degrees of freedom of the reacting molecules. At the present time, relatively little information is available on state-to-state rate coefficients for transitions between defined internal energy states of reactant and product molecules, and on the dependence of the detailed rate coefficients on internal energy, especially on the vibrational excitation of the reactants. In this section, we examine how the vibrational energy of the reacting molecules is utilized in overcoming the potential energy barrier of atom abstraction reactions.

Trajectory calculations greatly contributed to our understanding of the reaction dynamics and the enhancement of the reaction by vibrational excitation of the reactants.[84] The experimental information on the detailed rate coefficients for endoergic reactions has been derived mainly from infrared chemiluminescence study of the rate coefficients for the reverse exoergic reaction and the application of the principle of microscopic reversibility.[85-88]

All theoretical and experimental studies indicate that the efficiency of vibrational enhancement substantionally depends on the characteristics of the potential energy surface: endoergic, thermoneutral, or exoergic nature of the reaction, location and height of the saddle point, etc. The enhancement of the reaction rate as a result of vibrational excitation in a few atom + diatomic molecule metathesis reactions is shown in Table 3. It is evident from the data included in the table that the influence of vibrational excitation on the rate is most important for endoergic reactions. Such reactions have ''late barriers'', i.e., they have the crest of the potential barrier located in the exit valley of the potential energy surface. For a reaction of the type A + BC → AB + C, this means that the barrier is along the coordinate of separation (corresponding to the extension of the splitting bond BC); consequently, vibrational excitation of molecule BC is expected to be more efficient than translational energy in enhancing the rate of reaction.[89] Trajectory calculations indicate that the enhancement of rate increases rapidly with the extent of vibrational excitation up to the energy equal to the endoergicity of the reaction. A further increase in the vibrational energy has much less influence on the rate.

Trajectory studies were carried out for hypothetical thermoneutral reactions with barriers of 29 kJ/mol height.[84,85] If the crest of barrier is displaced slightly into the exit valley of the potential energy surface, vibrational motion in BC is almost parallel to motion along the reaction coordinate, and thus vibration promotes reaction efficiently; on the other hand, translation results in motion perpendicular to the reaction coordinate and is consequently ineffective. The situation is different if the crest of barrier is displaced slightly into the entry valley. Now vibrational motion is perpendicular to the reaction coordinate and is inefficient; however, translation of the reactants gives rise to motion which promotes reaction most effectively.

A widely studied exothermic reaction is

$$OH(v) + H_2(v') \rightarrow H_2O + H \quad (41)$$

Excitation of OH or H_2 to $v = 1$ and $v' = 1$ state, respectively, supplies more than enough energy to exceed the barrier. In accordance with the experience obtained with three-atom systems, vibrational excitation of H_2 enhances the rate by a factor of about 10^2.[96,97] In contrast to this, only a small rate enhancement is caused by the vibrational excitation of the OH radical.[97,98] Time-resolved LIF study of OH(v) + HCl and OH(v) + HBr reactions also showed that these low-activation energy exothermic reactions are not greatly enhanced, if increased at all, by vibrational excitation of the OH radical.[99] The results are understandable if the reaction is entirely adiabatic with regard to the vibrational motion along the OH bond.

"Hot" radicals are often assumed to be formed, for instance, in short wavelength photolysis of alkyl halides. These hot radicals react with enhanced rates in hydrogen abstraction reactions. It is not known so far to what extent the vibrational energy of the attacking polyatomic radical contributes to the enhanced rate. Some information can be derived from the investigation of exoergic reactions of atoms and polyatomic molecules via the principle of detailed balance. The results indicate that very little of the energy released in atom + polyatomic molecule interaction appears as vibrational energy of the polyatomic free radical formed. Thus, vibrational excitation of the polyatomic radical is not expected to enhance significantly the rate of the reverse endoergic reaction.

The results on the effect of internal energy in the metathesis reaction of a polyatomic radical and a molecule are for the time being not easy to interpret. A detailed trajectory study of the reaction

$$CH_3(v) + H_2(v') \rightarrow CH_4 + H \quad (42)$$

was carried out by Chapman and Bunker.[100] Promotion of the reaction by H_2 excitation has been established. On the other hand, the effect of vibrational excitation of the methyl radical proved to be insignificant: in some calculations, enhancement; in others, suppression; and in again others, no measurable effect was found, depending on the mode of methyl vibration excited and the location of the potential energy barrier.

Kovalenko and Leone[101] determined absolute rate coefficients for $CH_3 + Cl_2$ and $CH_3 + Br_2$ reactions by time-resolved measurements of IR fluorescence from the C–H stretch of the methyl halide product.[101] The methyl radical was produced by photodissociation of CH_3I at 248 nm with considerable excess energy. This resulted in an enhancement of the rates of metathesis reactions. A collisional moderation study was carried out in order to determine whether the reaction enhancement was caused by vibrational or translational excitation of the methyl radical. Interpretation of the results indicated that enhancement was predominately due to translational excitation.

Metathesis reactions of vibrationally excited species are only moderately fast even if one vibrational quantum supplies enough energy to surmount the potential barrier of the reaction. The reaction rates show definite temperature dependence, i.e., activation energy is still required. This indicates that the vibrational energy of the reactants is in general only partly utilized for overcoming the barrier. A parameter (α) has been introduced[102-104] to characterize the efficiency of the internal energy of the reactants in surmounting the barrier. The parameter is defined by the equation:

$$E_A^* = E_A - \alpha E_{vib} \tag{43}$$

where E_A^* and E_A are the Arrhenius activation energies for reaction with excited and non-excited reactants, respectively, and E_{vib} is the vibrational energy. If the pre-exponential factor is independent of vibrational excitation (an assumption that contradicts certain observations and requires experimental justification), then:

$$\alpha = \frac{RT}{E_{vib}} \ln k^*/k \tag{44}$$

The α parameter usually has a value around or less than 0.6[100] (see also Table 3). Birely and Lyman[105] have shown that the α parameter is in no correlation with E_A, E_{vib}, the excess energy $E_{vib} - E_A$, or the reaction exothermicity; thus, it should be regarded as an empirical parameter characteristic for the vibrational acceleration of the reaction rate.

Trajectory studies and experimental investigations show major differences in reaction dynamics as well as in rate enhancement by vibrational excitation for reactions characterized by potential energy surfaces with "early barrier crests" and "late barrier crests". Such surfaces are often referred to as type I and type II surfaces, respectively.[95] Endoergic and some of the nearly thermoneutral reactions of type $A + BC \rightarrow AB + C$ proceed on type II surfaces. These have the most part of the potential barrier in the exit (product) channel along the coordinate of separation of AB from C. Thus, a given energy is more efficient in enhancing reaction rates when present as vibration energy of the bond under attack than when present as translation or rotation. On the other hand, exoergic reactions proceed typically on type I surfaces with barriers largely in the entrance (reactant) channel along the coordinate of approach of the reactants. Thus, relative translational energy of reactants A and BC is more effective than vibration in surmounting the barrier.

The above generalizations, which were derived mainly from investigations of simple three- or four-atomic systems, are probably also valid for more complex polyatomic systems. Nevertheless, further experimental and theoretical work on reactions with polyatomic reactants and products is required.

III. CALCULATION OF THE TEMPERATURE DEPENDENCE OF RATE COEFFICIENTS FOR METATHESIS REACTIONS

Hydrogen atom abstraction reactions played a fundamental role in the development and testing of theories of chemical reaction rates. A great deal of our knowledge on hydrogen abstraction reactions originates from investigations of the dynamics of simple metathesis reactions. Here, we give a brief overview of the most important dynamic techniques (from exact quantum mechanical treatments to less sophisticated theoretical approaches) which are considered to be relevant in a survey of theoretical determination of rate coefficients and Arrhenius parameters for metathesis reactions of small organic free radicals. (The reader interested in a detailed discussion should consult the excellent recent review articles referred to in the following sections.)

Since most theoretical studies published so far dealt with reactions between atoms and diatomic molecules, naturally the discussion that follows cannot be devoid of the fundamental discoveries made in the investigations of the dynamics of the simplest reacting systems.

An ideal theoretical study of the elementary reaction sets out from the determination of the potential energy surface, carried out by solving the electronic Schrödinger equation for the energy of the microscopic system as a function of nuclear coordinates. Given the potential energy surface, next the Schrödinger equation describing the dynamics of the nuclear motion on this surface has to be solved. However, both conceptual and computational difficulties hinder the exact solution of the system of equations for nuclear motion. Exact solutions are and will be confined within reasonable time to the simplest atom + diatomics reactions. Therefore, much research has been devoted to approximate quantum treatments. The following sections will deal briefly with both exact and approximate quantum reactive calculations.

The most widely used treatment of reaction dynamics is the quasiclassical trajectory method, in which the nuclear motions are treated classically. One of the following sections gives a short review of recent trajectory results relevant to the subject of our essay.

Finally, we devote more space to the discussion of the transition state theory (TST) which undoubtedly proved to be the most outstanding method in predicting kinetic parameters for metathesis reactions involving polyatomic species. Our discussion includes both theoretically important modifications and developments of the TST[106,107] (i.e., generalized transition state theory and quantum transition state theory) and practically significant semiempirical[65] and thermochemical-kinetic[108] versions of TST.

Space limitations do not allow us to deal with the information theory approach which is becoming more and more important. Instead, we make reference to the excellent reviews of Levine.[109,110]

A. Quantum Reactive Calculations

Quantum reactive scattering is an extremely vigorous field which has been reviewed frequently during the last 10 years. Recent very comprehensive texts are the review by Walker and Light[111] which appeared in the 1980 volume of the *Annual Review of Physical Chemistry* series, Bernstein's[112] comprehensive handbook, and a volume of symposium proceedings edited by Truhlar.[113]

The problem of solving the nuclear-motion Schrödinger equation may be treated at different levels of complexity. In the following sections, some results obtained from exact, fully three-dimensional quantum calculations, exact collinear calculations, and approximate quantum treatments of reactive scattering are presented.

1. Exact Quantum Mechanical Reactive Rate Calculations

a. Three-Dimensional Calculations

Owing to the difficulties involved in exact quantum scattering treatments, only a very limited number of converged, fully three-dimensional calculations have been made. The majority of these are confined to the simple $H + H_2$ reaction. In most of the successful accurate quantum calculations, a coupled-channel solution of the Schrödinger equation, written in so-called natural collision coordinates, was used. The theoretical methods available for the solution have been described by Wyatt and Elkowitz,[114,115] Schatz and Kuppermann,[116,117] and Light and co-workers.[118,119] Application of these methods to the $H + H_2$ reaction has provided the first converged, fully three-dimensional quantum results.[120,121]

In the three-dimensional calculations quoted above, the semiempirical potential energy surface of Porter and Karplus[122] (PK surface) was applied. However, reaction dynamics is sensitive to the characteristics of the surface, therefore, exact quantum scattering results obtained on accurate surfaces are required. Today, the very accurate *ab initio* H_3 surface of and Liu and Siegbahn[123,124] with the careful fit of Truhlar and Horowitz[125] (SLTH surface) is available for dynamic studies. This surface — which is believed to be of chemical accuracy — has been used by Walker et al.[126] in exact three-dimensional quantum calculations of the reaction probabilities in the H + H_2 atom transfer system. In the same work, three-dimensional reaction probabilities calculated on the SLTH surface were compared with those obtained by using the semiempirical PK surface.

Only very few attempts have been made to calculate three-dimensional reaction probabilities for systems other than H + H_2 (or isotopes). Redmon and Wyatt[127] studied the F + H_2 reaction; mainly, resonance behavior was dealt with. The H + O_2 reaction, so important in oxidation chemistry, was studied by Redmon, who succeeded in obtaining unconverged results.

The number of reactions for which exact three-dimensional calculations are available will probably be extended only very slowly for atom + diatomics reactions, and such studies cannot be expected within reasonable time for reactions of polyatomic free radicals with polyatomic molecules. (The computer time required in the solution of the coupled-channel solution increases enormously with the increase in complexity of the reacting system.) Nevertheless, exact three-dimensional results of even the simplest atom transfer reactions are of particular importance for all types of metathesis reactions, since they can provide guidelines for the development of approximate rate theories and may contribute to the solution of various fundamental questions of kinetics (e.g., the dynamics of elementary reactions, the energy distribution in reaction products, the effect of excitation on reaction rate, the functional form of the reaction cross section, the temperature dependence of the rate coefficient, and tunneling).

b. Collinear Calculations

An unambiguous test of the validity of a kinetic theory is obtained by the comparison of predictions derived from the theory with accurate quantum dynamic results. However, for well-known reasons, widespread use of accurate three-dimensional calculations is greatly restricted. Fortunately, collinear results can successfully replace three-dimensional considerations in many cases. In the one-dimensional world, the interacting nuclei are constrained to move on one straight line. This reduction in dimensionality makes the procedure inexpensive, and yet collinear calculations predict various quantitative kinetic features in good agreement with the accurate three-dimensional results.

A number of theoretical treatments of quantum reactive scattering have been introduced for collinear systems and a variety of numerical methods have been proposed for the solution of coupled-channel equations. (For a review of the subject, see two recent articles.[111,128]) With these advancements, collinear quantum calculations have now become routine techniques.

Collinear reaction probabilities have been used to obtain approximate three-dimensional kinetic results. For instance, the *ab initio* SLTH potential energy surface with an adiabatic treatment of the ground state bending motion was employed in the calculation of exact collinear quantum reaction probabilities for D + H_2 ($v = 0$) and D + H_2 ($v = 1$) reactions.[129] These reaction probabilities were then used to obtain approximate three-dimensional rate coefficients.[129] Furthermore, in similar studies of the collinear X + F_2 and X + Cl_2 reactions (where X = Mu, H, D, T), Connor and co-workers[130,131] calculated collinear reaction probabilities using extended LEPS surfaces. The one-dimensional reaction probabilities were then transformed into three-dimensional vibrational product distributions.

A frequently used and very successful application of the exact quantum collinear calculations is the test of validity of an approximate theory. In this application, the dynamic results derived from an approximate treatment are compared with the exact one-dimensional quantum calculations made for the same potential energy surface. A number of papers have been published dealing with the comparison of quasiclassical trajectory results and exact collinear quantum dynamics. However, the most popular subject in this respect undoubtedly is the test of validity of different versions of the TST compared with one-dimensional quantum reaction probabilities. Some of the results are dealt with in Section III.C.

The results of exact quantum calculations have been used to test the dynamic features of various potential energy surfaces by carrying out one-dimensional dynamic calculations for a given reaction system simultaneously on an *ab initio* and on semiempirical surfaces. A good example of this type of application is presented for reaction $O(^3P) + H_2$ by Bowman et al.[132,133]

Collinear quantum calculations available today are restricted to atom + diatomics reactions, and the significance of this method for more complex metathesis reactions makes its way through the various theories of chemical kinetics, as described at the end of Section III.A.1.a. However, since one-dimensional calculations are much more feasible than three-dimensional considerations, collinear treatments are expected to remain popular in the future as well.

2. Approximate Quantum Mechanical Rate Calculations

The enormous computational requirements incident to the exact three-dimensional quantum calculations have initiated considerable research dealing with the development of approximate quantum treatments which are less expensive than the three-dimensional methods and, nevertheless, accurate enough. The methodology has been reviewed recently.[111,113,134] The most important approximate treatments may be divided into two categories:

1. Reactive decoupling approximations, which reduce the dimensionality of the coupled equations in the exact description. Two of such approximations should be mentioned. The more accurate one is the centrifugal sudden approximation ("J_2-conserving approximation"), and another method is the infinite-order sudden approximation (IOS).
2. Distorted wave approximation which was developed in its present form by Choi and Tang, who performed certain integrals analytically, thereby significantly reducing the numerical effort required in the calculation. The treatment can be further simplified by invoking the Born approximation. The results obtained with the distorted wave approximation are encouraging. A favorable feature of the method is that it may be easily generalized. Therefore, one expects a wider application of the treatment in the future.

B. Trajectory Calculations of the Thermal Rate Coefficients

Classical trajectory methods have been used much more frequently than quantum mechanical treatments for studying reaction dynamics and determining reaction probabilities. The greater popularity of the classical approach may be attributed to its applicability to more types of reactions. Most of the published classical trajectory studies dealt with atom + diatomic molecule reactions; however, in addition there are a number of trajectory calculations known for metathesis reactions involving polyatomic reactants.

Today, classical trajectory calculations for atom + diatomics reactions are commonplace and the methodology is well documented,[135-137] but the treatment is not so straightforward for reactions of polyatomics where a number of questions still need to be solved.[111,138,139]

Most trajectory calculations were performed using the quasiclassical trajectory method. In this method, classical trajectories corresponding to quantized initial vibrational states are calculated, and a histogram method is usually used to analyze the calculated classical energy distributions in terms of the correct product quantum states. Thus, the three stages of the procedure are

1. Selection of the initial conditions for trajectory calculations
2. Integration of the classical equations of motion

3. Analysis of the final results of trajectory calculations and determination of product energy distribution, reaction cross section, and rate coefficient

Set-up and integration of the equations of motion is a straightforward precedure which is well understood from the beginning of the use of the trajectory method. However, this is not so for the selection of initial conditions and analysis of results. In quasiclassical trajectory calculations of atom + diatomics reactions, the initial rotational and vibrational states are selected in accordance with the correct quantum states of the reactants. However, selection of quantized initial states for polyatomic reactants, and conversion of the continuous distribution of the action variables obtained from classical trajectory calculations into discrete distribution of the final action variables, become more and more difficult with the increasing number of internal degrees of freedom of the reacting species. The problems inherent in classical trajectory calculations involving polyatomic species have been surveyed by Walker and Light.[111]

Quasiclassical trajectory calculations have been performed for various reasons. One of the most important and often-studied question was, of course, the accuracy and reliability of the classical approach in dynamic treatments. In such studies, often quasiclassical reaction probabilities or rate coefficients were compared with the results of exact quantum mechanical calculations using the same potential energy surface. Collinear quantum results were used most often in these tests. The numerous studies of Truhlar, Connor, and others indicate that the quasiclassical trajectory method is generally a reliable method for the calculation of thermal (overall) rate coefficients and activation energies and is somewhat less accurate for the determination of state-to-state reaction probabilities (e.g., cf. Reference 140).

The classical approach does not take into account tunneling. Therefore, comparison of the quasiclassical trajectory results with accurate quantum mechanical calculations (or with experimental results) can show the importance of tunneling in the determination of the rate coefficient and activation energy for metathesis reactions (e.g., cf. Reference 141).

Trajectory results for reactions involving polyatomics are scarce. Lutz and Andresen[142] studied the reactions of $O(^3P)$ with saturated hydrocarbons using a LEPS potential energy surface. Assuming a triatomic O–H–R model for the reactions, these investigators obtained excellent agreement between quasiclassical trajectory calculations and experimental kinetic results. The good agreement was interpreted as a confirmation of the fact that reaction dynamics is dominated by a triatomic interaction at a single C–H bond. If this conclusion could be generalized, it would support the use of a simple triatomic approach in trajectory studies of metathesis reactions of polyatomic reactants.

Dynamic results are sensitive to the characteristics of the potential energy surface used in the calculations. The trajectory method is especially suited for the study of the influence of the features of the potential energy surfaces on the reaction dynamics (e.g., cf. References 133 and 143 to 146).

C. TST Calculations of the Temperature Dependence of the Rate Coefficients

TST, which dates back to more than 50 years, is beyond doubt the most remarkable theory in chemical kinetics. Its classical version has been used frequently and extensively to both interpret experimental results and predict kinetic data for metathesis reactions. The theory owes this unprecedented success to the fact that it incorporates the basic factors which control the rates of the chemical reactions. As is known, however, classical TST with unit transmission coefficient has two basic deficiencies:

1. Classical TST provides an upper bound on classical reaction rate, which follows from the fundamental dynamic criterion of the theory (see below). Violation of this dynamic criterion leads to an overestimation of the rate at high temperatures.

2. The neglect of quantum effects (and especially the omission of quantum mechanical tunneling) results in a calculated rate for H-atom transfer reactions which is too low at low temperatures.

Much of the recent theoretical research in chemical kinetics was directed to overcome these deficiencies of the theory and led to very significant progress in the formulation of the TST. Many excellent reviews dealt with the subject, among which the more notable recent ones are the concise summary by Walker and Light,[111] the excellent reviews of Pechukas,[107,147] and the enlightening papers of Truhlar and Garrett.[106,148,149]

Here, we discuss briefly the essence of variational TST and the development of quantum TST. Then we comment on the accuracy of various versions of TST by comparing TST results to accurate quantum calculations and quasiclassical trajectory calculations. Further on, the theoretical methods used most extensively in rate coefficient and activation energy determinations for metathesis reactions are discussed. These considerations include the discussion of semiempirical schemes, e.g., the well-known BEBO treatment, as well as a short survey of thermochemical kinetic estimation methods.

1. Classical TST, Variational TST, Quantum TST, and Other Theoretical Approaches

a. Classical Mechanical Rate Theories

TST is a classical theory based on a fundamental dynamical criterion which can be formulated unequivocally only in terms of classical mechanics. This fundamental assumption states[150] that the rate of forward reaction at equilibrium equals the equilibrium flux of classical trajectories through the transition state from the reactant to the product side. The microcanonical TST (μTST) rate coefficient is given by

$$k^{\ddagger}(\epsilon) = \frac{N^{\ddagger}(\epsilon)}{h\,\rho(\epsilon)} \tag{45}$$

where $N^{\ddagger}(\epsilon)$ is the number of internal energy levels of the transition state with energy less than or equal to ϵ, $\rho(\epsilon)$ is the density of states of reactants A + BC per unit energy and volume, and h designates the Planck's constant. The thermal equilibrium rate coefficient, i.e., the rate coefficient for a canonical ensemble of reactant molecules with energy distribution given by the Boltzmann law at a single temperature, may be written as an average over the microcanonical rate coefficients. Thus, the canonical TST (CTST) rate coefficient is

$$k(T) = \frac{\displaystyle\int_0^{\infty} k^{\ddagger}(\epsilon)\,\rho(\epsilon)\exp(-\epsilon/k_BT)\,d\epsilon}{\displaystyle\int_0^{\infty} \rho(\epsilon)\exp(-\epsilon/k_BT)\,d\epsilon} \tag{46}$$

where the denominator is the partition function for reactants A + BC per unit volume. The fundamental dynamical assumption of TST is justified if and only if trajectories do not recross the transition state represented by a dividing surface in the phase space. The μTST is exact only if no trajectory of a given energy crosses the transition state dividing surface more than once; the CTST is exact only if no trajectory of any kind of energy crosses the transition state dividing surface more than once. Otherwise, the theory overestimates the rate coefficient.

In classical TST, the transition state is a dividing surface in configuration space that separates reactant from product region and "cuts" through the saddle point of the potential energy surface. On the other hand, surfaces in the configuration space dividing reactants from products but not necessarily passing through the saddle point are designated generalized transition states. The classical transition state dividing surface passing through the saddle point, i.e., the highest point along the minimum energy path (MEP), is not necessarily the best dividing surface with the least flux of trajectories. Variation of the dividing surface in phase space and selection of the best generalized transition state gives the so-called variational transition-state theory (VTST).

Two notable basic versions of VTST have been worked out by Garrett and Truhlar:[148] the canonical variational theory (CVT) and the microcanonical variational theory (μVT). In μVT, the dividing surface is assumed to depend on energy and coordinates, and the best dividing surface (i.e., the surface minimizing the number of internal states of the generalized transition state) is searched for at each energy. From the microcanonical rate coefficients calculated at different energies, the thermal rate coefficient $k^{\mu VT}$ (T) is obtained by integration in accordance with Equation 46. In CVT, the dividing surface depends only on the coordinates, and the best compromise surface (i.e., the surface minimizing the partition function of the generalized transition state) is selected for a given temperature in order to obtain directly the rate coefficient k^{CVT} (T). Finally, one improved version of this latter theory exists which is called improved canonical variational theory[151] (ICVT). In this case, again a single best dividing surface is chosen for a given temperature, but in the selection of the best compromise dividing surface, only contributions from those trajectories which have energies exceeding a critical value characteristic for the reaction are taken into account. The designation of the rate coefficient derived from this theory is k^{ICVT} (T).

Various versions of the VTST give results which deviate more or less from the exact classical limit, depending on the extent they violate the basic TST assumptions. A great number of papers have been published in the past years which have dealt with the accuracy of VTST (e.g., see References 148, 151, and 152.). TST results were usually compared with exact classical trajectory results. Naturally, most of these investigations were carried out for atom + diatomics interactions, but some attempts have been made to reveal the accuracy of VTST methods for metathesis reactions of polyatomic species using a three-mass-point model. A representative set of collinear reactions is presented after Truhlar and Garrett[106] in Table 4. It may be seen from the table that for reactions of atoms with diatomic molecules, all versions of the theory, including conventional TST (designated TST in the table), are accurate enough at low energies which are represented by the 300 K results. However, as temperature is increased, the error in conventional TST becomes substantional and generalized transition state theories represent definite improvements. The best results are supplied by μVT, but ICVT results are almost as good.

In the last reaction presented in the table, a light hydrogen atom is transferred from a heavy atom to another heavy atom (^{57}C designates an atom of mass 57). The reaction may be regarded as a three-body model of reaction C_4H_9 + H–C_4H_9 → C_4H_9–H + C_4H_9. The error for this reaction in conventional TST calculations appears to be large, however, significant improvement can be attained by using variational optimization of the dividing surface.

The above results and other previous studies[153,154] have shown that considerable breakdown of the no-recrossing assumption may occur in the case of metathesis reactions in which a light atom is transferred between two heavy radicals, especially if the latter ones have similar masses. This conclusion is of particular interest in our present discussion where we are concerned with metathesis reactions involving polyatomic species. Accordingly, one expects for this type of reactions significant improvement in accuracy of TST rate calculations as a consequence of variational optimization of the dividing surface.

The above expectations gain some support from the model calculations carried out by Garrett and Truhlar. These authors[151] studied the collinear:

Table 4
RATIOS OF GENERALIZED TRANSITION STATE THEORY RATE COEFFICIENTS TO EXACT CLASSICAL DYNAMICAL ONES[106]

Reaction	TST	CVT	ICVT	μVT
$H + H_2$	1.0	1.0	1.0	1.0
	1.5	1.5	1.3	1.2
$Cl + HD$	1.1	1.1	1.1	1.0
	5.0	2.8	2.7	2.5
$F + H_2$	1.0	1.0	1.0	1.0
	2.0	1.8	1.8	1.8
$Br + H_2$	1.1	1.1	1.1	1.1
	4.2	3.9	3.9	3.7
$I + H_2$	1.7	1.0	1.0	1.0
	2.1	1.0	1.0	1.0
$H + Cl_2$	1.0	1.0	1.0	1.0
	1.3	1.3	1.3	1.1
$^{57}C + H^{57}C$	4.3	2.0	1.4	1.3
	11.4	2.0	1.9	1.7

Note: T = 300 K (upper entry) and 2400 K (lower entry).

Table 5
COLLINEAR THERMAL RATE COEFFICIENTS[151] FOR REACTION $^{57}C + H^{57}C \rightarrow {}^{57}CH + {}^{57}C$

			cm molecule^{-1} sec^{-1}		
T/K	$k_C^{TST}(T)$	$k_C^{CVT}(T)$	$k_C^{ICVT}(T)$	$k_C^{\mu VT}(T)$	$k_C(T)^a$
200	1.27×10^0	1.02×10^0	5.27×10^{-1}	4.77×10^{-1}	3.75×10^{-1}
300	6.74×10^1	3.16×10^1	2.21×10^1	2.10×10^1	1.57×10^1
400	5.13×10^2	1.83×10^2	1.46×10^2	1.42×10^2	1.01×10^2
600	4.13×10^3	1.12×10^3	9.99×10^2	9.78×10^2	6.68×10^2
1000	2.41×10^4	8.10×10^3	7.78×10^3	4.74×10^3	3.09×10^3
1500	6.27×10^4	1.24×10^4	1.19×10^4	1.09×10^4	6.82×10^3
2400	1.4×10^5	2.41×10^4	2.29×10^4	2.11×10^4	1.23×10^4
4000	2.6×10^5	3.88×10^4	3.70×10^4	3.48×10^4	1.72×10^4

[a] Results of accurate classical dynamics with estimated uncertainty <1% (up to 1000 K), around 10% (at 1500 to 2400 K), and 38% (at 4000 K), respectively.

$$^{57}C + H^{57}C \rightarrow {}^{57}CH + {}^{57}C \quad (47)$$

reaction (a reaction model for butyl radical attack on butane) as a function of temperature using an extended LEPS surface with Morse parameters assumed in accordance with those corresponding to the reaction $CH_3 + H - CH_3$. The thermal rate coefficients calculated by conventional TST and various VTST treatments are presented in Table 5. It can be seen

Table 6
COLLINEAR THERMAL RATE COEFFICIENTS[151] FOR HYPOTHETICAL PROCESSES REGARDED AS THREE-BODY MODELS OF ALKYL + ALKANE REACTIONS

	cm molecule^{-1} sec^{-1}				
Masses	$k_C^{TST}(T)$	$k_C^{CVT}(T)$	$k_C^{ICVT}(T)$	$k_C^{\mu VT}(T)$	$k_C^{US}(T)$
$^{15}C + H^{15}C$	1.4×10^5	4.65×10^4	4.41×10^4	4.09×10^4	2.42×10^4
$^{57}C + H^{57}C$	1.4×10^5	3.66×10^4	3.47×10^4	3.20×10^4	1.85×10^4
$^{42}C + H^{72}C$	1.40×10^5	2.49×10^4	2.37×10^4	2.16×10^4	1.17×10^4
$^{141}C + H^{57}C$	1.39×10^5	2.02×10^4	1.91×10^4	1.75×10^4	9.34×10^3

Note: T = 2400 K.

that conventional TST results are badly in error at all temperatures, and inaccuracy increases with temperature. The temperature dependence of the errors yields false activation energies if calculated by conventional TST. VTST results show better agreement with accurate classical trajectory calculations. As expected, agreement is best with the μVT kinetic coefficients. An important observation is that percentage errors appear to be constant and independent of the temperature.

Similar calculations also were carried out for isotopic analog reactions using extended C–H–C LEPS potential energy surfaces. The results are presented in Table 6. VTST results again show significant decrease compared to conventional TST rate coefficients. The difference is especially remarkable for the 141-1-57 mass combination.

In Table 6, the results appearing in the last column were obtained from the so-called unified statistical (US) model. This is a theory, proposed by Miller[155] and rederived by Pollak and Pechukas,[156] which interpolates between the TST expressions applicable to "direct" reactions and the statistical rate expressions valid for "complex" reactions characterized by long-lived complexes. It has been shown[151] that, in case of reaction $^{57}C + H^{57}C$, the US theory is very successful for energies 8 to 40 kJ/mol above the saddle point; at excess energies of 2 to 8 kJ/mol, US and μVT are comparable; and at even lower energies, μVT appears to be exact.

b. Quantum Effects

The second deficiency of the classical TST is the neglect of quantum effects. This deficiency is usually compensated for by replacing the classical partition functions in the rate equation by quantum mechanical partition functions, and by introducing in the TST equation a transmission coefficient to correct for nonclassical motion along the reaction coordinate. The problem in this treatment comes from the convenient but disputable approximation of separability of motion along the reaction coordinte. However, Garrett and Truhlar have shown that accurate quantum TST rate coefficients can be obtained even if the convenient assumption of reaction coordinate separability is retained provided that the location of the dividing surface is determined by variational optimization and quantal correction for the motion along the reaction coordinate is obtained by an appropriate treatment of tunneling (e.g., see References 106, 107, 147, 149, and 157).

One should use a separate tunneling correction for each internal state which contributes to the reaction. By taking into account the fact that quantum effects are most important at low temperature where the contribution of the ground state is predominant, Garrett and co-workers[157] recommended the use of an adiabatic ground state transmission coefficient. The vibrationally adiabatic ground state (VAG) transmission coefficient is designated κ^{VAG}, and its semiclassical counterpart (where tunneling is calculated semiclassically) is κ^{SAG}.

In the determination of tunneling correction there is a choice of tunneling path (for details, see References 106, 147, 157, ad 158). A simple approximation for the transmission coefficient is the SAG model based on tunneling along the MEP; this approximation is designated MEPSAG. A tunneling path proposed by Marcus and Coltrin (MCP) was found very successful and has been adapted frequently in the calculation of MCPSAG and MCPVAG transmission coefficients. Truhlar and Garrett found ICVT with MCPVAG or MCPSAG transmission coefficient to be the best practical theory of TST rate coefficient calculations;[106] such calculations reproduced very accurate quantal results for many simple atom + diatomics reactions.

The transfer of a light atom from a heavy particle to another heavy one represents a special type of metathesis reactions. These heavy-light-heavy atom transfers display small "skew angles" between the reactant and product valleys of the potential energy surface represented in terms of mass-weighted coordinates; the reaction paths show large curvatures. Hydrogen atom transfers occurring in the attack of a polyatomic free radical on a polyatomic molecule (i.e., the reaction types discussed in the present review) belong to this reaction category. Garrett et al.[158] and Koeppl[159] recognized that the MCP and other tunneling path models, which have been used successfully in TST rate coefficient calculations for various reactions, prove to be inappropriate in the calculation of tunneling corrections for heavy-light-heavy type atom transfer reactions. For such reactions, they proposed the use of the large-curvature ground state (LCG) model in the determination of transmission coefficients for generalized TST calculations.[158,159] The method is based on a transfer of a small mass on a straight path from one heavy particle to another; the light atom motion is treated adiabatically with respect to the heavy atom motion. Improved VTST with LCG transmission coefficients (ICVT/LCG) was found[158,159] to be accurate within a factor of 1.7 for reaction Cl + HCl over a wide range of experimental conditions. Model calculations of this type for metathesis reactions of polyatomic reactants have not been published so far.

c. TST Calculations of the Rate Coefficients and Activation Energies Using Ab Initio Potential Energy Surfaces

In the development and test of different versions of VTST, results of calculations representing various degrees of sophistication but using the same potential energy surface were usually compared. These considerations often employed realisic semiempirical or empirical potential surfaces, which perfectly served the purpose. However, if one compares calculated kinetic data with experiment, which tests at the same time the kinetic theory and the potential energy surface, one naturally needs to use accurate *ab initio* surfaces. Unfortunately, accurate *ab initio* calculations are restricted today to simple systems and, consequently, truly reliable comparisons of theory and experiments are not numerous.

Accurate and detailed calculations are available for the simple H + H_2 reaction and its isotopic analogs, like H + D_2, D + H_2, and T + HD.[159-161] These studies used the very accurate *ab initio* potential energy calculations of Liu and Siegbahn,[123,124] or the accurate analytic fit to this surface by Truhlar and Horowitz.[125] The calculated rate coefficients showed remarkably good agreement with experiment, and the activation energy was found to increase strongly with temperature.[161] TST calculations were carried out for the $O(^3P) + H_2 \rightarrow OH + H$ reaction[162,163] and for the $OH + H_2 \rightarrow H_2O + H$ reaction[164,165] using accurate *ab initio* POL-CI surfaces. Rate coefficient calculations, including a Wigner tunneling correction, were in excellent agreement with experiment over the temperature range 300 to 2000 K.

Determinations of *ab initio* potential energy surfaces and calculations of accurate rate coefficients for more complicated systems were attempted for reactions of the type:

$$R + H_2 \rightarrow RH + H \quad (48)$$

or the reverse processes:

$$H + RH \rightarrow H_2 + R \quad (49)$$

where R designates a simple polyatomic free radical. *Ab initio* TST rate coefficients and transition state properties were calculated for the $C_2H + H_2 \rightarrow C_2H_2 + H$ reaction[166] and the $H + H_2CO \rightarrow H_2 + HCO$ reaction[167] based on large-scale configuration interaction potential energy determinations in the reagent and saddle point regions. Potential energy barriers obtained from POL-CI calculations appeared to be too high. A lowering of the calculated $V^{\ddagger}$ was needed to achieve agreement between calculated TST rate coefficient and experiment.

The metathesis reaction:

$$CH_3 + H_2 \rightarrow CH_4 + H \quad (50)$$

and its reverse process:

$$H + CH_4 \rightarrow H_2 + CH_3 \quad (51)$$

may count on particular interest for more reasons. These reactions and their isotopic counterparts have played an important role in the development of modern chemical kinetics. Furthermore, in our present discussion, reaction Equations 50 and 51 are of underlined importance because the CH_5 six-atom system is among those relatively complicated molecular systems for which very sophisticated *ab initio* calculations have been carried out, supplying kinetic results which are believed to be of chemical accuracy.

In a recent paper, Walch[168] presented a potential energy surface based on accurate *ab initio* POL-CI calculations, and Schatz et al.[169,170] determined transition state properties and TST rate coefficients for Equations 50 and 51 and their numerous deuterium isotope variants using this *ab initio* surface and some new configuration interaction results. The rate coefficients were obtained from the conventional TST formulae with Wigner tunneling correction factors. For reactions $CH_3 + H_2$ and $CH_3 + D_2$, the agreement between calculated rate coefficients and experiment was good over the 900 K wide temperature range, as indicated in Figure 4. This agreement suggests that the theoretical POL-CI potential barrier of 45 kJ/mol is accurate to within 2 kJ/mol. For reaction $H + CH_4$ and its isotopic analogs, the TST rate coefficients were calculated with the POL-CI barrier $V^{\ddagger} = 56$ kJ/mol and an alternative barrier $V^{\ddagger} = 52$ kJ/mol. Better agreement between theory and experiment was obtained with the smaller barrier, especially at low temperatures. With the barriers of Reactions 50 and 51, 7 kJ/mol was obtained for the exoergicity of $CH_3 + H_2 \rightarrow CH_4 + H$, in good agreement with the experimental ΔE value.[170] Kinetic isotope effects in Reactions 50 and 51 were also investigated by Schatz et al.,[170] and their results seemed to be suitable for a study of the question if TST is adequate to describe quantitatively isotope effects.

Primary isotope effects for the deuterated analogs of $CH_3 + H_2$ were found to be correct in magnitude at high temperature, but with a weaker temperature dependence than experiment. The calculated temperature dependence of the secondary isotope effects also appeared to be less than the experimental one. A possible origin of the errors in the TST isotope effects calculated by Schatz et al. may be related to the use of conventional TST which does not take into account recrossing of the dividing surface. Furthermore, Wigner's expression for the tunneling factor may be insufficient and may have led to the underestimation of the rate of H-atom transfer at low temperature.

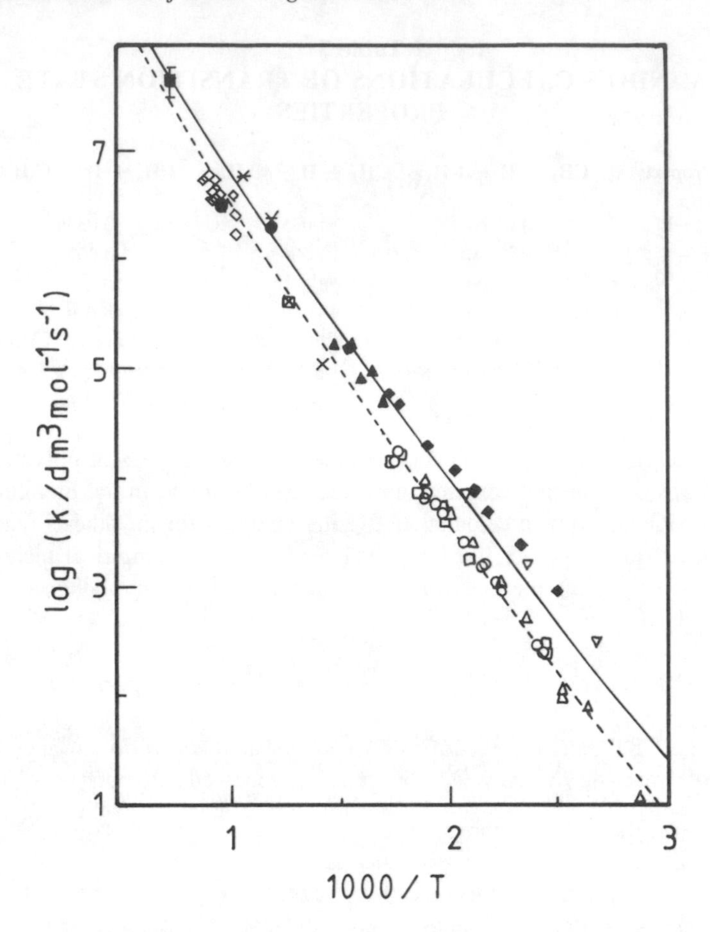

FIGURE 4. Calculated and experimental rate coefficients vs. 1/T for $CH_3 + H_2$ (——) and $CH_3 + D_2$ (- - -) reactions. Theoretical results are the TST rate coefficient calculations of Schatz et al.[170] Source of the experimental values is given in Figure 3 of Reference 170.

In order to be able to calculate reliable TST rate data, one requires a potential energy surface accurate to about ±3 kJ/mol. Unfortunately, this accuracy has not been achieved so far for atom transfer reactions involving two polyatomic reactants. Attempts of *ab initio* calculations for such systems indicated that, unless there is evidence for convergence of calculated values, only rough estimates of the activation energies can be expected from calculations involving relatively small basis sets.[171] An alternative to the *ab initio* calculations is the use of semiempirical procedures in the determination of potential energy surfaces. Dewar and Haselbach[172] have published a MINDO/2 study of the abstraction of a hydrogen atom from methane by a methyl radical and obtained an activation energy of 49 kJ/mol in reasonable agreement with experiment. However, a significantly smaller value was calculated in a later MINDO/2 investigation.[171]

In a very recent study,[173] the semiempirical MINDO/3 treatment has been used to calculate transition state geometries and potential barriers for methyl radical reactions with methane, ethane, and propane, respectively. Some of the results are given in Table 7. Comparison of the calculated barriers ($V^‡$) with the experimental activation energies (E_A) indicates that the MINDO/3 procedure underestimates the heights of the potential barriers, although the tendency in $V^‡$ along the reaction series is predicted correctly.

There has been much controversy regarding the value and applicability of the semiempirical

Table 7
MINDO/3 CALCULATIONS OF TRANSITION STATE PROPERTIES

Property	$CH_3 + H - CH_3$	$CH_3 + H - C_2H_5$	$CH_3 + H - C_3H_7$
$R^{\ddagger}_{AB} - R^{\circ}_{AB}$	0.0154	0.0351	0.0336
$R^{\ddagger}_{BC} - R^{\circ}_{BC}$	0.0154	0.0067	0.0079
$V^{\ddagger}$	43.4	28.3	20.8
E_A	59.9	48.1	48.6

Note: Units are nm for bond lengths, kJ for heights of potential barriers ($V^{\ddagger}$), and kJ/mol for experimental activation energies (E_A).

techniques in the calculation of molecular geometries and energies of polyatomic species. This may perhaps explain that very few examples can be found in the literature which use semiempirical MO methods to calculate activation energies for metathesis reactions. More experiences are required before the future value of such semiempirical techniques in the determination of activation energies for atom transfer reactions involving polyatomic reactants can be judged.

2. Calculation of Activation Energies and Transition State Properties by Semiempirical Treatments

Very accurate TST calculations based on sophisticated *ab initio* surfaces are confined today to simple reacting systems. For this reason, renewed attempts have been made to develop semiempirical treatments suitable for reliable calculation of the activation energies and transition state properties of metathesis reactions. Most of these treatments can be classified as follows:

1. Methods with adjustable parameters which are used to fit the treatment to representative kinetic data
2. Methods which use no adjustable parameters of kinetic nature, but are strongly based on empirical correlations from outside the field of chemical kinetics

Both approaches are represented below among the semiempirical methods developed for the study of atom transfer reaction:

$$A + BC \rightarrow AB + C \qquad (52)$$

The LEPS treatment represents a method belonging to category 1, while the equibonding method, BEBO, and BSBL schemes represent category 2.

a. LEPS Method

This treatment is based to a great extent on quantum mechanical results and yields the whole potential energy surface of the system, and (via TST principles) activation energies as well as transition state properties.

The LEPS method originates from the London equation[174] (disregarding the overlap integral):

$$V = Q_{AB} + Q_{BC} + Q_{CA} - \left\{\frac{1}{2}\left[(J_{AB} - J_{BC})^2 + (J_{BC} - J_{CA})^2 + (J_{CA} - J_{AB})^2\right]\right\}^{1/2} \qquad (53)$$

where Q and J designate pairwise Coulomb and exchange integrals, respectively. Eyring and Polanyi assumed[175] that $Q_{AB} + J_{AB}$, $Q_{BC} + J_{BC}$, and $Q_{CA} + J_{CA}$ can be calculated from the Morse function and the ratios $Q_{AB}/(Q_{AB} + J_{AB})$, etc., may be taken as constant (with a value of about 0.14), independent of interatomic distances. These assumptions allowed the calculation of separate values for Q and J at various interatomic distances and, by use of Equation 53, supplied the London-Eyring-Polanyi (LEP) potential energy surface.

Sato[176] proposed an improvement and modification of the treatment. He introduced a $1/(1 + \Delta)$ factor in the potential function:

$$V = \frac{1}{1 + \Delta}\left[(Q_{AB} + Q_{BC} + Q_{CA}) - \left\{\frac{1}{2}\left[(J_{AB} - J_{BC})^2 + (J_{BC} - J_{CA})^2 + (J_{CA} - J_{AB})^2\right]\right\}^{1/2}\right] \tag{54}$$

The new quantity, Δ, is treated in the LEPS approximation as an adjustable parameter. Furthermore, Sato suggested a new way for the determination of separate Q and J values. He obtained $(Q_{AB} + J_{AB})/(1 + \Delta)$, etc, from the Morse function (Equation 55), and derived $(Q_{AB} - J_{AB})/(1 - \Delta)$, etc, from the anti-Morse function (Equation 56):

$$^1E = D_e[\{1 - \exp(-\beta(R - R_o))\}^2 - 1] \tag{55}$$

$$^3E = 0.5\ D_e[\{1 + \exp(-\beta(R - R_o))\}^2 - 1] \tag{56}$$

In Equations 55 and 56, ^1E and ^3E are the ground and repulsive triplet state energies, respectively, of the diatomic molecule, while D_e denotes the spectroscopic dissociation energy defined as the sum of the observed dissociation energy and zero-point energy (zpe) i.e., $D_e = = D_o + zpe$. Furthermore, R_o and $(R - R_o)$ are the equilibrium interatomic distance and the deviation from this value, respectively, and β designates the Morse parameter. The use of Equations 55 and 56 allows calculation of separate Q and J values for various interatomic distances and, therefore, construction of the potential energy surface and the derivation of transition state properties. Although the activation energies derived from the LEPS treatment are not substantially better than the LEP values, however, the LEPS method seems to give much more realistic surfaces.

Several other modifications of the LEP treatment have been published. The best known among these is perhaps the often-cited method used by Porter and Karplus[122] in the study of the $H + H_2$ reaction. In their treatment, a number of contributions to the potential energy of the system were calculated theoretically and others were estimated semiempirically. The resulting surface was used in many dynamic studies.

The predictive power of LEP-type methods in determining activation energies and activated complex properties is not so significant because of the adjustable parameter occurring in the treatment. However, the simple LEPS potential energy surfaces have often been used and certainly will be used in the future in the investigations of the reaction dynamics.

b. Equibonding Method

The equibonding method of Zavitsas and Melikian[177,178] is not related directly to quantum mechanical results, and does not supply the entire potential energy surface or even the reaction profile. The method assumes a reasonable three-mass-point transition state model and estimates the energy of the transition complex by using empirical relations taken from domains outside chemical kinetics.

In this approximation, the transition state of the hydrogen abstraction reaction:

$$XH + Y \rightarrow YH + X \tag{57}$$

is assumed to be linear and is described in terms of a resonance hybrid:

$$
XH + Y \rightarrow
\begin{array}{ll}
X\cdot\cdot\cdot H\cdot\cdot\cdot Y & (I) \\
X : H \quad Y & (II) \\
X \quad H : Y & (III) \\
\dot{X} \quad \dot{H} \quad \dot{Y} & (IV)
\end{array}
\rightarrow X + HY \tag{58}
$$

The fundamental assumption of the method is that, in the transition state, partial bonding in X–H is equal to bonding in H–Y, i.e., the two canonical forms II and III are of equal energy. The geometry required by the equibonding criterion is at the same time the configuration resulting in maximum resonance stabilization for the odd electron.

The total energy of the system in the transition state has three energy terms:

$$
E^{\ddagger}_{tot} = \frac{1}{2}\left({}^{1}E^{\ddagger}_{XH} + {}^{1}E^{\ddagger}_{HY}\right) + E^{\ddagger}_{R} + {}^{3}E^{\ddagger}_{XY} \tag{59}
$$

The first term is the average bonding energy of canonical structures II and III which can be replaced (as a result of the equibonding criterion) by the bonding energy of either of these, e.g., $1/2({}^{1}E^{\ddagger}_{XH} + {}^{1}E^{\ddagger}_{HY}) = {}^{1}E^{\ddagger}_{XH}$. The bonding energies are estimated using the Morse equation (Equation 55).

The second term is the resonance energy of delocalization of one odd electron over three atoms, $E^{\ddagger}_{R}$. This energy is assumed to be similar to the resonance energy of the allyl radical and a value of 44.4 kJ/mol is used, independent of the type of X and Y atoms involved.

The last term in Equation 59 appears because simultaneous bonding of the H atom to both X and Y (structure I) requires either ↑ ↓ ↑ or ↓ ↑ ↓ spin orientation in structure IV. This results in repulsion between the terminal atoms X and Y. The antibonding energy ${}^{3}E^{\ddagger}_{XY}$ is estimated from the anti-Morse function (Equation 56).

The collinear transition state geometry is searched for by evaluating the total energy of the system at various R_{XH} distances. Starting at a small value of R_{XH} (e.g., $R_{XH} = R^{o}_{XH} +$ 0.001 nm), first ${}^{1}E_{XH}$ is calculated. Then, the corresponding R_{HY} distance is obtained in accordance with the equibonding criterion that requires ${}^{1}E_{XH} = {}^{1}E_{HY}$ to be satisfied. Next, the antibonding energy ${}^{3}E_{XY}$ is obtained at $R_{XY} = R_{XH} + R_{HY}$ interatomic distance. Finally, the total energy is derived from Equation 59. The whole procedure is repeated at 0.001-nm increments of R_{XH} until a minimum value for $E^{\ddagger}_{tot}$ is attained. The most stable arrangement satisfying the equibonding criterion is accepted in this approach as corresponding to the transition state. The appropriate activation energy can be given as the difference between calculated $E^{\ddagger}_{tot}$ and $D_e(X - H)$. A small correction may be applied for zero-point-energy (zpe) effects in the transition state according to Equation 60:

$$
zpe^{\ddagger} = \frac{1}{2}\left(zpe_{XH} + zpe_{HY}\right) \tag{60}
$$

Application of the method requires input data for those quantities which appear in the Morse functions: bond dissociation energies, stretching frequencies, bond lengths, and masses. Using literature data for the input parameters, Zavitsas[177] calculated activation energies for more than 100 hydrogen atom abstraction reactions[178] and found generally good agreement between calculation and experiment. Examples taken from the paper of Zavitsas and Melikian[178] are presented in Table 8. Notable deviation of the predictions from experiment was observed only for some F-atom, Cl-atom, and C_6H_5 radical reactions. The equibonding method overestimates the activation energies for these reactions.

Table 8
ACTIVATION ENERGIES FOR HYDROGEN ABSTRACTION REACTIONS CALCULATED BY THE EQUIBONDING METHOD[178]

Reaction	Calculated E*[a] (kJ/mol)	Experimental E_A[b] (kJ/mol)
$H + CH_3CH_3$	32	38, 28, 38, 50, 51, 38, 41
$H + (CH_3)_3CH$	17	31
$F + CH_3CH_3$	18	1
$Cl + CH_4$	54	16
$Cl + CHCl_3$	27	14, 27, 14
$Br + CH_3CH_3$	62	57, 56, 59
$Br + (CH_3)_3CH$	34	31
$I + CH_3CH_3$	125	117
$I + (CH_3)_3CH$	93	90
$OH + CH_3CH_3$	25	31, 15
$OH + (CH_3)_3CH$	11	12
$CH_3O + CH_3CH_3$	32	30
$CH_3O + (CH_3)_3CH$	23	17
$CH_3O + HCHO$	14	~17
$(CH_3)_3CO + CH_3CH_3$	31	27
$(CH_3)_3CO + (CH_3)_3CH$	22	17, 20
$CH_3 + H_2$	49	41, 42, 54, 55, 42, 43, 51, 46
$CH_3 + CH_3CH_3$	44	44, 41, 48, 47, 47, 48, 50
$CH_3 + (CH_3)_3CH$	33	32, 31, 28, 38, 32, 34
$CH_3 + CH_2{=}CHCH_3$	31	32, 31, 33, 34
$CH_3 + HCHO$	15	26, 28
$CH_3CH_2 + CH_3CH_3$	56	59, 53
$CH_3CH_2 + (CH_3)_3CH$	37	37
$CH_3CH_2 + CH_2{=}CHCH_3$	37	32, 36
$C_6H_5 + H_2$	41	27
$CF_3 + H_2$	44	37, 40, 46
$CF_3 + CH_3CH_3$	30	31, 35
$CF_3 + (CH_3)_3CH$	13	20, 13

Note: Values refer to the breaking of the weakest bond in the substrate

[a] Calculated values with zpe correction.
[b] Source of data given in Reference 178.

Reasonable good agreement between calculated and experimental values was found for *t*-butoxy radical reactions with carbonyl compounds[179] and for *t*-butoxy + paraffin reactions.[180] However, a study of H-atom abstraction from unsaturated molecules by *t*-butoxy radicals showed considerable discrepancies.[180] It appears that a more sophisticated procedure for the calculation of delocalization energy, $E_R^{\ddagger}$, would be required to improve the accuracy of atom abstractions from unsaturated molecules.

Studies made with the equibonding method have revealed some of the major factors which determine the activation energy of hydrogen atom transfer reactions. Considerations of nearly thermoneutral reactions proved especially instructive in this respect. Results obtained for such reactions,[178] arranged in order of increasing energy of activation, are presented in Table 9. There is no correlation at all between the activation energies and the strengths of bond broken. The activation energies follow the trend of increasing repulsion energy ${}^3E_{XY}^{\ddagger}$. There-

Table 9
ACTIVATION ENERGIES FOR NEARLY THERMONEUTRAL REACTIONS DETERMINED BY THE EQUIBONDING METHOD[178]

Reaction	E*/kJ/mol	$^3E^{\ddagger}_{XY}$	BDE(X–H)/kJ/mol
F + HF	0	36	569
$(CH_3)_3CO + (CH_3)_3COH$	6	39	439
Br + HBr	13	46	366
Cl + HCl	21	52	431
$CH_3O + H_2$	23	54	436
$OH + H_2O$	26	58	498
$H + H_2$	39	66	436
$CH_3 + NH_3$	41	68	439
$CF_3 + H_2$	44	74	436
$CH_3O + CH_4$	45	69	438
$CF_3 + CH_4$	45	72	438
$CH_3 + H_2$	49	76	436
$CH_3 + CH_4$	59	80	438

Note: Values refer to the breaking of the weakest bond in the substrate

fore, all factors that increase $^3E^{\ddagger}_{XY}$ also augment E_A. Thus, a stronger X–Y bond or lower stretching frequency results in a higher activation energy.

The results presented in Table 9 suggest that the activation energy serves to overcome triplet repulsion between X and Y. Comparison of E* and $^3E^{\ddagger}_{XY}$ indicates that repulsion energies are systematically higher than activation energies. Thus, repulsion energy is compensated for partly by the resonance energy due to delocalization of the odd electron, and partly by the activation energy. The question of how general these conclusions are, is discussed subsequently.

c. BEBO Method

The BEBO method is one of the most popular semiempirical schemes developed so far for the calculation of activation energies and rate coefficients of hydrogen atom transfer reactions. Extending the ideas formulated by Johnston and Goldfinger,[78,181] the BEBO method was developed by Johnston and Parr in 1963. The treatment includes no adjustable parameters and requires thermochemical and spectroscopic quantities, such as bond dissociation energies, bond lengths, and vibrational frequencies, as input data.

Considering atom transfer reactions (Equation 52), the method[65,182] sets out from the study of three-body systems assumed to be linear (in accordance with the results of quantum mechanical calculations for simple atom transfer reactions) and regards the complex as a superposition of three two-body systems:

$$\overset{\xleftrightarrow{\quad R_3 \quad}}{A \quad B \quad C}$$
$$\xleftrightarrow{R_1}\xleftrightarrow{R_2}$$

Recognition of the fact that the splitting of bond 2 is assisted by the formation of bond 1 has led the authors to the conclusion that there must be very strong correlation between changes in properties of the two bonds during reaction. This correlation is formulated in quantitative terms as the concept of conservation of bond order which implies that, along the minimum energy path from reactants to products, the sum of bond orders of breaking and forming bonds is unity:

$$n_1 + n_2 = 1 \tag{61}$$

For the bond order, the technical definition:

$$R = R_s - \lambda \ln n \tag{62}$$

originally proposed by Pauling[183] is used, where R is the length of the bond of order n, and R_s is the equilibrium bond length. The parameter λ in Equation 62 was chosen to be $\lambda = 0.26$ by Pauling, and Johnston and Parr adopted this value in the BEBO procedure.

As a next step, knowledge of the partial bonds (breaking and forming bonds) is required. Johnston found an empirical relationship between bond energy and bond length, analogous to Badger's rule for force constants, and combining this with Pauling's equation (Equation 62), he proposed the following relationship between bond energy and bond order:[65,78]

$$E = E_s n^p \tag{63}$$

where E_s and E are bond energies for a single bond and a bond of order n, respectively, and p is the bond index which depends on the atoms in the bond. Since partial bonds are of interest in the procedure, Johnston examined how bond energy vs. bond length and bond energy vs. bond order relationships can be extended to stretched bonds. It was postulated that "noble gas clusters" have bond orders equal to zero, and bond energy relationships were supposed to extrapolate to the corresponding noble gas cluster (i.e., to Ne–He for a C–H, N–H, O–H, or F–H bond; to Ar–He for a S–H or Cl–H bond, etc.). Thus, the bond index p is given by

$$p = \frac{\ln(E_s/E_x)}{R_x - R_s} \tag{64}$$

where R_x designates the interatomic distance in the noble gas diatomic cluster and E_x is the depth of the attractive well. It is immediately evident from Equation 64 that the value of p depends on the energy and length of the single bond in question and of the corresponding noble gas cluster. A few values, calculated by Johnston and Parr,[182] are given as examples: $p = 1.041$ for H–H, $p = 1.087$ for H–CH_3, $p = 1.032$ for H–NH_2, $p = 1.028$ for H–OH, $p = 1.036$ for H–F, and $p = 0.914$ for H–Cl. However, it should be remembered that p also depends on the value of λ. If a different value is used instead of $\lambda = 0.26$, this changes the estimated energy of the partial bond (See Equation 63).

Apart from the two bonding interactions A–B and B–C, a third interaction between A and C needs to be considered. If the transferred atom B is simultaneously bound in the complex to the other two atoms, the unpaired electrons of A and C atoms must have parallel spins; therefore, they repel each other. This triplet repulsion energy has been calculated in the original version of the BEBO procedure by the anti-Morse function (Equation 56). More recently, modifications of the triplet functions were suggested and are discussed below.

We now have all the information required for defining the locus of the line of minimum energy between reactants and products in the $R_{AB} - R_{BC}$ plane, and to calculate the energy along this reaction path:

$$V = E_{1s} - E_{1s}n^{p_1} - E_{2s}(1 - n)^{p_2} + V_{tr} \tag{65}$$

In this equation, the origin of the energy is chosen to be at the equilibrium potential energy of reactant B–C, and the energy of the system is given in terms of the progress variable n, which is the bond order of the bond being broken in the reaction. The designations E_{1s} and p_1 are the single bond energy and the bond index for the forming bond, respectively, E_{2s} and p_2 are the same quantities for the breaking bond, and V_{tr} is the triplet repulsion energy in the A and C interaction calculated by the anti-Morse function. The function V may be calculated from Equation 65 point by point for values of n varying from 1 to 0 and the maximum value, V*, is the potential energy of activation (i.e., height of the potential barrier). Alternatively, V* may be obtained directly by differentiating function V with respect to n and calculating the value of the energy where dV/dn = 0.

The Arrhenius activation energy E_A differs from the "classical" activation energy V* by the difference in both zero-point energy and thermal excitation energy between reactants and complex. The appropriate determination of the Arrhenius activation energy and the calculation of the rate coefficient requires a complete vibrational analysis of the transition state. This presumes the use of a kinetic theory, preferably TST. However, the BEBO method does not supply the complete potential energy surface, but only a cross section of the surface along the line of constant bond order. Thus, elaboration of the method necessitates introduction of further basic assumptions. Of these assumptions, we make here only a few remarks; for detailed discussion, see the original literature.[65,78,181,182]

The distance along the line of constant bond order is the progress variable ρ, and the second derivative of V with respect to ρ at the saddle point (at the point where $V = V_{max} = V^*$) is the force constant $\overset{*}{F_\rho} = \left(\frac{\partial^2 V}{\partial \rho^2}\right)_*$. This latter equation can be obtained from an analytic expression derived simply from Equation 65. On the other hand, small displacements perpendicular to the line of constant bond order define a progress variable σ and thereby a force constant $\overset{*}{F_\sigma} = \left(\frac{\partial^2 V}{\partial \sigma}\right)_*$. This force constant is approximated according to the BEBO procedure by applying a modified Badger's rule, as described by Johnston.[65,181] Finally, the streching force constants are given in terms of $\overset{*}{F_\rho}$ and $\overset{*}{F_\sigma}$:

$$F_{11} = \frac{F_\rho^*(1 - n_*)^2 + F_\sigma^* n_*^2}{n_*^2 + (1 - n_*)^2}$$

$$F_{22} = \frac{F_\rho^* n_*^2 + F_\sigma^*(1 - n_*)^2}{n_*^2 + (1 - n_*)^2}$$

$$F_{12} = \frac{(-F_\rho^* + F_\sigma^*)(1 - n_*)\, n_*}{n_*^2 + (1 - n_*)^2} \tag{66}$$

The stretching force constants F_{11}, F_{22}, and F_{12} together with the bending force constant F_ϕ (calculated from the first derivative of V with respect to R_3; see References 65 and 182) are used to carry out a frequency analysis for the linear three-mass-point transition state model. (These force constants are also used in analogous BEBO procedures based on four- or five-mass-point transition state models.) Knowing the energy, the interatomic distances, force constants, and vibrational frequencies of the transition state, we are now able to calculate the rate coefficients from simple separable TST. Frequently, partition functions are expressed in terms of local valence-bond-angle properties.[65,184] All information required for calculation of the Arrhenius activation energy is also available:

$$E_A = V^* + \Theta_B RT \tag{67}$$

where the temperature correction function θ_B can be expressed in terms of the temperature-dependent preexponential factor B(T) (see Equation 8):

$$\Theta_B = \frac{d \ln B(T)}{d \ln T} \quad (68)$$

Correction of the classical activation energy V* for the difference between E_A and V* by the procedure described above is somewhat lengthy and may not be justified if high precision is not required. Often a much simpler correction for the zpe effect appears to be sufficient:

$$E_A \simeq V^* + zpe^* - zpe_{BC} \quad (69)$$

where zpe_{BC} and zpe* destignate zero-point energies for the reactant BC and transition complex, respectively. The zpe* can be estimated by the average of the zpe of AB and BC molecules as proposed by Zavitsas (see Equation 60), or the equation suggested by Gilliom[185] can be used:

$$zpe^* = n_* \, zpe_{BC} + (1 - n_*) \, zpe_{AB} \quad (70)$$

The two estimations of zpe* are similar for nearly thermoneutral reactions, but Equation 70 is expected to be a better approximation for very exo- or endotermic reactions where bond order n_* may differ considerably from $^1/_2$.

The BEBO procedure has been carefully tested and applied in the calculation of rate coefficients and Arrhenius parameters for a wide variety of hydrogen atom transfer reactions of various free radical species. BEBO minimum-energy paths, activated complex geometries, and force constants were tested[65,185-187] against the results of *ab initio* calculations and remarkably good agreement was found. Naturally, the principle and thorough verification of the method should be based on comparison between calculated kinetic parameters and experimental data.

Systematic comparison has been made and published in the first report on the BEBO method,[182] where calculated results for about 130 reactions of hydrogen and halogen atoms, OH, CH_3, CF_3, C_2H_5, i-C_3H_7, and t-C_4H_9 radicals with various substrates are given. The method has been claimed to supply results which agree with experimental data within 8 kJ/mol for most cases. It has been found that the simple BEBO procedure is able to reproduce the temperature dependence of the rate coefficient in a wide range of temperature, including cases where deviations from the Arrhenius equation occur.[35,65,182] Successful BEBO calculations of the kinetic isotope effect in metathesis reactions have also been reported.[65,78,188,189] It appears that for reactions where there is agreement between calculated and experimental activation energies, kinetic isotope effects are also predicted correctly.[188]

In addition to the numerous successful applications of the BEBO scheme, there are, however, certain classes of reactions where the method appears to fail. Thus, metathesis reactions of halogen atoms belong to the problematic reactions. Calculated activation energies for fluorine and chlorine atom attacks on hydrocarbons were found to be higher than the observed values,[182] where the deviation was occasionally as great as 20 to 30 kJ/mol. Even more serious failures to the method were found for halogen atom reactions with hydrogen halides:[69,185] the BEBO method underestimated the activation energies, and the transition state geometries were miscalculated in some cases.[190] The BEBO activation energies for CF_3 radical reactions were found to be higher than experimental values.[191-193] Similarly, overestimations of the activation energies were observed[191,193] for hydrogen atom transfers from or to a multiple-bonded (double-bonded, triple-bonded, or aromatic) C atom.

BEBO calculations as well as equibonding results indicate that the activation energy is often mainly and sometimes entirely controlled by triplet repulsion. Hence, one may assume that the inadequacy of the anti-Morse function used to calculate repulsion energies in the BEBO procedure can be responsible for the failing of the method when applied to the reactions discussed above. Indeed, several modifications of the BEBO procedure have been suggested which employed improved functions to estimate the repulsion energy arising from the interaction of end groups A and C. Thus, Mayer and Schieler[194] doubled the triplet energy function (Equation 56) for certain types of reactions. On the other hand, Arthur et al.[192,195,196] found that a modified BEBO method, in which the Sato antibonding equation is replaced by a function fitted to the triplet H_2 ($^3\Sigma$) energy values of Hirschfelder and Linnett, gave more satisfactory correlation with the experimental Arrhenius parameters for metathesis reactions of alkyl and fluoro-alkyl radicals than obtained by the original BEBO method. Finally, another attempt[191] made to formulate a more accurate triplet repulsion function[197] applicable in a modified BEBO treatment should be mentioned. Although these modifications appear to have improved somewhat the outcome of the method, they have not changed the foundations of the BEBO procedure.

However, Jordan and Kaufman[198] reevaluated the BEBO method and found the predictive power of the BEBO procedure to be substantially lower when rare gas intermolecular potentials from crossed beam scattering measurements by Lee and co-workers[199,200] are used in place of earlier estimates based on Lennard-Jones parameters for noble gas "molecules". Jordan and Kaufman have shown that the use of the beam scattering results instead of earlier estimates of noble gas molecule properties in the calculation of bond indexes according to Equation 64 yields lower p values, and this substantially decreases the energy contribution of the fractional bonds. Consequently, the calculated activation energies become significantly lower than the earlier estimates and disagree with the experimental values.

In keeping with the modernization of the BEBO method initiated by Jordan and Kaufman, Gilliom[185,201] found that the predictive ability of the BEBO procedure can be restored by chosing a new value for the λ parameter in Pauling's relationship (Equation 62) in accordance with more recent and accurate data for interatomic distances. That is, using the consistent set of bond lengths given by Lide[202] for C–C, C=C and C≡C, the parameter in Pauling's bond length-bond order relationship can be reevaluated as $\lambda = 0.28$ compared with the original value of 0.26. With this λ value and the beam scattering noble gas molecule properties, bond indexes (p values) which are very similar to the old values (used in the original BEBO method) are obtained. Consequently, the modified BEBO procedure gives almost the same activation energies and kinetic parameters as the old BEBO method proposed by Johnston and Parr. In addition, Marschoff and Jatem[203] came to the conclusion that the original formulation of the BEBO method is applicable even when the new values for noble gas molecule properties are employed, provided that an appropriate choice of the λ parameter in Pauling's relation is made.

Gilliom[185] has made a systematic comparison of the activation energies estimated by the original BEBO procedure of Johnston and Parr and the new version of the method using beam scattering results for noble gas intermolecular potentials [replacing factor 0.5 in the anti-Morse function (Equation 56) with 0.45]. Comparison of old and new values with each other as well as with experimental activation energies was made for 97 metathesis reactions. The original method gave an average error of 6.5 kJ/mol with a standard deviation of 8.5 kJ/mol; the modified method afforded an average error of 5.4 kJ/mol and a standard deviation of 7.8 kJ/mol. Activation energies of the same reactions also were calculated by the equibonding method of Zavitsas and Melikian which gave an average error of 6.1 kJ/mol with a standard deviation of 10.0 kJ/mol. These figures, as well as the examples taken from the paper of Gilliom[185] and presented in Table 10, show that the predictive power of the simple BEBO method can be maintained even if recent structural and energetic data are taken into account; thus, we do not have to abandon the appealing notion of constant bond order or the use of rare gas "molecules" as zero-bond species.

Gilliom's modification of the BEBO method has been used widely[193,204-206] since its initiation in 1977 for the calculation of activation energies and estimation of kinetic isotope effects for various types of hydrogen atom transfer reactions. In general, reasonably good agreement was found with experiment and, where all the results were available, Gilliom's modification usually proved to be superior to the original BEBO method of Johnston and Parr and the equibonding method of Zavitsas and Melikian. However, the foundations of Gilliom's method and the original BEBO method are identical; thus, the new version also shows the same deficiencies as the old method, i.e., faulty prediction of the activation energies of fluorine and chlorine atom reactions and overestimation of E_A for CF_3 radical reactions and for hydrogen atom transfers from or to a multiple-bonded C atom. (Some examples of such reactions can be found in Table 10.) Since these failings appear to arise from the very essence of the BEBO treatment, no significant improvement can be expected from simple modifications; therefore, fundamentally different schemes are required. This became particularly obvious from the study of Brown and Laufer,[193] who investigated metathesis reactions of the unsaturated radical C_2H and found serious contradiction between the results of BEBO calculation and experiment. Calculated activation energies were on average 38 kJ/mol higher than the experimental data. We revert to this problem in the next section.

The BEBO procedure has been used for more than 20 years as a method for the calculation of activation energies and other kinetic parameters of hydrogen atom abstractions by atoms and free radicals. In addition to these major applications of the method, some special ones also have been presented. One of the most important special applications of the BEBO procedure is calculation of the activation energy for metathesis reactions of triplet species. As discussed in Section III.C.2.a, the original BEBO method — developed for the study of hydrogen atom abstractions by atoms or radicals in the doublet electronic state — describes the potential energy along the path of constant bond order by Equation 65, which contains an antibonding term. This term is meant to take into account the triplet repulsion between terminal groups A and C caused by parallel spins in the transition state. However, if the attacking species has more than one unpaired electron (e.g., the metathesis reactions of triplet oxygen atom, triplet CH_2, or any triplet molecule), increased repulsion in the transition state may be expected. Mayer and Schieler[194] suggested doubling the triplet repulsion term in case the attacking species is in a triplet electronic state. (Since the original procedure uses the Sato function with a 0.5 factor, doubling restores the unmodified function.) Activation energies for $O(^3P)$ atom reactions were calculated[194,206-208] with the BEBO method using doubled antibonding energy term. In most cases, only moderate agreement has been found with experiment. The exact reasons of the inconsistencies are not certain,[206] but it appears that handling of terminal atom interactions in the modified BEBO treatment is inadequate for these reactions.

Quantitative kinetic information on hydrogen abstraction reactions of ground-state triplet methylene is scarce in the literature, which renders the computational results provided by simple semiempirical schemes as the BEBO method even more valuable. Carr[209] has calculated activation energies and rate coefficients for $CH_2(^3B_1)$ reactions with hydrocarbons and hydrogen-containing substrates using the standard BEBO procedure with doubled triplet repulsion term. BEBO calculations predict that hydrogen abstractions by triplet methylene occur as spin-allowed processes with relatively low rates. The low reactivity can be ascribed, according to computational results, to the appreciable activation energies. When comparing metathesis reactions of CH_2 and CH_3, the more exothermic abstraction by triplet methylene always had the higher activation energy. Due to lack of reliable experimental results in 1972, Carr was not able to check these conclusions or test the applicability of the BEBO method

Table 10
ACTIVATION ENERGIES (kJ/mol) FOR HYDROGEN ABSTRACTION REACTIONS[a] CALCULATED BY THE BEBO METHOD[185]

Reaction	Calculated E*		Experimental E_A values[d]
	BEBO original[b]	BEBO modified[c]	
$H + CH_4$	54	55	39
$H + CH_3CH_3$	36	38	41
$F + CH_3CH_3$	11	8	1
$F + (CH_3)_2CH_2$	8	7	0
$Cl + CH_4$	44	43	16
$Cl + HCl$	1	2	25
$Cl + CCl_3H$	11	13	18
$Br + CH_4$	86	84	75
$Br + CH_3CH_3$	61	59	57
$I + CH_3CH_3$	125	123	117
$I + (CH_3)_3CH$	94	91	90
$OH + CH_3CH_3$	20	17	23
$OH + (CH_3)_3CH$	10	8	12
$CH_3O + CH_3CH_3$	31	28	30
$CH_3O + (CH_3)_3CH$	18	16	17
$CH_3O + HCHO$	12	10	17
$(CH_3)_3CO + CH_3CH_3$	30	27	27
$(CH_3)_3CO + (CH_3)_3CH$	18	15	18
$CH_3 + H_2$	52	52	47
$CH_3 + CH_3CH_3$	51	46	46
$CH_3 + (CH_3)_3CH$	38	33	33
$CH_3 + CH_2{:}CHCH_3$	36	31	33
$CH_3 + HCHO$	26	22	27
$CH_3 + HCl$	36	36	12
$CH_3CH_2 + CH_3CH_3$	62	56	56
$CH_3CH_2 + (CH_3)_3CH$	43	38	37
$C_6H_5 + H_2$	44	44	27
$CF_3 + H_2$	49	50	41
$CF_3 + CH_3CH_3$	43	38	33
$CF_3 + (CH_3)_3CH$	28	24	16
$CF_3 + HCl$	38	37	22

[a] Activation energies refer to breaking of the weakest bond of the substrate.
[b] Original method of Johnston and Parr.[182]
[c] Procedure modified by Gilliom.[185]
[d] Taken from Table 2 of Reference 185.

to hydrogen transfer reactions of triplet methylene. Very recently, however, Wagner and co-workers[210,211] carried out direct studies of CH_2 ($\tilde{X}^3B_1$) reactions with hydrocarbons using a discharge flow system with LMR detection and determined accurate activation energies for six metathesis reactions of triplet methylene. Comparison of the theoretical and measured values showed the BEBO activation energies to be higher by about a factor of two than the experimental results. The discrepancy may reflect problems in handling terminal atom interactions by the method, as already suggested in the discussion of triplet O-atom reactions.

A novel application of the BEBO method has been presented by Previtali and Scaiano,[212-214] who modified the BEBO procedure to provide a way of predicting activation energies and rate data for hydrogen abstractions by carbonyl triplets. This appears to be the first attempt to develop a semiempirical method for dealing with reactions of excited species. The modification sets out from the observation that the $n\pi^*$ states of the carbonyl compounds react like radicals, particularly like alkoxy radicals. The presence of two unpaired electrons instead of one in the abstracting species was taken into account by introducing a second repulsive term (V'_{tr}) into the potential energy function. Furthermore, it is assumed that the triplet state energy (E_t) provides the energy difference between double and single C–O bonds (E_d). The potential energy along the MEP is

$$V = E_{1s} - E_{1s}n_{p1} - E_{2s}(1 - n)_{p2} - (E_t - E_d)(1 - n) + V_{tr} + V'_{tr} \quad (71)$$

where the designations are the same as in Equation 65.

This modified treatment has been used[214] to calculate kinetic parameters for a number of intermolecular abstraction reactions of the triplet state of carbonyl compounds (acetophenone and benzophenone). The calculated activation energies and rate coefficients were compared with experimental values measured in nonpolar solvents and, in general, very good agreement was found.

The BEBO method has been applied to the abstraction of atoms other than hydrogen. The most important of these studies is the extension of the method to the calculation of activation energies for the transfer of multivalent atoms, i.e., the abstraction of oxygen atom. These modifications, which fall outside the scope of the present essay, have been described by Mayer.[215,216]

Finally, reference should be made to the application of the BEBO method to the study of alkyl radical disproportionation reactions.[217,218]

d. Bond Strength-Bond Length (BSBL) Method

A new method has been proposed recently by Bérces and Dombi[68] which endeavors to preserve the appealing characteristics of the BEBO method, but without the shortcomings of previous similar semiempirical treatments. The BSBL method sets out from structural and energetic considerations of postulated linear transition states of metathesis reactions (Equation 52) and calculates the "reaction profiles" along the MEP using empiricism entirely outside the field of chemical kinetics.

The MEP from reactant to product is given in the BSBL procedure by

$$\exp(-2\beta_{AB} X_{AB}) + \exp(-2\beta_{BC} X_{BC}) = 1 \quad (72)$$

In Equation 72, β_{ij} is the Morse parameter for bond between atoms i and j, $X_{ij} = R_{ij} - R^o_{ij}$ where R_{ij} is the interatomic distance of atoms i and j, and R^o_{ij} is its equilibrium value. Equation 72 allows for the particular characteristics of reaction paths of various atom transfer reactions by using the values for reactant A–B and product B–C molecules.

The method assumes that the energy of the triatomic complex A — B — C can be disintegrated into the following contributions:

1. Contributions of the bonding energies in structures:

   ·A B:C and A:B C·

   (I) (II)

2. Stabilization contribution resulting from delocalization of the odd electron over the three atoms of the complex

3. Repulsion between A and C caused by antibonding interaction due to parallel spins of electrons on the end atoms

Instead of attempting separate estimations of the stabilization and repulsion energies, the difference of these two quantities is evaluated in the BSBL method. This combined stabilization and repulsion energy is called the end-group contribution.

The bonding energy of the complex A — B — C along the MEP is given as a weighted average of $-V_{BC}$ and $-V_{AB}$, the bonding energies in structures I and II, respectively:

$$V_{bond} = -g_{BC}V_{BC} - g_{AB}V_{AB} \tag{73}$$

The energies $-V_{BC}$ and $-V_{AB}$ are obtained from the Morse function:

$$-V_{ij} = V^{\circ}_{ij}\{[1 - \exp(-\beta_{ij}X_{ij})]^2 - 1\} \tag{74}$$

where V°_{ij} is the potential energy of dissociation for bond i–j, defined as the sum of the observed dissociation energy and zero-point energy, and all other symbols have the meanings already given above. The weighting factors g_{BC} and g_{AB} are expressed as

$$g_{ij} = \exp(-2\beta_{ij}X_{ij}) \tag{75}$$

The end-group contribution, which allows for delocalization and repulsion interactions, is expected to depend[68] on the potential energy of dissociation of bond A–C in molecule AC (designated V°_{AC}), the interatomic distance of A and C, and the tendency of groups A and C to attract the odd electron. Thus, end-group contribution is given in the BSBL procedure by the term:

$$V_{endgr} = A_{AC}\, g_{AC}\, V^{\circ}_{AC} \tag{76}$$

Here V°_{AC} is the equilibrium dissociation energy of bond A–C and the factor g_{AC} expressed according to Equation 75 comprises the dependence on the distance between A and C. Alfassi and Benson[219] realized that here is a correlation between the intrinsic activation energy of an atom transfer reaction and the electron affinities of A and C; accordingly, a dependence on the electron affinities is taken into account in A_{AC} which allows for the electron-attracting tendency of end groups A and C. Considering symmetrical reactions for which the bonding contributions in the complex A — B — C are similar, A_{AC} is approximated[68] by

$$A_{AC} = 2\{\exp[0.75(1 - \overline{EA})] - 1\} \tag{77}$$

where EA is the average of the electron affinities of A and C expressed in eV:

$$\overline{EA} = \frac{1}{2}[EA(A) + EA(C)] \tag{78}$$

Depending on the electron affinities of A and C, the factor A_{AC} and consequently the end-group contribution may be positive, zero, or negative.

The energy of the complex A — B — C is given as the sum of bonding energy terms and end-group contribution: thus, the BSBL potential energy function along the MEP is

$$V = V^{\circ}_{BC} - g_{BC}V_{BC} - g_{AB}V_{AB} + A_{AC}g_{AC}V^{\circ}_{AC} \tag{79}$$

The reaction profile can be calculated according to Equation 79 point by point along the reaction path (Equation 72). The maximum of the reaction profile is identified as the potential energy of activation, $V^‡$, and the configuration of the A — B — C complex at this point is assumed to correspond to that of the transition state.

Since a reaction profile and not a complete potential energy surface is obtained from the BSBL procedure, a vibrational analysis for the activated complex can be made only at the expense of further assumptions (similar to the BEBO procedure). The BSBL method expresses the stretching force constants for the linear three-mass-point activated complex, in terms of the force constants $F^‡_\rho = (\partial^2V/\partial\rho^2)_‡$ and $F^‡_\sigma = (\partial^2V/\partial\sigma^2)_‡$ and the slope of the reaction path $S_‡ = (dR_{BC}/dR_{AB})$:

$$F^‡_{AB} = \frac{F^‡_\rho + F^‡_\sigma S^2_‡}{1 + S^2_‡} \qquad F^‡_{BC} = \frac{F^‡_\rho S^2_‡ + F^‡_\sigma}{1 + S^2_‡} \qquad F^‡_{AC} = \frac{F^‡_\rho - F^‡_\sigma}{1 + S^2_‡} \tag{80}$$

$F^‡_\rho$, $F^‡_\sigma$, and $S_‡$ can be calculated from analytical expressions given in the original literature,[68] and hence the stretching force constants are obtained. These, together with the bending force constant $F^‡_b$, supply the additional information required for the calculation of the frequencies of the linear three-mass-point activated complex. Finally, simple separable TST is used to determine the rate constant and obtain the activation energy from $V^‡$:

$$E_A = V^‡ + \Theta_B RT \tag{81}$$

where Θ_B is defined by Equation 68.

An approximate procedure may be used to obtain the Arrhenius activation energy from the BSBL potential energy of activation. As in the case of the BEBO procedure (see Equation 69), one simply corrects for the zero-point effect:

$$E_A = V^‡ + zpe^‡ - zpe_{BC} \tag{82}$$

The zpe of the complex may be approximated as the weighted average of the zpe of reactant BC and product AB:

$$zpe^‡ = g^‡_{AB}\, zpe_{AB} + g^‡_{BC}\, zpe_{BC} \tag{83}$$

where the weighting factors are the values of the progress variables g_{AB} and g_{BC} defined by Equation 75. Since $g_{AB} + g_{BC} = 1$ along the MEP, one obtains from Equations 82 and 83:

$$\begin{aligned} E_A &= V^‡ + g^‡_{AB}(zpe_{AB} - zpe_{BC}) \\ &= V^‡ + \exp(-2\beta_{AB}X^‡_{AB})\,(zpe_{AB} - zpe_{BC}) \end{aligned} \tag{84}$$

The BSBL method has been tested[69] carefully against both quantum mechanical calculations and experimental results. The BSBL reaction profile was found to be strikingly accurate when compared[69] with Liu's H_3 *ab initio* surface (standard deviation = 1.0 kJ/mol). Further H_3-activated complex properties (barrier height, bond lengths, force constants, and vibrational frequencies) were also in excellent agreement[69] with the best *ab initio* results.

Comparison of calculated and experimental activation energies for metathesis reactions of a number of atoms and hydrocarbon radicals is presented in Table 11. Reactions of oxygen-containing free radicals and miscellaneous radical reactions are given in Table 12. The calculated results agree with the experimental data generally within the limits of experimental error. Serious failings of the method have not been experienced. The method has

Table 11
ACTIVATION ENERGIES (IN kJ/mol AT 550 K) OF ATOM AND HYDROCARBON RADICAL REACTIONS CALCULATED BY THE BSBL METHOD

Reaction	Calculated E_A	Experimental E_A
H + H–CH_3	46	49
H + H–C_2H_5	34	38
H + H–$C(CH_3)_3$	27	31
H + H–$CH_2CH{:}CH_2$	22	21
F + H–CH_3	10	8
F + H–C_2H_5	3	2
Cl + H–CH_3	14	16
Cl + H–CCl_3	8	14
Cl + H–Cl	22	28
Br + H–CH_4	72	78
Br + H–C_2H_5	48	55
I + H–C_2H_5	111	111
I + H–$C(CH_3)_3$	89	90
H_3C + H–H	53	51
H_3C + H–C_2H_5	45	48
H_3C + H–$C(CH_3)_3$	37	40
H_3C + H–C_2H_3	48	42
H_3C + H–$CH_2CH{:}CH_2$	28	30
H_3C + H–C(O)H	21	26
H_3C + H–CF_3	41	43
H_3C + H–Cl	23	19
H_3C + H–OH	92	94
F_3C + H–H	35	39
F_3C + H–C_2H_5	28	31
F_3C + H–$C(CH_3)_3$	20	20
F_3C + H–C_6H_5	35	38
F_3C + H–$C(O)CH_3$	16	18
F_3C + H–Cl	14	22
Cl_3C + H–Cl	50	47
C_2H_5 + H–$C(CH_3)_3$	37	37
$(CH_3)_2CH$ + H–CH_3	82	82
$CH_2{:}CH$ + H–H	34	31
C_6H_5 + H–H	33	27
C_6H_5 + H–CF_3	18	22

been tested[69] for 128 reactions representing all the important types of hydrogen atom transfer processes. Standard deviation and average error in these tests were found to be 5.9 and 4.7 kJ/mol, respectively. The BSBL method gives more satisfactory results than the previous treatments. Reasonable activation energies are obtained from the BSBL method for reactions where other procedures, e.g., BEBO and equibonding methods, have failed: F- and Cl-atom reactions, CF_3 radical reactions, hydrogen atom abstraction from unsaturated C atoms, and metathesis reactions of unsaturated free radicals.

The BSBL method has been applied recently with success in the calculation of activation energies for various metathesis reactions, especially for reactions where hydrogen atoms are transferred from or to double-bonded C atoms. Thus, vinyl and substituted vinyl radical reactions were investigated.[220,221] A detailed study of the activation energies for reactions of ethynyl (HC≡c) radicals with H_2, CH_4, and C_2H_6 and for reactions of N≡c radicals with H_2 and CH_4 was made by Brown and Laufer[193] using both the BEBO and BSBL methods. It was found that the BSBL technique accurately predicted the activation energies for these

Table 12
ACTIVATION ENERGIES (IN kJ/mol AT 550 K) FOR REACTIONS OF O-CENTERED FREE RADICALS AND MISCELLANEOUS RADICAL REACTIONS

Reaction	Calculated E_A	Experimental E_A
$HO + H–C_2H_5$	19	15
$HO + H–CH(CH_3)_2$	17	12
$CH_3O + H–C_2H_5$	26	30
$CH_3O + H–C(CH_3)_3$	20	17
$CH_3O + H–C(O)H$	17	13
$(CH_3)_3CO + H–C(CH_3)_3$	18	20
$HOO + H–C_2H_5$	59	59
$HOO + H–C(O)H$	24	27
$CH_3CO + H–I$	7	6
$H_3Si + H–CH_3$	69	67
$NC + H–H$	28	29
$HS + H–CH_3$	70	65

reactions, while the BEBO method yielded energies averaging 38 kJ/mol higher than those observed. The BSBL method has been used to predict activation energies for hydrogen abstraction by triplet vinylidine ($^3B_2$, $H_2C{=}C$) radicals[222,223] and in considerations of hydrogen transfers from and to C atoms in aromatic molecules.[224,225] Another type of application of the results of BSBL calculations to the kinetics of unsaturated species has been the formation and reactions of the resonance-stabilized cyanomethyl ($cH_2C{\equiv}N$) radical.[226]

The greater success of the BSBL technique compared with the BEBO method in estimating kinetic parameters for metathesis reactions involving unsaturated reaction centers may be due to various reasons, however, one of these is certainly the appropriate handling of the end-group interaction in the BSBL procedure. In the evaluation of end-group contribution, which allows for the delocalization and A $\leftrightarrow$ C repulsion interaction in the A — B — C complex, the BSBL technique utilizes the idea[219] that the electron-attracting tendency of A and C lowers the energy of the transition complex. Improved handling of end-group interaction together with other properly chosen approximations appears to result is a reliable technique for predicting activation energies for metathesis reactions of all kinds.

Thorough examination of the results of BSBL calculations has been carried out[69] comparing calculated rate coefficients with experimental data over a range of temperatures. Calculated rate coefficients as a function of temperature are shown in Figures 5 and 6 for reactions of CH_3 and CF_3, respectively, with H_2 and hydrocarbons. The BSBL rate coefficients (estimated without tunneling correction) show good agreement with the experimental data. It can be seen from the results covering a wide temperature range that the BSBL method predicts deviations from the Arrhenius law. In the case of reaction $CH_3 + C_2H_6$, calculated activation energies increase from 40 up to 70 kJ/mol in the temperature range of 300 to 1000 K.

Although the BSBL method has been developed to predict kinetic properties for hydrogen atom transfer reactions, a few attempts have been made to apply the technique to abstraction of atoms other than hydrogen.[69] Calculated and experimental results were found to agree reasonably; no serious failings of the method have been experienced. These encouraging preliminary results appear to indicate that applicability of the procedure is not restricted to hydrogen atom transfer reactions.

The BSBL reaction profile is obtained from the potential energy function (Equation 79) in which terms $\{V^o_{BC} - g_{BC}V_{BC} - g_{AB}\,V_{AB}\}$ represent the bonding contribution without the stabilization energy resulting from delocalization of an odd electron, while the last term,

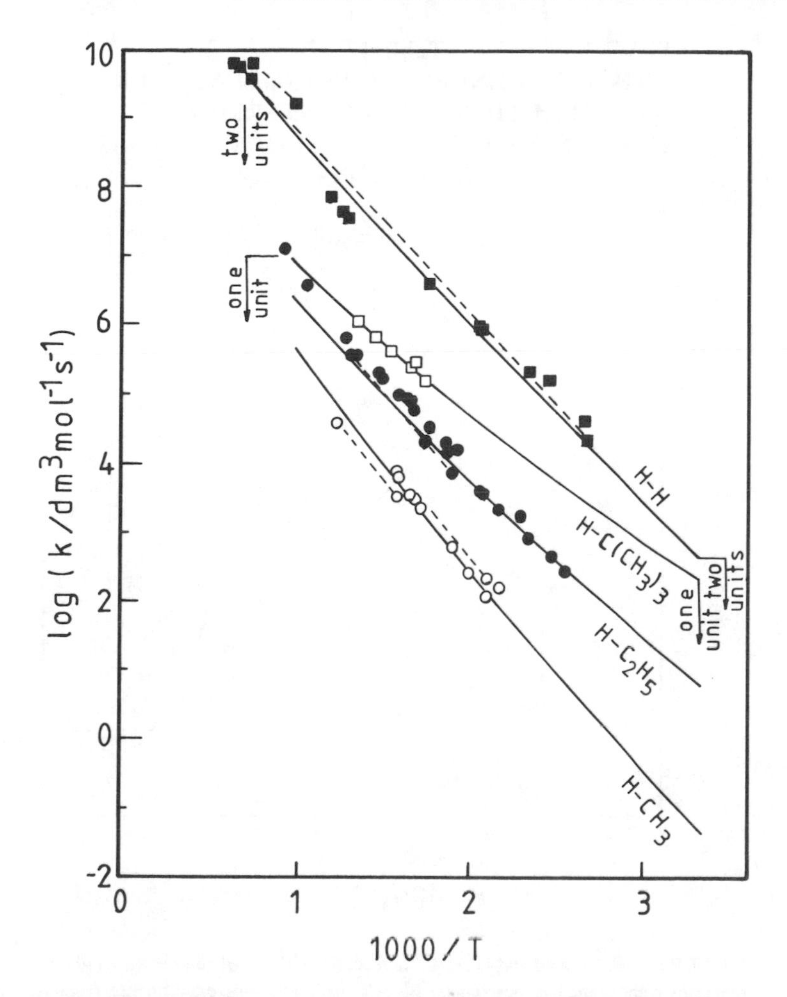

FIGURE 5. Calculated and observed rate coefficients for methyl radical attack on H_2 and hydrocarbons.[69] BSBL calculations (——); experimental results for H_3C + + H − H (■), H_3C + H − CH_3 (○), H_3C + H − C_2H_5 (●), and H_3C + + H − $C(CH_3)_3$ (□).

$A_{AC}g_{AC}V^{\circ}_{AC}$, known as end-group contribution, stands for the combined repulsion and stabilization energy. If one wants to decompose the potential energy of activation, $V^{\ddagger}$, calculated by the BSBL method, into separate total bonding contribution, $V^{\ddagger}_{bond}$, and triplet repulsion energy, $V^{\ddagger}_{rep}$, an independent estimation for either of them is required. A rough estimate of the repulsion energy can be obtained from the anti-Morse function (Equation 56) or the following repulsion function:[197]

$$V_{rep} = \frac{1}{\alpha_{AC}} V^{\circ}_{AC} \exp(-2\beta_{AC}X_{AC})[1 + (\beta_{AC}X_{AC})^{2\alpha_{AC}}] \tag{85}$$

where the α_{AC} parameter is the ratio of the singlet and triplet energies at the equilibrium interatomic distances, $\alpha_{AC} = {}^1V^{\circ}_{AC}/{}^3V^{\circ}_{AC}$, and other designations have their usual meaning. With the value of V_{rep}, the total bonding energy in the transition state is obtained simply from $V^{\ddagger}_{bond} = = V^{\ddagger} - V^{\ddagger}_{rep}$.

The above-described procedure was used[227] to study the factors predominant in the determination of the activation barrier. Three series of A + BC → AB + C type atom transfer

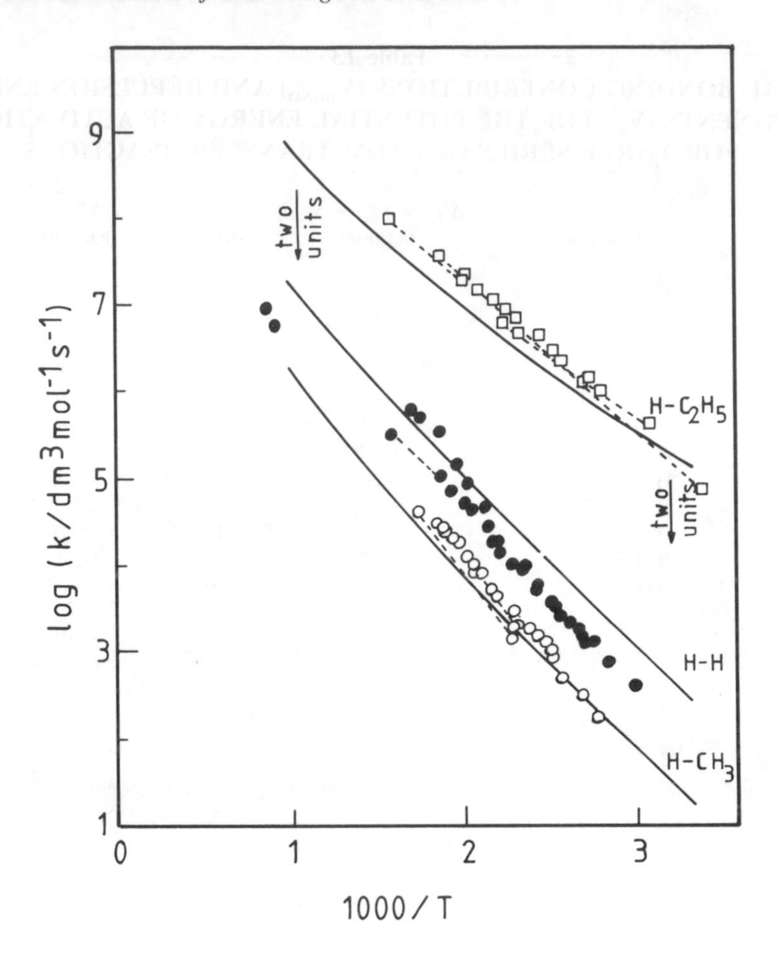

FIGURE 6. Calculated and observed rate coefficients for trifluoromethyl radical reactions with H_2 and hydrocarbons.[69] BSBL calculations (—); experimental results for $F_3C + H - H$ (●), $F_3C + H - CH_3$(○), and $F_3C + H - C_2H_5$ (□).

reactions were considered. Representatives of each series are presented in Table 13. The first two series consist of hydrogen atom transfer reactions in which the attacking species A and the remaining species C, respectively, vary, while the third series is composed of halogen transfer reactions where the transferred atom B changes.

It may be seen from the table that for hydrogen atom transfer reactions, the triplet repulsion between end groups A and C determines the height of the potential barrier. The total bonding contributions are small negative values, thus bond formation partly compensates for anti-bonding. Consequently, BSBL calculations support the conclusion derived from both BEBO results[65] and equibonding considerations;[178] triplet repulsion is the predominant factor determining the activation energy of hydrogen atom transfer reactions.

The situation is different for halogen atom transfer reactions. The bonding contributions are high positive values exceeding repulsion energies. Thus, the activation energy of the investigated halogen transfer reactions serves partly for overcoming triplet repulsion and partly for partially breaking the B–C bond. Investigations of other halogen transfer reactions and especially studies of metathesis reactions comprising the transfer of atoms other than hydrogen or halogen are required to check these conclusions on the nature of the activation energy of metathesis reactions.

3. Thermochemical-Kinetic Estimation of Rate Coefficients as a Function of Temperature

There is an increasing demand of kinetic parameters in science and technology for ana-

Table 13
TOTAL BONDING CONTRIBUTIONS ($V^{\ddagger}_{BOND}$) AND REPULSION ENERGY COMPONENTS ($V^{\ddagger}_{REP}$) OF THE POTENTIAL ENERGY OF ACTIVATION ($V^{\ddagger}$) FOR THREE SERIES OF ATOM TRANSFER REACTIONS

Series	Reaction	$\Delta V^{o} = V^{o}_{BC} - V^{o}_{AB}$ (kJ/mol)	$V^{\ddagger}$ (kJ/mol)	$V^{\ddagger}_{bond}$ (kJ/mol)	$V^{\ddagger}_{rep}$ (kJ/mol)
X + H–I	I + H–I	0.0	12.6	−2.5	22.1
	Br + H–I	−69.9	8.8	−3.8	12.5
	Cl + H–I	−137.4	5.0	−5.3	10.3
	F + H–I	− 281.0	1.6	−6.9	8.5
H + H–R	H + H –CH_3	−5.4	45.9	2.9	43.1
	H + H–C_2H_5	−30.5	32.5	−3.0	35.5
	H + H–$CH(CH_3)_2$	−47.2	27.0	−3.7	30.7
	H + H–$C(CH_3)_3$	−53.1	25.3	−3.8	29.1
H + X–CH_3	H + I–CH_3	−74.1	7.9	7.8	0.1
	H + Br–CH_3	−86.3	10.3	10.0	0.3
	H + Cl–CH_3	−95.9	12.9	12.1	0.7
	H + F–CH_3	−131.5	17.4	8.7	8.7

lyzing and controlling complex chemical systems of various types (e.g., atmospheric processes, high-temperature oxidations, and pyrolyses). These complex reactions occur under very different circumstances. In kinetic analysis of complex systems, one would naturally prefer to use rate coefficients and kinetic parameters of elementary reactions determined under conditions (temperature, pressure, and composition) similar to those prevailing in the system to be studied. This is, however, often not practical and sometimes even impossible. Thus, kinetic data measured at laboratory conditions need to be extrapolated, which may introduce significant uncertainties.

Thermochemical-kinetic formulation of TST is the best method for the estimation of Arrhenius parameters and extrapolation of the rate coefficients to different temperatures. The rate coefficient expression for a bimolecular reaction is

$$k(T) = \frac{c^2 k_B T}{h} \exp(\Delta S^{\ddagger}_c/R) \exp(-E_A/RT) \tag{86}$$

Determination of the A factor is related to the calculation of the activation entropy and thereby the estimation of the entropy of the transition state. The activated complex entropy, $S^{\ddagger}$, can be given by means of contributions from electronic, translational, rotational, and internal degrees of freedom:

$$S^{\ddagger} \simeq S^{\ddagger}_{elec} + S^{\ddagger}_{trans} + S^{\ddagger}_{i.rot} + S^{\ddagger}_{vib} \tag{87}$$

Benson[108] has proposed a thermochemical-kinetic method for the estimation of the entropy contributions. (The method is described in detail in Reference 108 and is not reproduced here.) By assumption of a tight transition complex for a metathesis reaction, the A factor can be estimated. With the A factor and an experimental rate coefficient at a single temperature, the activation energy can be obtained.

Estimation of the transition state entropy can be simplified and made more accurate if an appropriate model compound of known entropy which resembles the transition complex is selected.[108,228,229] The basis of this estimation method arises from the recognition[108,228] that molar entropies are relatively insensitive to structural details and to the number of H atoms in the species considered. Thus, the molar entropies and heat capacities for structurally similar species are similar. According to Benson's method, the known entropy of the model compound (which is usually the transition complex minus the transferred atom) is corrected for the differences in electronic degeneracy, symmetry, and rotational, internal rotational, and vibrational characteristics between the model compound and the activated complex. Often group additivity methods[108,230] are used to estimate the entropy difference between model and complex. Thus, $\Delta S^{\ddagger}$ and an A factor are obtained which, again with a known k value, yield the activation energy for the reaction.

The above-described thermochemical-kinetics method has been used to estimate Arrhenius parameters for a number of metathesis reactions of atoms and di- and polyatomic radicals.[108,228,229,231] However, knowledge of A and E_A alone is not sufficient for a long temperature range extrapolation of the kinetic data if significant deviations occur from the Arrhenius law. Cohen[229] "calibrated" the transition state models of OH + alkane reactions by adapting them to the experimental rate coefficients at 300 K, and then used conventional TST to extrapolate the rate coefficients to high temperatures. Calculated rate data were found to agree well with experiment in a wide temperature range.

A different approach focusing on the problem of extrapolation of rate coefficients has been proposed by Benson et al[232] and Shaw.[233,234] The procedure, which is also based on TST, is essentially empirical. The principle of the method can be outlined as follows. One starts from the thermodynamic formulation of the fundamental TST equation:

$$k(T) = \frac{k_B T}{h} R'T \exp(-\Delta G^{\ddagger}_T/RT) \tag{88}$$

where $R' = 0.082\ dm^3\ mol^{-1}\ atm\ K^{-1}$. The standard free energy of activation (standard state of 1 atm) at temperature T, $\Delta G^{\ddagger}_T$, can be expressed by means of the enthalpy and entropy of activation at 298 K ($\Delta H^{\ddagger}_{298}$ and $\Delta S^{\ddagger}_{298}$, respectively) and the heat capacity of activation ($\Delta C^{\ddagger}_p$):

$$\Delta G^{\ddagger}_T = \Delta H^{\ddagger}_{298} + \int_{298}^{T} \Delta C^{\ddagger}_p\, dt - T\Delta S^{\ddagger}_{298} - T\int_{298}^{T} \Delta C^{\ddagger}_p\, d\ln T \tag{89}$$

Assuming that $\Delta C^{\ddagger}_p$, the heat capacity of activation, is a constant, we may easily perform the integrations indicated in Equation 89. Substitution in Equation 88 and comparison with the three-parameter equation:

$$k(T) = B'T^m \exp(-E_B/RT) \tag{90}$$

yields the following expressions for the parameters:

$$B'/(dm^3\ mol^{-1}\ sec^{-1}) = \frac{R'k_B}{h} 298 - (\Delta C^{\ddagger}_p/R) \exp(\Delta S^{\ddagger}_{298} - \Delta C^{\ddagger}_p)/R \tag{91}$$

$$m = 2 + (\Delta C^{\ddagger}_p/R) \tag{92}$$

$$E_B/(kJ\ mol^{-1}) = \Delta H^{\ddagger}_{298} - 298\ \Delta C^{\ddagger}_p \tag{93}$$

It is obvious from the equations that the form of the extrapolation function (Equation 90) and the extent of non-Arrhenius behavior are essentially determined by the heat capacity of activation. Equation 90 reduces to the Arrhenius function if $(\Delta C^{\ddagger}_p/R) = -2$.

The value of $\Delta C^{\ddagger}_p$ does depend on the type of reaction considered. Benson et al.[232] have studied metathesis reactions of type:

$$X + YZ \rightarrow XY + Z \tag{94}$$

where X, Y, and Z are atoms H, N, and O. These authors estimated $(\Delta C_p^{\ddagger}/R) = -1.5 \pm 0.5$ for Equation 94 type reactions and could describe the temperature dependence of the rate coefficient for a number of atom + diatomics reactions. Shaw[233] extended the estimation technique to reactions:

$$W + XYZ \rightarrow WZ + YZ \tag{95}$$

where W, X, Y, and Z are H and O atoms. From reasonable assumptions and standard thermodynamics, Shaw obtained $(\Delta C_p^{\ddagger}/R) = -1.25 \pm 0.75$ and was able to represent successfully the temperature dependence of the rate coefficients for 7 atom transfer reactions of type Equation 95. In a latter communication, Shaw[234] studied H-atom abstractions from methane:

$$X + CH_4 \rightarrow XH + CH_3 \tag{96a}$$

and

$$XY + CH_4 \rightarrow XYH + CH_3 \tag{96b}$$

and found that the choice of m = 2 [i.e., $(\Delta C_p^{\ddagger}/R) = 0$] fits best the temperature dependence of the rate coefficients for Equation 96 type reactions. It has also been mentioned that Equation 90 with m = 2 may be a better representation for reaction Equations 94 and 95 as well.

The information available so far appears to indicate that m depends on the complexity of the reactants. The values of m proposed for Equations 94 to 96 are 0.5 , 0.75, and 1.0 or 2.0, respectively. This seems to indicate that the non-Arrhenius behavior increases as the reacting species grow larger. However, more evidence is required to support this conclusion and to be able to predict the form of temperature dependence of metathesis reactions. A simple and reliable semiempirical extrapolation technique would be of great value since there is a considerable and continually increasing demand for kinetic data under uncommon and often extreme conditions.

D. Estimation of Activation Energies by Using Empirical Schemes

In this section, we deal with the estimation of the activation energies by using empirical schemes. The treatments which belong to this category are either based on empirical correlations between activation energies and chemical or physical properties of separated reactant and product molecules, or assume additivity for some molecular properties.

1. Estimation Methods Based on Bond Dissociation Energy Correlations

Estimation methods of this kind make use of empirical correlations between activation energies and dissociation energies of bonds broken or formed, or are based on relationships between activation energies of metathesis reactions and reaction heats.

For homologous series of exoergic atom transfer reactions A + BC → AB + C, the activation energy often has been found to decrease with increasing exoergicity of the reaction. Evans and Polanyi[235,236] studied the relation between activation energy and reaction heat on sodium atom reactions with a series of alkyl halides and discussed their observations in terms of potential energy profiles. They suggested a correlation which has been used widely since its recognition:

$$E_A = a\,\Delta H + b \tag{97}$$

where E_A is the experimental activation energy, ΔH is the heat of reaction ($-\Delta H = Q$ is the heat release in the reaction), and a and b are assumed to be constants[240] for series of reactions in which structurally similar substrates are involved.

The Evans-Polanyi equation has been applied successfully[229,230,237,238] to a number of homologous series in which an atom or free radical reacted with substrates of the same type. For such series, the same bond is formed, therefore, the Evans-Polanyi relationship requires the activation energy to be proportional to the dissociation energy of the rupturing bond:

$$E_A = a\,D(B - C) + b' \tag{97a}$$

Equation 97a has been used occasionally to determine bond dissociation energies.

A special application of the Polanyi-type Equation 97a was suggested by Grenier,[239] who found that a plot of the activation energies for hydrogen atom abstractions from alkanes by OH radicals vs. the activation energies of CH_3 radical reactions with the same alkanes gave an excellent straight line which could be described by

$$E_A(OH) = 0.77\,E_A(CH_3) - 23.8\text{ kJ mol}^{-1} \tag{98}$$

It can be shown easily that the two parameters in Equation 98 (namely, $\alpha = 0.77$ and $\beta = 23.8$ kJ/mol) may be given in terms of the a and b′ paramaters of Equation 97a:

$$\alpha = \frac{a_{OH}}{a_{CH_3}} \quad\text{and}\quad -\beta = b'_{OH} - \frac{a_{OH}}{a_{CH_3}}\,b'_{CH_3} \tag{99}$$

where the subscripts refer to the series of OH and CH_3 radical reactions, respectively. The significance of correlations of Equation 98 type equations is that they allow the estimation of activation energies for reactions of a free radical from values for reactions of another free radical, and vice versa. Moreover, activation energies are directly compared and thereby the use of bond dissociation energies (which are often insufficiently known) can be avoided.

Semenov[241] generalized the activation energy vs. reaction heat correlation suggesting that the activation energy (kJ/mol) of practically all exoergic reactions can be obtained from the equation with fixed parameters:

$$E_A = 0.25\,\Delta H + 46 \tag{100}$$

and E_A for endoergic reactions from $E_A = 0.75\,\Delta H + 46$; i.e., the Semenov-Polanyi relationship requires that the correlation between activation energy and reaction heat be roughly independent of the type of metathesis reactions. However, use of the generalized Equation 100 appears to decrease the accuracy of the Equation 97 type equations.

Although the simple exoergicity rule that is expressed in the Evans-Polanyi and Semenov-Polanyi equations seems to be valid for many reaction series and various reaction groups, there are a number of exceptions known, and some objections to the rule can be made. In the first place, one observes that Equations 97 and 100 predict negative activation energies for highly exothermic reactions. One would expect rather a gradual decrease to zero of the activation energies of metathesis reactions with increasing exoergicity. Indeed, this is what follows from more sophisticated treatments. Thus, for a series of reactions of a free radical A with a number of substrates BC of similar structure, the relationship:

$$E_A = [D^\circ(A-B) + \Delta H^\circ]\left\{1 - \left[1 + \left(\frac{D^\circ(A-B) + \Delta H^\circ}{D^\circ(A-B)}\right)^{1/(p-1)}\right]^{(1-p)}\right\} \tag{101}$$

is obtained from the BEBO formalism.[65] In Equation 101, D° (A–B) and ΔH° (= −Q) designate the dissociation energy of the bond formed and the reaction heat, respectively, and p is a bond parameter which is treated preferably as an empirical parameter with a value adjusted to one activation energy in the reaction series. Equation 101 predicts nonlinear relationship between E_A and ΔH°, and approaches zero for highly exothermic reactions. The nonlinearity is more definite for reaction series characterized by p values not significantly different from 1.0. Thus, it can be shown[65] that for CH_3 radical reactions with RH (where p = 1.17), only small deviations from the Polanyi-type correlations are expected which hardly exceed the error limits, while for Cl atom reactions with RH (where p = 1.05), significant curvature occurs.

Several extensions of Evans-Polanyi-type relations have been published in the literature. Most of these propose correlations between the activation energy and dissociation energies of broken and formed bonds rather than between E_A and the exoergicity of the reaction. The empirical schemes developed by Kagiya et al.[242,243] belong to such types of estimation methods. Kagiya et al.[242] extended the approach originally initiated by Ogg and Polanyi[244] by assuming that the potential energy of the system along the reaction path of a metathesis reaction can be expressed by two Morse functions. The product of the Morse parameters (a′) and the deviation (Δr*) of the interatomic distance from its equilibrium value in the transition state, i.e., the a′Δr* product, is evaluated from plots against the reaction heat. Finally, the following expression of the activation energy is obtained:

$$E_A = \frac{D^\circ(B-C)[(1 - 2\alpha)\, D^\circ(A-B) + \alpha^2 D^\circ(B-C)]^2}{[D^\circ(A-B) - \alpha^2 D^\circ(B-C)]^2} \tag{102}$$

where α = exp(−0.019 ΔH) and other designations have their usual meaning. A very detailed test including various metathesis reactions of atoms and alkyl radicals and a few miscellaneous radicals showed Equation 102 to supply data which agreed in most cases within 8 kJ/mol with the experimental results. However, systematic deviations also occurred, especially in case of halogen atom transfer reactions. Moreover, the basic equation (i.e., Equation 102) is rather complicated, not promoting the application of the method.

In order to simplify the estimation procedure, Kagiya et al.[243] introduced some approximations and finally obtained:

$$E_A = D^\circ(B-C) - \epsilon\, D^\circ(A-B)[1 + \delta D^\circ(B-C)] \tag{103}$$

where ε and δ are empirical parameters with approximate values of ε = 0.0850 and δ = 0.0944. The activation energies estimated according to Equation 103 are in good agreement with experiment.[243] Calculated and measured data are generally within 6 to 8 kJ/mol, but the method appears to fail[219] for reactions X + RH → HX + R and R + HX → RH + X (where X designates a halogen atom).

By substitution of ΔH° = D°(B–C) − D°(A–B), Equation 103 can be rearranged to

$$E_A = [1 - \epsilon\delta D^\circ(A-B)]\, \Delta H^\circ + [1 - \epsilon - \epsilon\delta D^\circ(A-B)]\, D^\circ(A-B) \tag{104}$$

From Equations 104 and 97 one finds that the two parameters in the Polanyi equation are given by

$$\begin{aligned} a &= 1 - \epsilon\delta D^\circ(A-B) \\ b &= [1 - \epsilon - \epsilon\delta D^\circ(A-B)]\, D^\circ(A-B) \end{aligned} \tag{105}$$

This indicates that the Polanyi-type correlations are expected to hold in homologous series of reactions of a free radical with structurally similar substrates, while the relation should fail for a series of reactions of different radicals with a given substrate. (Note that, according to Equation 105, both a and b depend on the energy of the forming bond.)

A relationship similar to Eq. (103) of Kagiya et al. has been proposed by Szabó[245,246] which also takes into account the dissociation energies of both breaking and forming bonds. For atom transfer reactions, the equation takes the form:

$$E_A = D^\circ(B-C) - \alpha D^\circ(A-B) \quad (106)$$

where α is a parameter which depends on the structure of the transition state and the exo- or endoergic nature of the reactions considered. The recommended value of α is 0.83 and 0.96 for exo- and endoergic metathesis reactions, respectively. Equation 106 was found[246,247] to be obeyed somewhat better than the two-parameter Evans-Polanyi-type equations.

2. Estimation Methods Based on Correlations with Nonthermochemical Molecular Properties of Reactants and Products

Polanyi-type correlations were found to be obeyed in many series of similar reactions, and Equation 97 was used many times with success in estimating activation energies for atom transfer reactions. However, a number of cases are known where the simple exoergicity rule appears to fail. The best documented exceptions are found[219,248] among halogen atom transfer reactions, e.g., reactions $H + X_2 \rightarrow HX + X$, $H + CH_3X \rightarrow HX + CH_3$, and $CF_3 + RX \rightarrow CF_3X + R$ (where X is a halogen atom and R designates an alkyl group). There is some evidence that molecular properties other than bond dissociation energies must also be taken into account, at least in those empirical estimation schemes of activation energies which are intended for more general applicability.

An early attempt correlates the activation energies with molecular properties of reactants and products and was made by Spirin,[249] who proposed a relationship between E_A, on the one hand, and the reaction heat and polarizabilities of the participating reactive species, on the other hand. His considerations set out from the London equation[174] and yield the following expression for the activation energy:

$$E_A = d(0.75\, D^\circ_{AC} + \Delta H^\circ)\left(\frac{1}{p_A} + \frac{1}{p_C}\right) \quad (107)$$

where d is an empirical parameter (with a value of about 0.06), D°_{AC} stands for the dissociation energy of bond A–C in molecule AC, ΔH° is the heat of reaction,and p_A and p_C designate the polarizabilities of reactant and product radicals (or atoms), respectively. The equation allows for the intuitively expected effect that greater polarizability of the reactants is associated with less repulsion between A and C and thereby lower activation energy of the reaction $A + BC \rightarrow AB + C$. Spirin studied various metathesis reactions of atoms and simple free radicals (CH_3, CF_3, CCl_3, etc). Most of the reactions were hydrogen atom transfers, but a number of halogen atom transfers were also considered. The activation energies calculated from Equation 107 were found to agree with experiment better than those obtained from the Evans-Polanyi relationship. Greater deviations occurred only in case of formyl hydrogen transfers, for hydrogen abstractions from substrates containing several chlorine atoms, and in reactions where polar effects occur.

The importance of polarizability of the reactants is manifested in the simple correlation used by Krech and McFadden[248] in which an inverse proportionality between E_A and the mean polarizability of reactant BC (i.e., p_{BC}) is assumed:

$$E_A = \frac{c}{p_{BC}} \quad (108)$$

Activation energies for hydrogen abstractions by a given free radical from structurally similar compounds, RH, have also been correlated[240,250] with the ionization potentials of free radicals, IP(R). It has been suggested[240] that the correlation between E_A and IP(R) should be preferred to the conventional Evans-Polanyi-type relations representing E_A in terms of ΔH° or $D^\circ(RH)$. As it has been pointed out, conventional Evans-Polanyi-type relations are not sensitive to the structural variation of R and thus these correlations are not very informative.

Surprisingly successful simple empirical methods for the estimation of activation energies in radical molecule metathesis reactions of A + BC → AB + C type have been suggested by Alfassi and Benson.[219] The development of these methods was inspired by the recognition of the importance of contribution from end groups A and C. Considering the transition state as consisting of two one-electron bonds, A·B·C, and one antibonding electron shared by A and C, an opposite tendency is expected between the activation energy and the electron affinities of A and C. Indeed, a correlation of the type:

$$E_A = (a - bI)/(1 + c\Delta H^\circ) \qquad (109)$$

was shown to be reasonably obeyed, where I is the sum of electron affinities of A and C, and a, b, and c are positive constants.

The idea of end-group contributions was exploited in the estimation of the activation energy by the additivity scheme:[219]

$$E_A = X_A + X_C \qquad (110)$$

Here, X_A and X_C are empirical increments characteristic of groups or atoms A and C, respectively, which were determined by a multilinear regression method from known activation energies. Since some of the end-group contributions are negative numbers, negative E_A values may be derived from Equation 110; for these cases, E_A has taken zero arbitrarily. Activation energies for about 50 reactions, representing 15 different end groups, were calculated by this method and compared with measured values. The standard deviation of the estimations was 3.4 kJ/mol and the average error was 4.3 kJ/mol. A few representative results are given in Table 14.

Alfassi and Benson[219] also proposed another simple estimation method which eliminates the problem of calculating negative activation energies by using the product of characteristic end-group contributions F_A and F_C:

$$E_A = F_A F_C \qquad (111)$$

Comparison of calculated activation energies with measured data is given for a few representative cases in Table 14. A detailed test of the method including more than 50 reactions gave an average error of 3.4 kJ/mol and a standard deviation of 4.6 kJ/mol.

An important factor which determines the activation energy of metathesis reactions is the interaction of end groups A and C. Thus, the success of the simple estimation methods of Alfassi and Benson, which are based on end-group contributions, is really not unexpected. Naturally, further terms may be added to the expressions which make allowance for the effect of other molecular properties of the reactants and products. First of all, the dependence on the dissociation energy of the newly formed bond and the bond broken in the reaction should be considered:[219]

$$E_A = X_A + X_C + \alpha D^\circ(B-C) - \beta D^\circ(A-B) \qquad (112)$$

Such an extension may improve the predictive power of the method for such reactions as,

Table 14
ACTIVATION ENERGIES (kJ/mol) CALCULATED BY THE END-GROUP CONTRIBUTION METHODS OF ALFASSI AND BENSON[219]

Reaction	Addition (Equation 110)	Multiplication (Equation 111)	Experimental value
$H + CH_4 \rightarrow H_2 + CH_3$	45.6	43.9	51.0
$H + C_2H_6 \rightarrow H_2 + C_2H_5$	37.2	36.0	40.2
$H + HCl \rightarrow H_2 + Cl$	10.0	7.1	18.0
$F + CH_4 \rightarrow HF + CH_3$	10.0	5.0	6.3
$Cl + C_2H_6 \rightarrow HCl + C_2H_5$	8.0	6.7	4.2
$OH + CH_4 \rightarrow H_2O + CH_3$	25.1	19.3	21.8
$OH + C_2H_6 \rightarrow H_2O + C_2H_5$	16.7	15.5	15.1
$CH_3O + CH_4 \rightarrow CH_3OH + CH_3$	38.9	36.8	35.6
$CH_3 + CH_4 \rightarrow CH_4 + CH_3$	51.9	51.1	61.1
$CH_3 + C_2H_6 \rightarrow CH_4 + C_2H_5$	43.5	41.8	48.1
$CH_3 + HCHO \rightarrow CH_4 + CHO$	27.2	22.6	25.9
$CH_3 + HCl \rightarrow CH_4 + Cl$	16.3	8.4	12.1
$CH_3CH_2 + H_2O_2 \rightarrow LC_2H_6 + HO_2$	22.2	20.9	25.1
$CF_3 + H_2 \rightarrow CF_3H + H$	36.8	37.2	39.3
$CF_3 + CH_4 \rightarrow CF_3H + CH_3$	43.1	43.1	42.7
$CF_3 + C_2H_6 \rightarrow CF_3H + C_2H_5$	34.7	35.2	34.7
$CF_3CF_2 + C_2H_6 \rightarrow C_2F_5H + C_2H_5$	42.7	43.1	38.1
$Cf_3CF_2 + HCl \rightarrow C_2F_5H + Cl$	15.5	8.8	19.3

e.g., hydrogen abstractions from double-bonded C atoms where the original method appears to fail.[251]

3. Bond Energy Additivity Methods

Additivity laws are used widely in chemistry. One of the most frequently employed additivity rules states that the heat of atomization of a compound can be given as the sum of bond energy terms corresponding to the number and type of bonds constituting the compound. The bond energy terms are by definition constants, and the validity of additivity approximation depends on the assumption that the interatomic forces are of very short range which makes interactions between nonbonded atoms negligible. Bond energy parameters are used in calculations of heats of atomization for stable compounds and thereby in the estimation of heats of formation.

Moin[252] assumed that the additivity rule of bond energy terms can be extended to activated complexes. He proposed that the energy of dissociation of the activated complex ABC:

$$A \ldots B \ldots C \rightarrow A + B + C \tag{113}$$

be given by the sum of the bond energies of the two partial bonds:

$$\Sigma Q^{\ddagger} = D_{A \cdots B \cdots C} = Q_{A \cdots B} + Q_{B \cdots C} \tag{114}$$

In the course of formation of the activated complex, the bond B–C of the reactant is broken (requiring Q_{B-C} energy) and two partial bonds are formed (resulting in the release of $\Sigma Q^{\ddagger}$ energy). Hence, the activation energy of the atom transfer reaction $A + BC \rightarrow AB + C$ is given by

$$E_A = Q_{B-C} - (Q_{A \cdots B} + Q_{B \cdots C}) \tag{115}$$

Table 15
BOND ENERGY PARAMETERS

Bond X–Y	B(X–Y)[a] (kJ/mol)	D(X–Y)[b] (kJ/mol)	Bond X–Y	B(X–Y)[a] (kJ/mol)	D(X–Y)[b] (kJ/mol)
C–H[c]	410.8—420.7	413.4	H–Cl[c]	444.8	431.8
C–H[d]	420.7		H–Cl[d]	25.9	
C–F[c]	504.2	441.0	F–F[d]	291.6	
C–Cl[d]	336.7		Cl–Cl[d]	−20.5	
H–F[d]	−140.2		Cl–F[d]	44.7	

[a] Bond energy parameters.[254,255]
[b] Conventional bond energies.[257]
[c] Bonding interactions.
[d] Nonbonding interactions.

The bond energy terms for the partial bonds are obtained from known bond energies of single bonds and experimental activation energies of the appropriate atom transfer reactions.[251,252] The method has been used by Moin[252] to calculate the activation energies for hydrogen atom abstractions (mainly from paraffins) by atoms and alkyl (including CF_3) radicals. Good agreement was found with experiment. Pivovarov and Stepukhovich[251] applied the procedure to hydrogen abstractions from double-bonded carbon atoms and found reasonably good results even when other empirical methods failed. It should be mentioned, however, that a very thorough test of the method has not been made so far.

The concepts of bond index and bond energy have been exploited in the procedure developed and applied recently by Bell and Perkins.[253-256] The bond indexes, derived from the Coulson density matrix elements, correlate with the chemist's idea of bond multiplicity. These indexes reflect even minor variations of bond properties (dissociation energies and lengths) caused by changes in environment, hence Bell and Perkins assumed that the use of bond indexes could promote the definition of an improved concept for bond energies and thereby lead to an improved calculation of the heats of atomization.

According to the proposal, standard state heat of atomization of a compound is given by the sum of products of bond indexes and bond energy terms:

$$\Delta H^{\circ}_{atom} = \Sigma m_{X-Y} B(X-Y) \tag{116}$$

where summation comprises each (bonding or nonbonding) interaction between X and Y in the molecule. In Equation 116, B(X–Y) designates the bond-energy-like term called bond energy parameter and the weighting factor m_{X-Y} is the bond index for interaction X–Y. Similar equations can be written for other molecules. With theoretical values used for bond indexes, the solution of the system of simultaneous equations of the Equation 116 type supplies values for the bond energy parameters. Such values are given in Table 15. For comparison, some conventional bond energies, as reported by Pauling,[257] are also indicated in the table.

The above-described procedure for the calculation of heats of atomization of molecules was extended to species with bonds of nonequilibrium lengths.[253,254] Consequently, a method has been provided for the computation of the heats of atomization of systems A···B···C having extended bonds. As a first step, heats of atomization of species in the standard state and at 298 K at various positions in the coordinate space are determined so that the minimum enthalpy path for the reaction can be identified. Then, $\Delta H^{\ddagger}_{298}$ is obtained as the difference between the heats of atomization in the initial state and the transition state (i.e., at saddle point geometry). With $\Delta H^{\ddagger}_{298}$, the value for $\Delta H^{\ddagger}_{T}$ is derived from the integrated form of

Kirchoff's law, taking C_p° for the reactants from the literature and estimating specific heat for the transition state from group contributions or by statistical thermodynamics.[108] Finally, the Arrhenius activation energy at temperature T is obtained in accordance with TST from Equation 19.

In calculation of the heat of atomization of the complex A···B···C, certain problems arise which have to be dealt with. These can be described best by considering an example. Let us choose reaction $CH_3 + CH_4$:

$$CH_3 + CH_4 \rightarrow H\underset{H}{\overset{H}{-}}C\text{–}H\text{–}C\underset{H}{\overset{H}{-}}H \rightarrow CH_4 + CH_3 \qquad (117)$$

In the initial state, there are two types of C–H bonds, the "normal" C–H bond in CH_4 and the C–H bond in the methyl radical. Accordingly, two different bond energy parameters are needed. B(C–H) for CH_3 may be derived[254] by using the known heat of formation of CH_3radical. In this way, B(C–H) = 410.8 kJ/mol was obtained which is to be compared with B(C–H) = 420.7 kJ/mol, the value appropriate for the methane C–H bond. In the course of the reaction, all C–H bonds change character, since CH_3 is converted into CH_4 and the original CH_4 changes into CH_3 in the reaction. Thus, in addition to using two different C–H bond energy parameters, they have to be considered a function of the reaction coordinate. A further problem occurs because the planar CH_3 radical takes on a pyramidal shape as the two reactants approach each other. The authors have developed[254] a sophisticated procedure to handle all these problems, and presented graphically as well as in tabular form the variation of the bond energy parameters in the course of the reaction (i.e., as the length of the newly formed C–H bond changes).

The method of Bell and Perkins was applied for the calculation of E_A (and A factors) for hydrogen abstraction reactions of CH_3 and CF_3 radicals from CH_4, H_2, and HCl substrates[254] for Cl-atom reactions with chlorinated methanes[255] and for H and Cl abstractions by Cl atoms from C_2H_6 and chlorinated ethanes. The calculated activation energies agreed with the experimental values generally within the estimated error limits. The uncertainties of the experimental data of H-atom abstractions from more common substrates are reasonably small, however, in the case of highly chlorinated substrates and especially for chlorine atom transfer reactions, experimental errors are substantional. Thus, the reactions studied are not particularly good choices for determination of the predictive power of the method. Apart from this, it has to be stated that this additivity method is very complicated; therefore, the extra effort made in applying the procedure would be justified only by highly accurate results.

E. Empirical Correlations of Barrier Properties in Homologous Reaction Series

Discussion in this section concentrates on structure-reactivity type relationships in homologous reaction series. The question briefly considered is how height and location of potential energy barrier change with systematic variation of the reactants in a series of related atom transfer reactions A + BC → AB + C.

1. Correlations of Barrier Heights and Barrier Locations Along the Reaction Coordinate

It is known for some time that transition state properties (i.e., heights and locations of the potential energy barriers) change with variation of the structure of reactants. Among the earliest findings were the Evans-Polanyi relationship and the Hammond rule which correlated reaction heat in a group of related reactions with activation energy (barrier height) and the barrier location, respectively. Polanyi-type relationships have been discussed in detail in Section III.D.1. The Hammond rule[258] is concerned with the position of the transition state along the reaction coordinate, and predicts early saddle points and transition states resembling reactants for highly exothermic reactions, while late saddle points and transition states resembling products are predicted for endothermic reactions.

Numerous attempts have been made to derive potential barrier heights and barrier locations from empirical correlations valid in homologous reaction series. Such empirical relationships were formulated most often by generalizing the results of theoretical or semiempirical calculations for prototypical series of homologous reactions. Thus, Mok and Polanyi[259] examined related families of hydrogen and halogen atom reactions with hydrogen halides in the LEPS and BEBO approximations. These authors found that the barrier for substantially exothermic reactions is in the entry valley and for substantially endothermic reactions the barrier is in the exit valley of the potential energy surface. For decreasing barrier height in related exothermic reactions, the barrier moves to successively earlier positions along the entry valley. For increasing barrier height in related endothermic reactions, the barrier moves to successively later positions along the exit valley. In a quantitative form:

$$\Delta\log V^{\ddagger} = \alpha\Delta(V^{\circ}_{BC} - V^{\circ}_{AB}) \tag{118}$$

$$\Delta\log V^{\ddagger} = -\beta\Delta X^{\ddagger}_{AB} \tag{119}$$

$$\Delta(V^{\circ}_{BC} - V^{\circ}_{AB}) = -(\beta/\alpha)\,\Delta X^{\ddagger}_{AB} \tag{120}$$

where α and β are constants within a reaction series, $(V^{\circ}_{BC} - V^{\circ}_{AB})$ is the difference in product and reactant energies (not including zpes), $V^{\ddagger}$ designates the classical barrier height, and $X^{\ddagger}_{AB} = R^{\ddagger}_{AB} - R^{\circ}_{AB}$ is the extension of the A–B bond at the saddle point.

In a recent similar study, Bérces et al.[227] used the BSBL method for the investigation of location and height of activation barriers in various homologous series of A + BC → AB + C atom transfer reactions. For all reaction series investigated, the barrier shifted to an earlier position along the reaction coordinate as the ratio $V^{\circ}_{BC}/V^{\circ}_{AB}$ decreased (where V°_{BC} and V°_{AB} designate the energies of the splitting and forming bonds, respectively):

$$\ln(RC)_{\ddagger} = a_1 \frac{V^{\circ}_{BC}}{V^{\circ}_{AB}} - b_1 \tag{121}$$

Here, a_1 and b_1 are constants and $(RC)_{\ddagger}$ designates the reaction coordinate at the saddle point [i.e., $(RC_{\ddagger} = \exp(-2\beta_{AB}X^{\ddagger}_{AB})$; see Reference 227]. Correlations between barrier height and $V^{\circ}_{BC}/V^{\circ}_{AB}$, as well as between barrier height and barrier location were found valid for series of metathesis reactions in which either the attacking species A or the leaving species C was varied:

$$\ln V^{\ddagger} = a_2 \frac{V^{\circ}_{BC}}{V^{\circ}_{AB}} - b_2 \tag{122}$$

and

$$V^{\ddagger} \simeq a_3(RC)_{\ddagger} \tag{123}$$

However, Equations 122 and 123 failed for groups of halogen atom transfer reactions (in which the transferred atom B was varied). The breakdown of the barrier height correlations for groups in which the transferred atom was varied may be a result of neglecting end-group interactions in these simple empirical correlations, i.e., in Equations 118 and 119 as well as Equations 122 and 123.

Dunning[260] used generalized valence bond and configuration interaction methods for the calculation of reaction energies, barrier heights, and saddle point geometries for $H + HX \rightarrow H_2 + X$ (where X = halogen atom) hydrogen abstraction reactions. A remarkably good representation of the calculated barrier heights was obtained by a relationship equivalent to Equations 118 and 122.

Basilevsky et al.[261] developed a procedure for the estimation of the position of the transition state along the reaction coordinate in reaction series of various free radical metathesis reactions. Use of the method requires the knowledge of transition state properties for one reaction (the simplest reaction without substituents) in the series. Transition state properties for this standard reaction together with an experimental activation energy vs. reaction heat relationship are then used for the estimation of barrier locations for other members of the series:

$$\alpha^{\ddagger} = \alpha_o^{\ddagger} + \beta(Q - Q_o) \tag{124}$$

where $\alpha^{\ddagger}$ and Q are reaction coordinates for the transition state and reaction heat, respectively, and subscript o refers to the standard reaction. Coefficient β can be calculated from transition state properties of the standard reaction and from the experimental activation energy vs. reaction heat relationship.

Thermodynamic-like relations are used by Agmon and Levine[262] to interpret a quantitative version of the Hammond postulate and an Evans-Polanyi-type correlation. The procedure views the reaction as a continuous transformation from reactants to products, and introduces the concept of mixing entropy to account for an intrinsic barrier of the transformation. Explicit expressions are derived for the potential energy along the reaction coordinate and thereby for the location and height of the potential barrier.

A simple and remarkably successful relationship has been proposed by Miller[263] for predicting the barrier position ($X^{\ddagger}$) in terms of the barrier height ($\Delta E^{\ddagger}$) and difference between product and reactant energies (ΔE):

$$X^{\ddagger} = \frac{1}{2 - \Delta E/\Delta E^{\ddagger}} \tag{125}$$

A theoretical derivation of Equation 125 has been given by Miller in the original paper, and the criteria required to give the correlation were discussed.[263,264]

Finally, a free energy relationship proposed by Marcus[265,266] should be considered:

$$\Delta F^{*} = \Lambda[1 + (\Delta F^{\circ\prime}/4\,\Lambda)]^2 \tag{126}$$

where ΔF^{*} and $\Delta F^{\circ\prime}$ designate the free energy barrier and the "standard" free energy of the reaction, respectively, and Λ is the intrinsic barrier for the reaction (i.e., the barrier at $\Delta F^{\circ\prime} = 0$). The slope of the free energy plot is $\beta = d\Delta F^{*}/d\Delta F^{\circ\prime}$, which is given by

$$\beta = 0.5\left(1 + \frac{\Delta F^{\circ\prime}}{4\,\Lambda}\right) \tag{127}$$

when Λ is constant in the reaction series considered. For such systems, β reflects the position of the transition state along the reaction coordinate.

IV. SPECIFIC EFFECTS INFLUENCING THE RATE AND ACTIVATION ENERGY

A. Polar Effects

Discussion in Sections III.D and III.E revealed the importance of the strengths of breaking and forming bonds in the determination of the activation energies in metathesis reactions of free radicals. However, kinetic observations made from time to time contradicted predictions derived from thermochemical considerations. The cause of the unexpected behavior was often the so-called polar effect.

The polar effect can be demonstrated best by considering the reactions of methyl and halogenated methyl radicals with HCl and hydrocarbons, respectively.[267] Comparison of the activation energies given below (kJ/mol) for CH_3, CCl_3, and CF_3 radical reactions with CH_4 and HCl shows significant differences:

$$CH_3 + CH_4,\ E_A = 60.7; \qquad CH_3 + HCl,\ E_A = 9.6$$

$$CCl_3 + CH_4,\ E_A = 71.5; \qquad CCl_3 + HCl,\ E_A = 46.9$$

$$CF_3 + CH_4,\ E_A = 46.0; \qquad CF_3 + HCl,\ E_A = 21.6$$

Activation energies for reactions with HCl are consistently and significantly lower than those obtained for reactions of the same radicals with CH_4 or hydrocarbons. There is no great difference in the reaction heats of the reactions with CH_4 and HCl. The lower activation energies for the reactions with HCl can be rationalized in terms of attractive polar forces in the transition state. Difference in the activation energies is greatest for methyl radical reactions, in accordance with the great difference in electronegativities between attacking radical and substrate molecule.

Reaction CCl_3 + HCl is endothermic, thus it is not easily compared with the other two exothermic reactions. Among the latter ones, a higher activation energy is determined for the hydrogen abstraction from HCl by CF_3 than by CH_3, despite the greater exothermicity of the trifluoromethyl radical reaction. (It is to be noted that, usually, reactions of CH_3 have higher activation energies than reactions of CF_3 radicals.) The observed activation energy difference can be rationalized[268] by taking into account that the repulsion forces between the carbon atom of the polar CF_3 radicals and the polar HCl molecules will be greater than in the case of reaction CH_3 + HCl. The electronegative trifluoromethyl radical will resist the formation of a polar transition state, while the electropositive methyl radical will facilitate the formation of a polar transition state.[269]

Evidence for polar effects has been found[270] for hydrogen abstractions from silanes by CF_3 and CH_3 radicals. Polar substituents have opposite effects on the activation energies of CH_3 and CF_3 reactions, which confirms the role of polar forces in the transition states of these reactions.

B. Comparison of Activation Energies for Liquid- and Gas-Phase Reactions

The traditional ways of determination of kinetic parameters in the gas phase and in solution are different. The rate coefficients of gas-phase atom transfer reactions of polyatomic free radicals are normally measured against the specific rates of radical recombination reactions, while in liquid phase the rates of metathesis reactions are often compared with the rates of hydrogen abstractions from the solvent. Therefore, kinetic data for the same reaction determined under similar conditions are seldom available, which hinders direct comparison of gas- and liquid-phase results.

For a metathesis reaction free of polar effect, one expects similar rates and reactivities in the gas phase in nonpolar solvents, where solvation energy of the transition state is expected to be similar to the solvation energy of the reactants. Convincing support for this hypothesis has been received only recently in the study of hydrogen abstraction reactions of CCl_3 radicals by Katz et al.[271,272] and Alfassi and Feldman,[273,274] as well as in the investigation of $CHCl_2$ metathesis reactions.[275] These authors used the method of gamma-radiation-induced free radical chain reactions in solutions of CCl_4 (or $CHCl_3$) and hydrogen-containing substrates. The rate coefficients of hydrogen abstraction reactions were measured in competition with the recombination of CCl_3 radicals to form hexachloroethane. The investigations confirm the thesis that activation energies of radical metathesis reactions in the liquid phase are equal to their respective values in the gas phase.

Interpretation of liquid-phase kinetic results obtained in polar solvents is more complicated. The extent of solvation of the reactants and the transition state may vary in a wide range causing significant solvent effects. Thus, one expects different reactivities in the gas phase and in solution.

ACKNOWLEDGMENT

The authors would like to thank Dr. S. Dóbé and Dr. Gy. Lendvay for their comments on the manuscript.

REFERENCES

1. **Kondratiev, V. N.,** *Rate Constants of Gas Phase Reactions,* Fristrom, R. M., Ed., National Standard Reference Data Series, National Bureau of Standards, Washington, D.C., 1972.
2. **Trotman-Dickenson, A. F., and Milne, G. S.,** *Tables of Biomolecular Gas Reactions,* National Standard Reference Data Series, National Bureau of Standards, Washington, D.C., 1967, 9.
3. **Ratajczak, E. and Trotman-Dickenson, A. F.,** *Supplementary Tables of Bimolecular Gas Reactions,* University of Wales, Institute of Science and Technology, Cardiff, 1970.
4. **Kerr, J. A. and Ratajczak, E.,** *Second Supplementary Tables of Bimolecular Gas Reactions,* The University, Birmingham England, 1973.
5. **Kerr, J. A. and Ratajczak, E.,** *Third Supplementary Tables of Bimolecular Gas Reactions,* The University, Birmingham, England, 1977.
6. **Kerr, J. A. and Mose, S. J., Eds.,** *CRC Handbook of Bimolecular and Termolecular Gas Reactions,* Vol. 1, CRC Press, Boca Raton, Fla., 1981.
7. **Kerr, J. A. and Parsonage, M. J.,** *Evaluated Kinetic Data on Gas Phase Hydrogen Transfer Reactions of Methyl Radicals,* Butterworths, London, 1975.
8. **Allara, D. L. and Shaw, R.,** A compilation of kinetic parameters for the thermal degradation of n-alkane molecules, *J. Phys. Chem. Ref. Data,* 9, 523, 1980.
9. **Hoyerman, K. and Wagner, H. Gg.,** Elementary reactions in the high temperature oxidation of hydrocarbons, *Oxid. Comm.,* 2, 259, 1982.
10. **DeMore, W. B., Watson, R. T., Golden, D. M., Hampson, R. F., Kurylo, M., Howard, C. J., Molina, M. J., and Ravishankara, A. R.,** *Chemical Kinetics and Photochemical Data for* Use in *Stratospheric Modeling,* JPL Publ. No. 82-57, NASA and Jet Propulsion Laboratory, Pasadena, 1982.
11. **Atkinson, R. and Lloyd, A. C.,** Evaluation of kinetic and mechanistic data for modeling of photochemical smog, *J. Phys. Chem. Ref. Data,* 13, 315, 1984.
12. **Baulch, D. L., Cox, R. A., Crutzen, P. J., Hampson, R. F., Jr., Kerr, J. A., Troe, J., and Watson, R. T.,** Evaluated kinetic and photochemical data for atmospheric chemistry: supplement I, *J. Phys. Chem. Ref. Data,* 11, 327, 1982.
13. **Kerr, J. A.,** Rate processes in the gas phase, in *Free Radicals,* Vol. 1, Kochi, J. K., Ed., John Wiley & Sons, New York, 1973, chap. 1.
14. **Kerr, J. A.,** Metathetical reactions of atoms and radicals, in *Comprehensive Chemical Kinetics,* Vol. 18, Bamford, C. H. and Tipper, C. F. H., Eds., Elsevier, Amsterdam, 1976, chap. 2.
15. **Denisov, E. T.,** Konstanty skorosti gomoliticheskikh zhidkofaznykh reaktsii, *Izd. Nauka,* Moscow, 1971.
16. **Hendry, D. G., Mill, T. Piszkiewicz, L., Howard, J. A., and Eigenmann, H. K.,** A critical review of H-atom transfer in the liquid phase: chlorine atom, alkyl, trichloromethyl, alkoxy, and alkylperoxy radicals, *J. Phys. Chem. Ref. Data,* 3, 937, 1974.

17. **Fischer, H., Ed.,** *Landolt-Börnstein, Numerical Data and Functional Relationships in Science and Technology,* Vol. 13, (Subvols. b to d), Springer-Verlag, Basel, 1984.
18. **Ingold, K. U.,** Rate constants for free radical reactions in solution, in *Free Radicals,* Vol. 1, Kochi, J. K., Ed., John Wiley & Sons, New York, 1973, chap. 2.
19. **Pacey, P. D.,** Changing conceptions of activation energy, *J. Chem. Educ.,* 58, 612, 1981.
20. **Eliason, M. A. and Hirschfelder, J. O.,** General collision theory treatment for the rate of bimolecular gas phase reactions, *J. Chem. Phys.,* 30, 1426, 1959.
21. **Tolman, R. C.,** Statistical mechanics applied to chemical kinetics, *J. Am. Chem. Soc.,* 42, 2506, 1920.
22. **Karplus, M., Porter, R. N., and Sharma, R. D.,** Energy dependence of cross sections for hot tritium reactions with hydrogen and deuterium molecules, *J. Chem. Phys.,* 45, 3871, 1966.
23. **Smith, I. W. M.,** *Kinetics and Dynamics of Elementary Gas Reactions,* Butterworths, London, 1980.
24. **Heneghan, S. P., Knoot, P. A., and Benson, S. W.,** The temperature coefficient of the rates in the system $Cl + CH_4 \rightleftharpoons CH_3 + HCl$. Thermochemistry of the methyl radical, *Int. J. Chem. Kinet.,* 13, 677, 1981.
25. **Hulett, J. R.,** Deviations from the Arrhenius equation, *Q. Rev.,* 18, 227, 1964.
26. **Perlmutter-Hayman, B.,** The temperature dependence of E_a, in *Progress in Inorganic Chemistry,* Vol. 20, Lippard, S. J., Ed., John Wiley & Sons, New York, 1976, 229.
27. **Gardiner, W. C., Jr.,** Temperature dependence of bimolecular gas reaction rates, *Acc. Chem. Res.,* 10, 326, 1977.
28. **Zellner, R.,** Non-Arrhenius behavior in bimolecular reactions of the OH radical, *J. Phys. Chem.,* 83, 18, 1979.
29. **Fontijn, A. and Zellner, R.,** Influence of temperature on rate coefficients of bimolecular reactions, in *Reactions of Small Transient Species,* Fontijn, A. and Clynne, M. A. A., Eds., Academic Press, London, 1983, chap. 1.
30. **Clark, T. C., Dove, J. E., and Finkelman, M.,** Temperature dependence of rates of bimolecular elementary reactions, *Acta Astronaut.,* 6, 961, 1979.
31. **Moelwyn-Hughes, E. A.,** Über den Temperaturkoeffizienten der Rohrzuckerinwersion, *Z. Phys. Chem.,* B26, 281, 1934.
32. **Arrhenius, S.,** Über die Reaktionsgeschwindigkeit bei der Inersion von Rohrzucker durch Säuren, *Z. Phys. Chem.,* 4, 226, 1889.
33. **Westenberg, A. A. and deHaas, N.,** Atom-molecule kinetics using ESR detection. II. Results for $D + H_2 \rightarrow HD + H$ and $H + D_2 \rightarrow HD + D$, *J. Chem. Phys.,* 47, 1393, 1967.
34. **Blais, N. C., Truhlar, D. G., and Garrett, B. C.,** Temperature dependence of the activation energy: $D + H_2$, *J. Chem. Phys.,* 76, 2768, 1982.
35. **Clark, T. C. and Dove, J. E.,** Examination of possible non-Arrhenius behavior in the reactions $H + C_2H_6 \rightarrow H_2 + C_2H_5$, $H + CH_4 \rightarrow H_2 + CH_3$ and $CH_3 + C_2H_6 \rightarrow CH_4 + C_2H_5$, *Can. J. Chem.,* 51, 2147, 1973.
36. **Manning, R. G. and Kurylo, M. J.,** Flash photolysis resonance fluorescence investigation of the temperature dependencies of the reactions of $Cl(^2P)$ atoms with CH_4, CH_3Cl, CH_3F, CH_3F^* and C_2H_6, *J. Phys. Chem.,* 81, 291, 1977.
37. **Whytock, D. A., Lee, J. H., Michael, J. V., Payne, W. A., and Stief, L. J.,** Absolute rate of the reaction of $Cl(^2P)$ with methane from 200 to 500 K, *J. Chem. Phys.,* 66, 2690, 1977.
38. **Zahniser, M. S., Berquist, B. M., and Kaufman, F.,** Kinetics of the reaction $Cl + CH_4 \rightarrow CH_3 + HCl$ from 200 to 500 K, *Int. J. Chem. Kinet.,* 10, 15, 1978.
39. **Keyser, L. F.,** Absolute rate and temperature dependence of the reaction between chlorine (^2P) atoms and methane, *J. Chem. Phys.,* 69, 214, 1978.
40. **Clyne, M. A. A. and Walker, R. F.,** Absolute rate constants for elementary reactions in the chlorination of CH_4, CD_4, CH_3Cl, CH_2Cl_2, $CHCl_3$, $CDCl_3$ and $CBrCl_3$, *J. Chem. Soc. Faraday Trans. 1,* 1547, 1973.
41. **Ravishankara, A. R., Nicovich, J. M., Thompson, R. L., and Tully, F. P.,** Kinetic study of the reaction of OH with H_2 and D_2 from 250 to 1050 K, *J. Phys. Chem.,* 85, 2498, 1981.
42. **Tully, F. P. and Ravishankara, A. R.,** Flash photolysis-resonance fluorescence kinetic study of the reactions $OH + H_2 \rightarrow H_2O + H$ and $OH + CH_4 \rightarrow H_2O + CH_3$ from 298 to 1020 K, *J. Phys. Chem.,* 84, 3126, 1980.
43. **Jeong, K.-M. and Kaufman, F.,** Kinetics of the reaction of hydroxyl radical with methane and with nine Cl- and F-substituted methanes. I. Experimental results, comparisons, and applications, *J. Phys. Chem.,* 86, 1808, 1982.
44. **Jeong, K.-M., Hsu, K.-J., Jeffries, J. B., and Kaufman, F.,** Kinetics of the reactions of OH with C_2H_6, CH_3CCl_3, $CH_2ClCHCl_2$, $CH_2ClCClF_2$ and CH_2FCF_3, *J. Phys. Chem.,* 88, 1222, 1984.
45. **Kobrinsky, P. C. and Pacey, P. D.,** The reaction of methyl radicals with molecular hydrogen, *Can. J. Chem.,* 52, 3665, 1974.
46. **Clark, T. C. and Dove, J. E.,** The rate coefficient for $CH_3 + H_2 \rightarrow CH_4 + H$ measured in reflected shock waves; a non-Arrhenius reaction, *Can. J. Chem.,* 51, 2155, 1973.

47. **Pacey, P. D. and Purnell, J. M.,** Arrhenius parameters of the reaction $CH_3 + C_2H_6 \rightarrow CH_4 + C_2H_5$, *J. Chem. Soc. Faraday Trans. 1,* 68, 1462, 1972.
48. **Clark, T. C., Izod, T. P. J., and Kistiakowsky, G. B.,** Reaction of methyl radicals produced by pyrolysis of azomethane, or ethane in reflected shock waves, *J. Chem. Phys.,* 54, 1295, 1971.
49. **Rebbert, R. E. and Steacie, E. W. R.,** The photolysis of mercury dimethyl in presence of hydrocarbons, *J. Chem. Phys.,* 21, 1723, 1953.
50. **Trotman-Dickenson, A. F., Birchard, J. R., and Steacie, E. W. R.,** The reactions of methyl radicals, II. The abstraction of hydrogen atoms from paraffins, *J. Chem. Phys.,* 19, 163, 1951.
51. **Kerr, J. A. and Timlin, D.,** Hydrogen abstraction from neopentane by methyl radicals, *J. Chem. Soc. A,* 1241, 1969.
52. **Furimsky, E. and Laidler, K. J.,** Kinetics of the mercury-photosensitized decomposition of neopentane. I. The overall mechanism, *Can. J. Chem.,* 50, 1115, 1972.
53. **Pacey, P. D.,** The reaction of methyl radicals with neopentane, *Can. J. Chem.,* 51, 2145, 1973.
54. **Camilleri, P., Marshall, R. M., and Purnell, J. H.,** Arrhenius parameters for the unimolecular decompositions of azomethane and n-propyl and isopropyl radicals and for methyl radical attack on propane, *J. Chem. Soc. Faraday Trans. 1,* 71, 1491, 1975.
55. **Durban, P. C. and Marshall, R. M.,** Photolysis of azomethane-propane mixtures. Arrhenius parameters for the hydrogen abstraction reaction of methyl with azomethane and with propane, *Int. J. Chem. Kinet.,* 12, 1031, 1980.
56. **Pacey, P. D. and Purnell, J. H.,** Arrhenius parameters for the reactions $CH_3 + C_4H_{10} \rightarrow CH_4 + C_4H_9$ and $C_2H_5 + C_4H_{10} \rightarrow C_2H_6 + C_4H_9$, *Int. J. Chem. Kinet.,* 4, 657, 1972.
57. **Konar, R. S., Marshall, R. M., and Purnell, J. H.,** The self-inhibited pyrolysis of isobutane, *Int. J. Chem. Kinet.,* 5, 1007, 1973.
58. **Anastasi, C.,** Study of the methyl-isobutane reaction in the range $478 \leqq T/K \leqq 560$, *J. Chem. Soc. Faraday Trans. 1,* 79, 741, 1983.
59. **Larson, G. F. and Gilliom, R. D.,** Kinetic isotope effect in the homolytic abstraction of benzylic hydrogen by tert-butoxy radical, *J. Am. Chem. Soc.,* 97, 3444, 1975.
60. **Hay, J. M.,** Some features of radical reactivity, *J. Chem. Soc. B,* 1175, 1967.
61. **Lin, S. H., Lau, K. H., and Eyring, H.,** Absolute reaction rate constants and chemical reaction cross sections of bimolecular reactions, *J. Chem. Phys.,* 55, 5657, 1971.
62. **Lin, S. H., Lau, K. H., and Eyring, H.,** Absolute reaction rate constants and chemical reaction cross section of bimolecular reactions. II. Numerical results, *J. Chem. Phys.,* 58, 1261, 1973.
63. **Menzinger, M. und Wolfgang, R. L.,** Bedeutung und Anwendung der Arrhenius-Aktivierungsenergie, *Angew. Chem.,* 81, 446, 1969; *Angew. Chem. Int. Ed.,* 8, 438, 1969.
64. **LeRoy, R. L.,** Relationship between Arrhenius activation energies and excitation functions, *J. Phys. Chem.,* 73, 4338, 1969.
65. **Johnston, H. S.,** *Gas Phase Reaction Rate Theory,* The Ronald Press, New York, 1966.
66. **Bron, J.,** The importance of anharmonicity of the vibrational excited states in chemical kinetics, *Can. J. Chem.,* 53, 3069, 1975.
67. **Weston, R. E.,** H_3 activated complex and the rate of reaction of hydrogen atoms with hydrogen molecules, *J. Chem. Phys.,* 31, 892, 1959.
68. **Bérces, T. and Dombi, J.,** Evaluation of the rate coefficients and Arrhenius parameters of hydrogen atom transfer reactions. I. The method, *Int. J. Chem. Kinet.,* 12, 123, 1980.
69. **Bérces, T. and Dombi, J.,** Evaluation of the rate coefficients and Arrhenius parameters of hydrogen atom transfer reactions. II. Application of the method, *Int. J. Chem. Kinet.,* 12, 183, 1980.
70. **Cohen, N. and Westberg, K. R.,** Chemical kinetics data sheets for high-temperature chemical reactions, *J. Phys. Chem. Ref. Data,* 12, 531, 1983.
71. **Bell, R. P.,** The tunnel effect correction for parabolic potential barriers, *Trans. Faraday Soc.,* 55, 1, 1959.
72. **Eckart, C.,** The penetration of a potential barrier by electrons, *Phys. Rev.,* 35, 1303, 1930.
73. **Stern, M. J. and Weston, R. E., Jr.,** Phenomenological manifestations of quantum-mechanical tunneling. I. Curvature in Arrhenius plots, *J. Chem. Phys.,* 60, 2803, 1974.
74. **Pacey, P. D.,** Curvature of Arrhenius plots caused by tunneling through Eckart barriers, *J. Chem. Phys.,* 71, 2966, 1979.
75. **Caldin, E. F.,** Tunneling in proton-transfer reactions in solution, *Chem. Rev.,* 69, 135, 1969.
76. **Brunton, G., Griller, D., Barclay, L. R. C., and Ingold, K. U.,** Kinetic applications of electron paramagnetic resonance spectroscopy. XXVI. Quantum-mechanical tunneling in the isomerisation of sterically hindered aryl radicals, *J. Am. Chem. Soc.,* 98, 6803, 1976.
77. **Le Roy, D. J., Ridley, B. A., and Quickert, K. A.,** Kinetics of the four basic thermal three-centre exchange reactions of the hydrogen isotopes, *Disc. Faraday Soc.,* 44, 92, 1967.
78. **Johnston, H. S.,** Large tunneling corrections in chemical reaction rates, *Adv. Chem. Phys.,* 3, 131, 1961.
79. **Sprague, E. D. and Williams, F.,** Evidence for hydrogen atom abstraction by methyl radicals in the solid state at 77 K, *J. Am. Chem. Soc.,* 93, 787, 1971.

80. **Le Roy, R. J., Sprague, E. D., and Williams, F.,** Quantum mechanical tunneling in hydrogen atom abstraction from solid acetonitrile at 77 to 87 K, *J. Phys. Chem.*, 76, 546, 1972.
81. **Wang, J. T. and Williams, F.,** Hydrogen atom abstraction by methyl radicals in γ-irradiated crystalline methyl isocyanate at 77 to 125 K, *J. Am. Chem. Soc.*, 94, 2930, 1972.
82. **Campion, A. and Williams, F.,** Hydrogen atom abstraction by methyl radicals in methanol glasses at 67 to 77 K, *J. Am. Chem. Soc.*, 94, 7633, 1972.
83. **Toriyama, K., Nunome, K., and Iwasaki, M.,** Electron spin resonance evidence for tunneling hydrogen atom transfer reaction at 4.2 K in organic crystals, *J. Am. Chem. Soc.*, 99, 5823, 1977.
84. **Polanyi, J. C.,** Some concepts in reaction dynamics, *Acc. Chem. Res.*, 5, 161, 1972.
85. **Douglas, D. J., Polanyi, J. C., and Sloan, J. J.,** Effect of changing reagent energy on reaction dynamics. VI. Dependence of reaction rate on vibrational excitation in substantially endothermic reactions, $XH(v') + Y \rightarrow X + HY$, *Chem. Phys.*, 13, 15, 1976.
86. **Anlauf, K. G., Maylotte, D. H., Polanyi, J. C., and Bernstein, R. B.,** Rates of the endothermic reactions $HCl + X (X \equiv I, Cl)$ as a function of reagent vibration, rotation and translation, *J. Chem. Phys.*, 51, 5716, 1969.
87. **Anlauf, K. G., Charters, P. E., Horne, D. S., McDonald, R. G., Maylotte, D. H., Polanyi, J. C., Skrlac, W. J., Tardy, D. C., and Woodall, K. B.,** Translational energy-distribution in the products of some exothermic reactions, *J. Chem. Phys.*, 53, 4091, 1970.
88. **Perry, D. S. and Polanyi, J. C.,** Energy distribution among reaction products. IX. $F + H_2$, HD and D_2, *Chem. Phys.*, 12, 419, 1976.
89. **Polanyi, J. C., Sloan, J. J., and Wanner, J.,** The effect of changing reagent energy on reaction dynamics. V. Reagent translation and vibration in $H + F_2 \rightarrow HF + F$, *Chem. Phys.*, 13, 1, 1976.
90. **Arnoldi, D. and Wolfrum, J.,** The reaction of vibrationally excited HCl with oxygen and hydrogen atoms, *Chem. Phys., Lett.*, 24, 234, 1974.
91. **Heidner, R. F., III and Kasper, J. V. V.,** An experimental rate constant for $H + H_2 (v'' = 1) \rightarrow H + H_2 (v'' = 0)$, *Chem. Phys. Lett.*, 15, 179, 1972.
92. **Persky, A., Rubin, R., and Broida, M.,** Quasiclassical trajectory studies of the chlorine-hydrogen system. IV. The effect of reagent vibrational excitation and of the location of the potential energy barrier on the dynamics, *J. Chem. Phys.*, 79, 3279, 1983.
93. **Broida, M. and Persky, A.,** Quasiclassical trajectory study of the reaction $O(^3P) + H_2 \rightarrow OH + H$. The effects of the location of the potential energy barrier, vibrational excitation and isotopic substitution on the dynamics, *J. Chem. Phys.*, 80, 3687, 1984.
94. **Arnoldi, D., Kaufman, K., and Wolfrum, J.,** Chemical-laser-induced isotopically selective reaction of HCl, *Phys. Rev. Lett.*, 34, 1597, 1975.
95. **Polanyi, J. C. and Wong, W. H.,** Location of energy barriers. I. Effect on the dynamics of reaction A + BC, *J. Phys. Chem.*, 51, 1439, 1969.
96. **Zellner, R. and Steinert, W.,** Vibrational rate enchancement in the reaction $OH + H_2 (v = 1) \rightarrow H_2O + H$, *Chem. Phys. Lett.*, 81, 568, 1981.
97. **Truhlar, D. G. and Isaacson, A. D.,** Statistical-diabatic model for state-selected reaction rates. Theory and application of vibrational-mode correlation analysis to $OH(n_{OH}) + H_2(n_{HH}) \rightarrow H_2O + H$, *J. Chem. Phys.*, 77, 3516, 1982.
98. **Light, J. C. and Matsumoto, J. H.,** The effect of vibrational excitation in the reaction of OH with H_2, *Chem. Phys. Lett.*, 58, 578, 1978.
99. **Cannon, B. D., Roberthaw, J. S., Smith, I. W. M., and Williams, M. D.,** A time-resolved LIF study of the kinetics of OH (υ = O) and OH (υ = 1) with HCl and HBr, *Chem. Phys. Lett.*, 105, 380, 1984.
100. **Chapman, S. and Bunker, D. L.,** An exploratory study of reactant vibrational effects in $CH_3 + H_2$ and its isotopic variants, *J. Chem. Phys.*, 62, 2890, 1975.
101. **Kovalenko, L. J. and Leone, S. R.,** Laser studies of methyl radical reactions with Cl_2 and Br_2: absolute rate constants, product vibrational excitation, and hot radical reactions, *J. Chem. Phys.*, 80, 3656, 1984.
102. **Kurylo, M. J., Braun, W., Kaldor, A., Freund, S. M., and Wayne, R. P.,** Infra-red laser enhanced reactions: chemistry of vibrationally excited O_3 with NO and $O_2 (^1\Delta)^*$, *J. Photochem.*, 3, 71, 1974/75.
103. **Braun, W., Kurylo, M. J., Kaldor, A., and Wayne, R. P.,** Infrared laser enhanced reactions: spectral distribution of the No_2 chemiluminescence produced in the reaction of vibrationally excited O_3 with NO, *J. Chem. Phys.*, 61, 461, 1974.
104. **Kaldor, A., Braun, W., and Kurylo, M. J.,** Infrared laser enhanced reactions: $O_3 + SO$, *J. Chem. Phys.*, 61, 2496, 1974.
105. **Birely, J. H. and Lyman, J. L.,** The effect of reagent vibrational energy on measured reaction rate constants *J. Photochem.*, 4, 269, 1975.
106. **Truhlar, D. G. and Garrett, B. C.,** Variational transition-state theory, *Acc. Chem. Res.*, 13, 440, 1980.
107. **Pechukas, P.,** Transition state theory, *Ann. Rev. Phys. Chem.*, 32, 159, 1981.
108. **Benson, S. W.,** *Thermochemical Kinetics*, 2nd ed., John Wiley & Sons, New York, 1976.

109. **Levine, R. D.,** Information theory approach to molecular reaction dynamics, *Ann. Rev. Phys. Chem.*, 29, 59, 1978.
110. **Levine, R. D. and Kinsey, J. L.,** Information-theoretical approach: applied to molecular collisions, in *Atom-Molecule Collision Theory: A Guide for the Experimentalist,* Bernstein, R. B., Ed., Plenum Press, New York, 1979, chap. 22.
111. **Walker, R. B. and Light, J. C.,** Reactive molecular collisions, *Ann. Rev. Phys. Chem.*, 31, 401, 1980.
112. **Bernstein, R. B., Ed.,** *Atom-Molecule Collision Theory: A Guide for the Experimentalist,* Plenum Press, New York, 1979.
113. **Truhlar, D. G., Ed.,** *Potential Energy Surfaces and Dynamics Calculations,* Plenum Press, New York, 1981.
114. **Wyatt, R. E.,** Direct-mode chemical reactions. I. Methodology for accurate quantal calculations, in *Atom-Molecule Collision Theory: A Guide for the Experimentalist,* Bernstein, R. E., Ed., Plenum Press, New York, 1979, chap. 17.
115. **Elkowtiz, A. B. and Wyatt, R. E.,** Quantum mechanical cross sections for the three-dimensional exchange reaction, *J. Chem. Phys.*, 62, 2504, 1975.
116. **Kuppermann, A. and Schatz, G. C.,** Quantum mechanical reactive scattering: an accurate three-dimensional calculation, *J. Chem. Phys.*, 62, 2502, 1975.
117. **Schatz, G. C. and Kuppermann, A.,** Quantum mechanical reactive scattering for three-dimensional atom plus diatom system. I. Theory, *J. Chem. Phys.*, 65, 4642, 1976.
118. **Walker, R. B., Light, J. C., and Altenberger-Siczek, A.,** Chemical reaction theory for asymmetric atom-molecule collisions, *J. Chem. Phys.*, 64, 1166, 1976.
119. **Light, J. C. and Walker, R. B.,** An R matrix approach to the solution of coupled equations for atom-molecule reactive sacttering, *J. Chem. Phys.*, 65, 4272, 1976.
120. **Elkovitz, A. B. and Wyatt, R. E.,** Three-dimensional natural coordinate asymmetric top theory of reactions: applications to $H + H_2$, *J. Chem. Phys.*, 63, 702, 1975.
121. **Schatz, G. C. and Kuppermann, A.,** Quantum mechanical reactive scattering for three-dimensional atom plus diatomic systems. II. Accurate cross section for $H + H_2$, *J. Chem. Phys.*, 65, 4668, 1976.
122. **Porter, R. N. and Karplus, M.,** Potential energy surface for H_3, *J. Chem. Phys.*, 40, 1105, 1964.
123. **Liu, B,** *Ab initio* potential energy surface for linear H_3, *J. Chem. Phys.*, 58, 1924, 1973.
124. **Siegbahn, P. and Liu, B.,** An accurate three-dimensional potential energy surface for H_3, *J. Chem. Phys.*, 68, 2457, 1978.
125. **Truhlar, D. G. and Horowitz, C. J.,** Functional representation of Liu and Siegbahn's accurate *ab initio* potential energy calculations, *J. Chem. Phys.*, 68, 2466, 1978; *J. Chem. Phys.*, 71, 1514(E), 1979.
126. **Walker, R. B., Stechel, E. B., and Light, J. C.,** Accurate H_3 dynamics on an accurate H_3 potential surface, *J. Chem. Phys.*, 69, 2922, 1978.
127. **Redmon, M. J. and Wyatt, R. E.,** Quantal resonance structure in the three-dimensional $F + H_2$ reaction, *Chem. Phys. Lett.*, 63, 209, 1979.
128. **Connor, J. N. L.,** Reactive molecular collision calculations, *Comput. Phys. Commun.*, 17, 117, 1979.
129. **Bowman, J. M., Lee, K.-T., and Walker, R. B.,** Reduced dimensionality quantum rate constants for the $D + H_2$ ($\upsilon = 0$) and $D + H_2$ ($\upsilon = 1$) reactions on the LSTH surface, *J. Chem. Phys.*, 79, 3742, 1983.
130. **Connor, J. N. L., Jakubetz, W., and Manz, J.,** Isotope effects in the reaction $X + F_2 \rightarrow XF + F$ (X = Mu, H, D, T): a quantum mechanical and information theoretic investigation, *Chem. Phys.*, 28, 219, 1978.
131. **Connor, J. N. L., Jakubetz, W., Laganá, A., Manz, J., and Whitehead, J. C.,** The reaction $X + Cl_2 \rightarrow XCl + Cl$ (X = Mu, H, D,). II. Comparison of experimental data with theoretical results derived from a new potential energy surface, *Chem. Phys.*, 65, 29, 1982.
132. **Bowman, J. M. and Lee, K.-T.,** Exact quantum reaction probabilities for the collinear $O(^3P) + H_2$ reaction on an *ab initio* surface, in *Potential Energy Surfaces and Dynamics Calculations,* Truhlar, D. C., Ed., Plenum Press, New York, 1981, chap. 15.
133. **Lee, K.-T., Bowman, J. M., Wagner, A. F., and Schatz,G. C.,** A comparative study of the reaction dynamics of the $O(^3P) + H_2 \rightarrow OH + H$ reaction on several potential energy surfaces. III. Collinear exact quantum transmission coefficient correction to transition state theory, *J. Chem. Phys.*, 76, 3583, 1982.
134. **Wyatt, R. E.,** Reactive scattering cross sections. II. Approximate quantal treatments, in *Atom-Molecule Collision Theory: A Guide for the Experimentalist,* Bernstein, R. E., Ed., Plenum Press, New York, 1979, chap. 15.
135. **Truhlar, D. G. and Muckerman, J. T.,** Reactive scattering cross sections. III. Quasiclassical and semiclassical methods, in *Atom-Molecule Collision Theory: A Guide for the Experimentalist,* Bernstein, R. E., Ed., Plenum Press, New York, 1979, chap. 16.
136. **Porter, R. N. and Raff, L. M.,** Classical trajectory methods in molecular collisions, in *Dynamics of Molecular Collisions,* Part B, Miller, W. H., Ed., Plenum Press, New York, 1976, 1.

137. **Karplus, M., Porter, R. N., and Sharma, R. D.,** Exchange reactions with activation energy. I. Simple barrier potential for /H,H_2/, *J. Chem. Phys.*, 43, 3259, 1965.
138. **Hase, W. L., Wolf, R. J., and Sloane, C. S.,** Trajectory studies of the molecular dynamics of ethyl radical decomposition, *J. Chem. Phys.*, 71, 2911, 1979.
139. **Grant, E. R. and Bunker, D. L.,** Dynamical effects in unimolecular decomposition: a classical trajectory study of the dissociation of C_2H_6, *J. Chem. Phys.*, 68, 628, 1978.
140. **Truhlar, D. G., Merrick, J. A., and Duff, J. W.,** Comparison of trajectory calculations, transition state theory, quantum mechanical reaction probabilities, and rate constants for the collinear reaction H + Cl_2 → HCl + Cl, *J. Am. Chem. Soc.*, 98, 6771, 1976.
141. **Baer, M., Halavee, V., and Persky, A.,** The collinear Cl + XY system (X,Y = H,D,T). A comparison between quantum mechanical, classical, and transition state theory results, *J. Chem. Phys.*, 61, 5122, 1974.
142. **Lutz, A. C. and Andresen, P.,** The chemical dynamics of the reactions of O(^{3}P) with saturated hydrocarbons. II. Theoretical model, *J. Chem. Phys.*, 72, 5851, 1980.
143. **Polanyi, J. C. and Sathyamurty, N.,** Location of energy barriers. VII. Sudden and gradual late-energy-barriers, *Chem. Phys.*, 33, 287, 1978.
144. **Polanyi, J. C. and Sathyamurty, N.,** Location of energy barriers. VIII. Reagent → product energy conversion on surfaces with sudden or gradual late-barriers, *Chem. Phys.*, 37, 259, 1979.
145. **Persky, A., Rubin, R., and Broida, M.,** Quasiclassical trajectory studies of the chlorine-hydrogen system. IV. The effect of reagent vibrational excitation and of the location of the potential energy barrier on the dynamics, *J. Chem. Phys.*, 79, 3279, 1983.
146. **Broida, M. and Persky, A.,** Quasiclassical trajectory study of the reaction O(^{3}P) + H_2 → OH + H. The effects of the location of the potential energy barrier, vibrational excitation and isotopic substitution on the dynamics, *J. Chem. Phys.*, 80, 3687, 1984.
147. **Pechukas, P.,** Recent development in transition state theory, *Ber. Bunsenges. Phys. Chem.*, 86, 372, 1982.
148. **Garrett, B. C. and Truhlar, D. G.,** Generalized transition state theory. Classical mechanical theory and applications to collinear reactions of hydrogen molecules, *J. Phys. Chem.*, 83, 1052, 1979.
149. **Garrett, B. C. and Truhlar, D. G.,** Generalized transition state theory. Quantum effects for collinear reactions of hydrogen molecules and isotopically substituted hydrogen molecules, *J. Phys. Chem.*, 83, 1079, 1979.
150. **Wigner, E.,** The transition state method, *Trans. Faraday Soc.*, 34, 29, 1938.
151. **Garrett, B. C. and Truhlar, D. G.,** Improved canonical variational theory for chemical reaction rates. Classical mechanical theory and applications to collinear reactions, *J. Phys. Chem.*, 84, 805, 1980.
152. **Truhlar, D. G.,** Accuracy of trajectory calculations and transition state theory for thermal rate constants of atom transfer reactions, *J. Phys. Chem.*, 83, 188, 1979.
153. **Sverdlik, D. I. and Koeppl, G. W.,** An energy limit of transition state theory, *Chem. Phys. Lett.*, 59, 449, 1978.
154. **Sverdlik, D. I., Stein, G. P., and Koeppl, G. W.,** The accuracy of transition state theory in its absolute rate theory and variational formulations, *Chem. Phys. Lett.*, 67, 87, 1979.
155. **Miller, W. H.,** Unified statistical model for "complex" and "direct" reaction mechanisms, *J. Chem. Phys.*, 65, 2216, 1976.
156. **Pollak, E. and Pechukas, P.,** Unified statistical model for "complex" and "direct" reaction mechanisms: a test on the collinear H + H_2 exchange reaction, *J. Chem. Phys.*, 70, 325, 1979.
157. **Garrett, B. C., Truhlar, D. G., Grev, R. S., and Magnuson, A. W.,** Improved treatment of threshold contributions in variational transition state theory, *J. Phys. Chem.*, 84, 1730, 1980.
158. **Garrett, B. C., Truhlar, D. G., Wagner, A. F., and Dunning, T. H., Jr.,** Variational transition state theory and tunneling for a heavy-light-heavy reaction using an *ab initio* potential energy surface. ^{37}Cl + H(D) ^{35}Cl → H(D) ^{37}Cl + ^{35}Cl, *J. Chem. Phys.*, 78, 4400, 1983.
159. **Koeppl, G. W.,** Best *ab initio* surface transition state theory rate constants for the D + H_2 and H + D_2 reactions, *J. Chem. Phys.*, 59, 3425, 1973.
160. **Garrett, B. C. and Truhlar, D. G.,** Generalized transition state theory calculations for the reactions D + H_2 and H + D_2 using an accurate potential energy surface: explanation of the kinetic isotope effect, *J. Chem. Phys.*, 72, 3460, 1980.
161. **Garrett, B. C., Truhlar, D. G., Grev, R. S., and Walker, R. B.,** Comparison of variational transition state theory and the unified statistical model with vibrationally adiabatic transmission coefficients to accurate collinear rate constants for T + HD → TH + D, *J. Chem. Phys.*, 73, 235, 1980.
162. **Walch, S. P., Dunning, T. H., Jr., Raffenetti, R. C., and Bobrowicz, F. W.,** A theoretical study of the potential energy surface for O(^{3}P) + H_2, *J. Chem. Phys.*, 72, 406, 1980.
163. **Schatz, G. C., Wagner, A. F., Walch, S. P., and Bowman, J. M.,** A comparative study of the reaction dynamics of several potential energy surfaces of O(^{3}P) + H_2 → OH + H, *J. Chem. Phys.*, 74, 4984, 1981.
164. **Schatz, G. C. and Walch, S. P.,** An *ab initio* calculation of the rate constant for the OH + H_2 → H_2O + H reaction *J. Chem. Phys.*, 72, 776, 1980.

165. **Walch, S. P. and Dunning, T. H., Jr.,** A theoretical study of the potential energy surface of OH + H_2, *J. Chem. Phys.*, 72, 1303, 1980.
166. **Harding, L.B., Schatz, G. C., and Chiles, R. A.,** An *ab initio* determination of the rate constant for H_2 + $C_2H \rightarrow H + C_2H_2$, *J. Chem. Phys.*, 76, 5172, 1982.
167. **Harding, L. B. and Schatz, G. C.,** An *ab initio* determination of the rate constant for H + $H_2CO \rightarrow H_2$ + HCO, *J. Chem. Phys.*, 76, 4296, 1982.
168. **Walch, S. P.,** Calculated barriers to abstraction and exchange for CH_4 + H, *J. Chem. Phys.*, 72, 4932, 1980.
169. **Schatz, G. C., Walch, S. P., and Wagner, A. F.,** *Ab initio* calculation of transition state normal mode properties and rate constants for the H(T) + $CH_4(CD_4)$ abstraction and exchange reactions, *J. Chem. Phys.*, 73, 4536, 1980.
170. **Schatz, G. C., Wagner, A. F., and Dunning, T. H., Jr.,** A theoretical study of deuterium isotope effects in the reactions H_2 + CH_3 and H + CH_4, *J. Phys. Chem.*, 88, 221, 1984.
171. **Rayez-Meaume, M. T., Donnenberg, J. J., and Whitten, J. L.,** A theoretical study of the reaction of methane with methyl radical using several different *ab initio* and semiempirical methods, *J. Am. Chem. Soc*, 100, 747, 1978.
172. **Dewar, M. J. and Haselbach, E.,** Ground state of σ-bonded molecules. IX. the MINDO/2 method, *J. Am. Chem. Soc*, 92, 590, 1970.
173. **Bálint, I.,Révész, M., Bán, M. I., Márta, F., and Bérces, T.,** Calculation of activation energies and transition state geometries for hydrogen abstraction reactions by semiempirical methods, *Acta Chim. Hung.*, 120, 3, 1985.
174. **London, F.,** Quantenmechanische Deutung des Vorganges der Aktivierung, *Z. Elektrochem.*, 35, 552, 1929.
175. **Glasstone, S., Laidler, K. J., and Eyring, H.,** *The Theory of Rate Processes,* McGraw-Hill, New York, 1941.
176. **Sato, S.,** On a new method of drawing the potential energy surface, *J. Chem. Phys.*, 23, 592, 1955.
177. **Zavitsas, A. A.,** Activation energy requirements in hydrogen abstractions. Quantitative description of the causes in terms of bond energies and infrared frequencies, *J. Am. Chem. Soc.*, 94, 2779, 1972.
178. **Zavitsas, A. A. and Melikian, A. A.** Hydrogen abstraction by free radicals. Factors controlling reactivity, *J. Am. Chem. Soc.*, 97, 2757, 1975.
179. **Al Akeel, N. Y., Selby, K., and Waddington, D. J.,** Reactions of oxygenated radicals in the gas phase. VIII. Reactions of alkoxyl radicals with aldehydes and ketones, *J. Chem. Soc. Perkin Trans. 2,* 1036, 1981.
180. **Sway, M. I. and Waddington, D. J.,** Reactions of oxygenated radicals in the gas phase. XIII. Reactions of t-butoxyl radicals with alkenes and elkenes, *J. Chem. Soc. Perkin Trans. 2,* 1984, 63, 1984.
181. **Johnston, H. S. and Goldfinger, P.,** Theoretical interpretation of reactions occurring in photochlorination, *J. Chem. Phys.*, 37, 700, 1962.
182. **Johnston, H. S., and Parr, C.,** Activation energies from bond energies. I. Hydrogen transfer reactions, *J. Am. Chem. Soc.*, 85, 2544, 1963.
183. **Pauling, L.,** Atomic radii and interatomic distances in metals, *J. Am. Chem. Soc.*, 69, 542, 1947.
184. **Herschbach, D. R., Johnston, H. S., and Rapp, D.,** Molecular partition functions in terms of local properties, *J. Chem. Phys.*, 31, 1652, 1959.
185. **Gilliom, R. D.,** Activation energies from bond energies. A modification, *J. Am. Chem. Soc.*, 99, 8399, 1977.
186. **Truhlar, D. G.,** Test of bond-order methods for predicting the position of the minimum-energy path for hydrogen atom transfer reactions, *J. Am. Chem. Soc.*, 94, 7584, 1972.
187. **Bérces, T., and Dombi, J.,** Minimum energy path equation for A + BC $\rightarrow$ AB + C type hydrogen atom abstraction reactions, *React. Kinet. Catal. Lett.*, 5, 281, 1976.
188. **Arthur, N. L. and McDonell, J. A.,** BEBO calculations. I. Activation energies and kinetic isotope effects for the reactions of CH_3 and CF_3 radicals with HCl and H_2S, *J. Chem. Phys.*, 56, 3100, 1972.
189. **Arthur, N. L. and McDonell, J. A.,** BEBO calculations. II. Arrhenius parameters and kinetic isotope effects for the reactions of CH_3 and CF_3 radicals with NH_3, *J. Chem. Phys.*, 57, 3228, 1972.
190. **Endo, H. and Glass, G. P.,** Reactions of atomic hydrogen and deuterium with HBr and DBr, *J. Phys. Chem.*, 80, 1519, 1976.
191. **Bérces, T. and Dombi, J.,** Estimation of activation barriers for hydrogen abstraction reactions by a modified BEBO treatment, *React. Kinet. Catal. Lett.*, 9, 153, 1978.
192. **Arthur, N. L., Donchi, K. F., and McDonell, J. A.,** BEBO calculations. III. A new triplet repulsion energy term, *J. Chem. Phys.*, 62, 1585, 1975.
193. **Brown, R. L. and Laufer, A. H.,** Calculation of activation energies for hydrogen-atom abstractions by radicals containing carbon triple bonds, *J. Phys. Chem.*, 85, 3826, 1981.
194. **Mayer, S. W. and Schieler, L.,** Activation energies and rate constants computed for reactions of oxygen with hydrocarbons, *J. Phys. Chem.*, 72, 2628, 1968.

195. **Arthur, N. L., Donchi, K. F., and McDonell, J. A.,** BEBO calculations. IV. Arrhenius parameters and kinetic isotope effects for the reactions of CH_3 and CF_3 radicals with H_2 and D_2, *J. Chem. Soc. Faraday Trans. 1,* 71, 2431, 1975.
196. **Arthur, N. L., Donchi, K. F., and McDonell, J. A.,** BEBO calculations. V. Arrhenius parameters and kinetic isotope effects for the reactions of C_2H_5 and C_2F_5 radicals with H_2 and D_2, *J. Chem. Soc. Faraday Trans. 1,* 71, 2442, 1975.
197. **Bérces, T.,** Evaluation of triplet repulsion energies, *React. Kinet. Catal. Lett.,* 7, 379, 1977.
198. **Jordan, R. M. and Kaufman, F.,** Re-evaluation of the BEBO method, *J. Chem. Phys.,* 63, 1691, 1975.
199. **Farrar, J. M. and Lee, Y. T.,** Intermolecular potentials from crossed beam differential elastic scattering measurements. V. The attractive well of He_2, *J. Chem. Phys.,* 56, 5801, 1972.
200. **Chen, C. H., Siska, P. E., and Lee, Y. T.,** Intermolecular potentials from crossed beam differential elastic scattering measurements. VIII. He + Ne, He + Ar, He + Kr, and He + Xe, *J. Chem. Phys.,* 59, 601, 1973.
201. **Gilliom, R. D.,** Re-evaluation of the BEBO method, *J. Chem. Phys.,* 65, 5028, 1976.
202. **Lide, D. R., Jr.,** A survey of carbon-carbon bond lengths, *Tetrahedron,* 17, 125, 1962.
203. **Marschoff, C. M. and Jatem, A.,** On the validity of the BEBO method, *Chem. Phys. Lett.,* 56, 35, 1978.
204. **Park, C. R., Song, S. A., Lee, Y. E., and Choo, K. Y.,** Arrhenius parameters for the tert-butoxy radical reactions with trimethylsilane in the gas phase, *J. Am. Chem. Soc.,* 104, 6445, 1982.
205. **Löser, V. and Scherzer, K.,** BEBO calculations of kinetic isotope effects for methyl radical reactions with alkanes and ketones, *React. Kinet. Catal. Lett.,* 11, 155, 1979.
206. **Singleton, D. L. and Cvetanovic, R. J.,** Temperature dependence of rate constants for the reactions of oxygen atoms, $O(^3P)$ with HBr and HI, *Can. J. Chem.,* 56, 2934, 1978.
207. **Brown, R. D. H. and Smith, I. W. M.,** Absolute rate constants for the reactions $O(^3P)$ atoms with HCl and HBr, *Int. J. Chem. Kinet.,* 7, 301, 1975.
208. **Michael, J. V., Keil, D. G., and Klemm, R. B.,** Theoretical rate constant calculations for $O(^3P)$ with saturated hydrocarbons, *Int. J. Chem. Kinet.,* 15, 705, 1983.
209. **Carr, R. W., Jr.,** Predictions of the rates of hydrogen abstractions by $CH_2(^3B_1)$ by the bond-energy bond-order method, *J. Phys. Chem.,* 76, 1581, 1972.
210. **Dóbé, S., Böhland, T., Temps, F., and Wagner, H. Gg.,** A direct study of the reactions of CH_2 $(\tilde{X}^3B_1)$-radicals with selected hydrocarbons in the temperature range 296 K $\leq$ T $\leq$ 705 K, *Ber. Bunsenges. Phys. Chem.,* 89, 432, 1985.
211. **Böhland, T., Dóbé, S., Temps, F., and Wagner, H. Gg.,** Kinetics of the reactions between CH_2 $(\tilde{X}^3B_1)$-radicals and saturated hydrocarbons in the temperature range 296 K $\leq$ T $\leq$ 707 K, *Ber. Bunsenges. Phys. Chem.,* 89, 1110, 1985.
212. **Previtali, C. M. and Scaiano, J. C.,** A kinetic model for the intermolecular photoreduction of carbonyl compounds; a novel application of the bond-energy-bond-order method, *Chem. Comm.,* 1971, 1298, 1971.
213. **Previtali, C. M. and Scaiano, J. C.,** The kinetics of photochemical reactions. I. Applications of a modified bond-energy-bond-order method to the atom abstraction reactions of excited carbonyl compounds, *J. Chem. Soc. Perkin Trans. 2,* 1972, 1667, 1972.
214. **Previtali, C. M. and Scaiano, J. C.,** The kinetics of photochemical reactions. II. Calculation of kinetic parameters for the intermolecular hydrogen abstraction reactions of the triplet state of carbonyl compounds, *J. Chem. Soc. Perkin Trans. 2,* 1972, 1672, 1972.
215. **Mayer, S. W.,** Computed activation energies for bimolecular reactions of O_2, N_2, NO, N_2O, NO_2 and CO_2, *J. Phys. Chem.,* 71, 4159, 1967.
216. **Mayer, S. W.,** Estimation of activation energies for nitrous oxide, carbon dioxide, nitrogen dioxide, nitric oxide, oxygen and nitrogen reactions by a bond energy method, *J. Phys. Chem.,* 73, 3941, 1969.
217. **Lissi, E., Passeggi, M., and Previtali, C.,** Application del metodo semiempirico de Johnston-Parr a la desproporcionacion de radicales etilo, *An. Asoc. Quim. Argent.,* 58, 167, 1970.
218. **Thommarson, R. L.,** Alkyl radical disproportionation, *J. Phys. Chem.,* 74, 938, 1970.
219. **Alfassi, Z. B. and Benson, S. W.,** A simple empirical method for the estimation of activation energies in radical molecule metathesis reactions, *Int. J. Chem. Kinet.,* 5, 879, 1973.
220. **Löser, V., Scherzer, K., and Rochde, R.,** BEBO calculations of hydrogen abstraction reactions by alkenyl radicals: vinyl radicals, *Z. Chem.,* 25, 68, 1985.
221. **Scherzer, K., Claus, P., und Karwath, M.,** Untersuchungen zur Kinetik und zum Mechanismus der Addition von Methylradikalen an Vinylacetylen, *Z. Phys. Chem. (Leipzig),* 266, 321, 1985.
222. **Laufer, A. H. and Yung, Y. L.,** Equivalence of vinylidene and $C_2H_2^*$: calculated rate constant for vinylidene abstraction from CH_4, *J. Phys. Chem.,* 87, 181, 1983.
223. **Laufer, A. H., Gardner, E. P., Kwok, T. L., and Yung, Y. L.,** Computations and estimates of rate coefficients for hydrocarbon reactions of interest to the atmosphere of the outer solar system, *Icarus,* 56, 560, 1983.
224. **Louw, R., Dijks, H. M., and Mulder, P.,** Pyrolysis of benzene and chlorobenzene in an atmosphere of hydrogen; formation of methane, *Recl. Trav. Chim. Pays Bas,* 103, 271, 1984.

225. **Mulder, P. and Louw, R.,** Gas-phase Thermolysis of tert-butyl hydroperoxide with benzene and chlorobenzene in the temperature range 200 to 300°C, *Recl. Trav. Chim. Pays Bas,* 103, 282, 1984.
226. **Kurylo, M. J. and Knable, G. L.,** A kinetics investigation of the gas-phase reactions of $Cl(^2P)$ and $OH(X^2\pi)$ with CH_3CN: atmospheric significance and evidence for decreased reactivity between strong electrophiles, *J. Phys. Chem.,* 88, 3305, 1984.
227. **Bérces, T., László, B., and Márta, F.,** Location of the saddle point and height of the activation barrier in atom transfer reactions, *Acta Chim. Acad. Sci. Hung.,* 109, 363, 1982.
228. **Benson, S. W.,** Current status of methods for the estimation of rate parameters, *Int. J. Chem. Kinet.,* 7(Symp. 1), 359, 1975.
229. **Cohen, N.,** The use of transition-state theory to extrapolate rate coefficients for reactions of OH with alkenes, *Int. J. Chem. Kinet.,* 14, 1339, 1982.
230. **O'Neal, H. E. and Benson, S. W.,** Thermochemistry of free radicals, in *Free Radicals,* Vol. 2 (Part 3), Kochi, J. K., Ed., John Wiley & Sons, New York, 1973, chap. 17.
231. **Jeong, K.-M. and Kaufman, F.,** Kinetics of the reaction of hydroxyl radical with methane with nine Cl- and F-substituted methanes. II. Calculation of rate parameters as a test of transition-state theory, *J. Phys. Chem.,* 86, 1816, 1982.
232. **Benson, S. W., Golden, D. M., Lawrence, R. W., Shaw, R., and Woolfolk, R. W.,** Estimation of rate constants as a function of temperature for reactions $X + YZ \rightleftharpoons XY + Z$, $X + Y + M \rightleftharpoons XY + M$ and $X + YZ + M \rightleftharpoons XYZ + M$, where X, Y and Z are atoms H,N,O, *Int. J. Chem. Kinet.,* 7(Symp. 1), 399, 1975.
233. **Shaw, R.,** Estimation of rate constants as a function of temperature for the reactions $W + XYZ = WX + YZ$, where W,X,Y, and Z are H or O atoms, *Int. J. Chem. Kinet.,* 9, 929, 1977.
234. **Shaw, R.,** Semi-empirical extrapolation and estimation of rate constants for abstraction of H from methane by H, O, HO, and O_2, *J. Phys. Chem. Ref. Data,* 7, 1179, 1978.
235. **Evans, M. G. and Polanyi, M.,** Further considerations on the thermodynamics of chemical equilibria and reaction rates, *Trans. Faraday Soc.,* 32, 1333, 1936.
236. **Evans, M. G. and Polanyi, M.,** Inertia and driving force of chemical reactions, *Trans. Faraday Soc.,* 34, 11, 1938.
237. **Bagdasaryan, Kh. S.,** O svyazi mezhdu stroeniem vinilovykh soedinenij i ikh sposobnosti k polimerizatsii, *Zhur. Fiz. Khim.,* 23, 1375, 1949.
238. **Tikhomirova, N. N. and Voevodskii, V. V.,** O reaktsion-nosposobnosti svobodnikh alkil'nykh radikalov, *Dokl. Akad. Nauk. SSSR* 79, 993, 1951.
239. **Grenier, N. R.,** Comparison of the kinetics of alkane H-atom abstraction by methyl and hydroxyl radicals, *J. Chem. Phys.,* 53, 1285, 1970.
240. **Screttas, C. G.,** Equivalent or alternative forms of the Evans-Polanyi-type relations, *J. Org. Chem.,* 45, 1620, 1980.
241. **Semenov, N. N.,** *Some Problems in Chemical Kinetics and Reactivity,* Princeton University Press, Princeton, N. J., 1958.
242. **Kagiya, T., Sumida, Y., Inoue, T., and Dyachkovskii, F. S.,** Evaluation of the activation energies of radical substitution reaction in the gaseous phase. I. An empirical method employing the Morse function, *Bull. Chem. Soc. Jpn.,* 42, 1812, 1969.
243. **Kagiya, T., Sumida, Y., and Inoue, T.,** Evaluation of the activation energies of radical substitution reactions in the gaseous phase. II. An approximate formula with two constants, *Bull. Chem. Soc. Jpn.,* 42, 2422, 1969.
244. **Ogg, R. A., Jr. and Polanyi, M.,** Mechanism of ionic reactions, *Trans. Faraday Soc.,* 31, 604, 1935.
245. **Szabó, Z. G.,** Discussion on the paper "the transition state in radical reactions", in *The Transition State,* Spec. Publ. No. 16, Chemical Society of London, 1962, 113.
246. **Szabó, Z. G. and Bérces, T.,** The transition state and the Arrhenius parameters. II. Evaluation of the activation energy, *Z. Phys. Chem. N. F.,* 57, 113, 1968.
247. **Szabó, Z. G. and Konkoly Thege, I.,** The transition state and Arrhenius parameters. IV. On the "constants" of the Polanyi-Evans equation, *Z. Phys. Chem. N. F.,* 84, 62, 1973.
248. **Krech, R. H. and McFadden, D. L.,** An empirical correlation of activation energy with molecular polarizability for atom abstraction reaction, *J. Am. Chem. Soc.,* 99, 8402, 1977.
249. **Spirin, Yu. L.,** Ob energii aktivatsii radikal'nykh reaktsii, *Zhur. Fiz. Khim.,* 36, 1202, 1962.
250. **Screttas, C. G.,** Could ionization potentials of free radicals serve as alkyl inductive substituent constants?, *J. Org. Chem.,* 44, 1471, 1979.
251. **Pivovarov, V. V. and Stepukhovich, A. D.,** Otsenka energii aktivatsii radikal'nykh reaktsii H-otryva s uchastiem olefinov, *Zhur. Fiz. Kim.,* 57, 609, 1983.
252. **Moin, F. B.,** Raschet energii aktivatsii khimicheskikh reaktsii na osnove printsipa additivnosti, *Usp. Khim.,* 36, 1223, 1967.
253. **Bell, T. N. and Perkins, P. G.,** A novel method for the calculation of energies of activation for gas reactions, *Nature (London),* 256, 300, 1975.

254. **Bell, T. N. and Perkins, P. G.,** Calculation of the energies of activation for some gas-phase reactions, *J. Phys. Chem.*, 81, 2012, 1977.
255. **Bell, T. N., Perkins, K. A., and Perkins, P. G.,** The chlorination of paraffin hydrocarbons. Calculation of the activation energies and A factors for reactions in the total chlorinated methane, *J. Phys. Chem*, 81, 2610, 1977.
256. **Bell, T. N., Perkins, K. A., and Perkins, P. G.,** Chlorination of paraffin hydrocarbons. II. Ethanes and propanes, *J. Phys. Chem.*, 83, 2321, 1979.
257. **Pauling, L.,** *The Nature of the Chemical Bond,* 3rd ed., Cornell University Press, Ithaca, N. Y., 1960.
258. **Hammond, G. S.,** A correlation of reaction rates, *J. Am. Chem. Soc.*, 77, 334, 1955.
259. **Mok, M. H. and Polanyi, J. C.,** Location of energy barriers. II. Correlation with barrier height, *J. Chem. Phys.*, 51, 1451, 1969.
260. **Dunning, T. H., Jr.,** Theoretical studies of the energetics of the abstraction and exchange reactions in H + HX, with X = F – I, *J. Phys. Chem.*, 88, 2469, 1984.
261. **Basilevsky, M. V., Weinberg, N. N., and Zhulin, V. M.,** Estimating the position of transition states from experimental data, *Int. J. Chem. Kinet.*, 11, 853, 1979.
262. **Agmon, N. and Levine, R. D.,** Energy, entropy and the reaction coordinate: thermodynamic-like relations in chemical kinetics, *Chem. Phys. Lett.*, 52, 197, 1977.
263. **Miller, A. R.,** A theoretical relation for the position of the energy barrier between initial and final states of chemical reactions, *J. Am. Chem. Soc.*, 100, 1984, 1978.
264. **Murdoch, J. R.,** Barrier heights and the position of stationary points along the reaction coordinate, *J. Am. Chem. Soc.*, 105, 2667, 1983.
265. **Marcus, R. A.,** Theoretical relations among rate constants, barriers, and Brønsted slopes of chemical reactions, *J. Phys. Chem.*, 72, 891, 1968.
266. **Marcus, R. A.,** Unusual slopes of free energy plots in kinetics, *J. Am. Chem. Soc.*, 91, 7224, 1969.
267. **Matheson, I. A., Sidebottom, H. W., and Tedder, J. M.,** The reaction of trichloromethyl radicals with hydrogen chloride and a new estimation of the rate of combination of trichloromethyl radical, *Int. J. Chem. Kinet.*, 6, 493, 1974.
268. **Tucker, B. G. and Whittle, E.,** Reactions of trifluoromethyl radicals with iodine, bromine and hydrogen bromide, *Trans. Faraday Soc.*, 61, 866, 1965.
269. **Tedder, J. M.,** The importance of polarity, bond strength and steric effects in determining the site of attack and the rate of free radical substitution in aliphatic compounds, *Tetrahedron,* 38, 313, 1982.
270. **Kerr, J. A. and Timlin, D. M.,** Hydrogen abstraction from organosilicon compounds. The reactions of fluoromethyl radicals with tetramethylsilane. Polar effects in gas phase reactions, *Int. J. Chem. Kinet.*, 3, 69, 1971.
271. **Katz, M. G., Baruch, G., and Rajbenbach, L. A.,** Radiation-induced dechlorination of carbon tetrachloride in cyclohexane solutions. The kinetics and liquid-phase reactions of trichloromethyl radicals, *Int. J. Chem. Kinet.*, 8, 131, 1976.
272. **Katz, M. G., Baruch, G., and Rajbenbach, L. A.,** Radiation-induced dechlorination of carbon tetrachloride in cyclohexane and n-hexane mixtures. III. The kinetics of liquid-phase reactions of trichloromethyl radicals, *Int. J. Chem. Kinet.*, 10, 905, 1978.
273. **Alfassi, Z. B. and Feldman, L.,** Kinetics of radiation-induced abstraction of hydrogen atoms by CCl_3 radicals in the liquid phase: 2,3-dimethylbutane, *Int. J. Chem. Kinet.*, 12, 379, 1980.
274. **Alfassi, Z. B. and Feldman, L.,** The kinetics of radiation-induced hydrogen abstraction by CCl_3 radicals in the liquid phase: cyclanes, *Int. J. Chem. Kinet.*, 13, 517, 1981.
275. **Dickey, L. C. and Firestone, R. F.,** Radiolysis of chloroform vapor. Effects of phase on the Arrhenius parameters of the hydrogen-atom abstraction reaction of dichloromethyl radicals with chloroform, *J. Phys. Chem.*, 74, 4310, 1970.

Chapter 10

COMPUTER MODELING OF CHEMICAL REACTIONS

Jenn-Tai Hwang

TABLE OF CONTENTS

I. INTRODUCTION

The object of much experimentation in chemical kinetics is to build a mechanistic model of a reaction process, with the (usual) additional aim of estimating unknown rate constants, while the procedure of deducing the mathematical description of a process from experimental observations is modeling. Over the recent past, significant development in the computational tools has been made, and more and more complex numerical models, devised to reflect what actually happens at the molecular level, are being used as predictive tools and as aids for understanding the processes underlying observed chemical phenomena.

To start modeling a chemical phenomenon, the chemist must first work out a model. In the most general terms, a model is a functional relationship or relationships among state variables (concentrations, temperatures, pressures) and parameters (rate constants, activation energies, transport coefficients, etc). The relationships may be in the form of algebraic, differential, or integral equations. Of particular interest in the present chapter are models formulated in terms of differential equations, which are called dynamic models.

The second step in the modeling procedure is to fit the model predictions to the experimental data by adjusting the values of unknown or imprecisely known parameters. This is usually done by minimizing the sum of squares of the deviations of the predicted values from the corresponding experimental values.

Finally, the adequacy of the model in explaining the chemical phenomenon is examined. Any relationships between the deviations of the model predictions and the experimental variables can prove useful in pinpointing deficiencies in the model. The presence of systematic errors is an indication that the model is inadequate and the parameter estimates are invalid. Based on the adequacy test, new model(s) may be deduced, new experiments designed, and the whole modeling procedure may be iterated until the chemist is satisfied that the model is well developed for his purpose. Here, the phase "well developed" simply means that the model explains the set of experimental data to within deviations which can be ascribed to experimental error.

The goal of the present chapter is to provide an introduction to the computer modeling of chemical reactions that will stimulate interest. The key concepts and ideas of the parameter estimation problem are reviewed briefly in the next section. The third section discusses the equations for the dynamic models of both well-stirred homogeneous and inhomogeneous flow systems; the basics of numerical techniques for solving these equations are also given. The subject of sensitivity analysis is reviewed in the fourth section, which gives a thorough discussion on the proper usage of sensitivity information in unraveling dynamic details of a complex mechanistic model, and presents a newly developed method, the polynomial approximation method (PAM), for computing sensitivity coefficients. The fifth section concludes with a flow diagram showing the essential steps of a model-building process and the interrelations among the various topics discussed in the chapter.

II. PARAMETER ESTIMATION

Parameter estimation problems result when one attempts to match a model of known form to experimental data by optimal determination of unknown model parameters.[1-4] The exact nature of the parameter estimation problem depends on the type of mathematical model. Let us begin by considering algebraic models.

A. Algebraic Models

Let $\mathbf{k}$ = p-dimensional column vector of parameters (e.g., rate constants) whose numerical values are unknown, $[k_1, k_2, \ldots, k_p]^T$ (here the superscript T denotes transpose); $\mathbf{x}$ = m-dimensional column vector of states — these are the independent variables which are either fixed arbitrarily for each experiment or known precisely for each experimental observation, $[x_1,x_2,\ldots,x_m]^T$, e.g., initial species concentrations and reaction time; and $\mathbf{Y}$ = n-dimensional column vector of observed variables — these are the model variables which are actually measured in the experiments, $[y_1,y_2,\ldots,y_n]^T$ (note that $\mathbf{y}$ is *not* the actual experimental data, but only the dependent variables of the model; examples of $\mathbf{y}$ are species concentrations, optical absorbances, and exothermicity). In modeling, the term "vector" is customarily used to denote sets of functions, variables, or parameters.

A single experiment consists of the measurement of each of the m observed variables for a given set of values of the independent variables. The algebraic models are those in which the observed variables are related to the independent variables and parameters by an algebraic equation of the form:

$$\mathbf{y} = \mathbf{h}(\mathbf{x}, \mathbf{k}) \tag{1}$$

where $\mathbf{h}$ is an n-dimensional vector function of known form. In component form, Equation 1 reads:

$$y_i = h_i(x_1, x_2,\ldots, x_m, k_1, k_2,\ldots, k_p), \quad i = 1, 2,\ldots, n \tag{2}$$

As an example of Equation 1, consider the reaction $A \xrightarrow{k_1} B \xrightarrow{k_2}$ product. The species concentrations of this system are

$$C_A = C_A^{(0)} e^{-k_1 t}$$

$$C_B = C_B^{(0)} e^{-k_2 t} + \frac{k_1 C_A^{(0)}}{k_2 - k_1}(e^{-k_1 t} - e^{-k_2 t}) \tag{3}$$

where t is the reaction time and $C_A^{(0)}$ and $C_B^{(0)}$ are the initial concentrations of A and B, respectively. Suppose that both A and B are measured. Then, by letting $\mathbf{y} = [C_A, C_B]^T$, $\mathbf{x} = [t, C_A^{(0)}, C_B^{(0)}]^T$, and $\mathbf{k} = [k_1, k_2]^T$, Equation 3 is seen to be equivalent to Equation 1.

To determine the p unknown parameters, at least p different experiments must be performed. For the jth experiment, Equation 1 reads:

$$\mathbf{y}_j = \mathbf{h}(\mathbf{x}_j, \mathbf{k}), \quad j = 1, 2,\ldots, M \tag{4}$$

where we assume that M separate experiments have been performed. Now let $\boldsymbol{\eta}_j = [\eta_{1j},\eta_{2j},\ldots,\eta_{nj}]^T$, j = 1,2,...,M, be the actual data points, that is, experimentally measured values of $\mathbf{y}_j$. Then:

$$\boldsymbol{\eta}_j = \mathbf{h}(\mathbf{x}_j, \mathbf{k}) + \boldsymbol{\epsilon}_j, \quad j = 1, 2,\ldots, M \tag{5}$$

where $\boldsymbol{\epsilon}_j = [\epsilon_{1j},\epsilon_{2j},\ldots,\epsilon_{nj}]^T$ is the vector of errors, or residuals, between the observations and the predicted dependent variables. Parameter estimation tries to find a set of values for the parameters $\mathbf{k}$ such that some scalar function of the errors, $S(\mathbf{k})$, is minimized.

1. Least-Squares Estimation

In least-squares estimation, the objective function $S(\mathbf{k})$ is chosen to be a weighted sum of squares of the errors:

$$S(\mathbf{k}) = \sum_{j=1}^{M} \boldsymbol{\epsilon}_j^T \mathbf{Q}_j \boldsymbol{\epsilon}_j$$

$$= \sum_{j=1}^{M} \sum_{r=1}^{n} \sum_{s=1}^{n} [\eta_{rj} - h_r(\mathbf{x}_j, k)]\, q_{rsj}[\eta_{sj} - h_s(\mathbf{x}_j, \mathbf{k})] \qquad (6)$$

where $\mathbf{Q}_j$ is an $n \times n$ weighting matrix with elements q_{rsj}. The weighting matrix is used to reflect knowledge of the relative precision of the various measurements. If $q_{rsj} = \partial_{rs}$, where ∂_{rs} is a Kronecker delta, then Equation 6 is an *un*weighted least-squares criterion.

To find the minimum of S in the parameter space, either direct search or gradient methods may be used. Direct search methods are those that do not require explicit evaluation of the derivatives $(\partial S/\partial k_l)$, while gradient methods are those that do. The idea of direct search methods is appealing, with the obvious advantage of not requiring differentiation, and they seem to perform well on well-conditioned problems. In difficult problems, however, gradient methods tend to outperform direct search methods in both reliability and speed of convergence.[1,2]

The gradient methods are iterative in nature: one starts with an initial guess $\mathbf{k}^{(1)}$ of the parameters, and proceeds to generate a sequence of values $\mathbf{k}^{(2)}, \mathbf{k}^{(3)}, \ldots$, which one hopes converges to the optimum value $\hat{\mathbf{k}}$ at which S is minimum. In the class of gradient methods which have proved successful, the formula used for finding the sequence is

$$\mathbf{k}^{(i+1)} = \mathbf{k}^{(i)} - \lambda \mathbf{R}\mathbf{g}, \quad i = 1, 2, \ldots \qquad (7)$$

where the step length λ is a scalar, $\mathbf{R}$ is a matrix, and $\mathbf{g}$ is the gradient vector of S, $g_l = (\partial S/\partial k_l)$, $l = 1,2\ldots, p$. Different choice of $\mathbf{R}$ and λ results in different gradient methods. For example, the choice of $\mathbf{R} = \mathbf{I}$, where $I_{ij} = \delta_{ij}$, constitutes the method of steepest descent.[1,2,5] The choice of R as the inverse of the Hessian matrix $\mathbf{G}$, $G_{ij} = (\partial^2 S/\partial k_i \partial k_j)$, constitutes the Newton Raphson method,[1,2,5,6] while the choice of $\mathbf{R}$ as the inverse of $(\mathbf{G} + \nu\mathbf{I})$, where ν is some sufficiently large number, is the Marquardt method.[1,2,5-7]

The minimization procedure proceeds iteratively by computing successive correction, $-\lambda\mathbf{R}\mathbf{g}$, of an initial guess. In principle, the gradient $\mathbf{g}$ should vanish at the minimum of S. However, due to the round-off errors, this condition can never be attained precisely, and in practice the iteration is terminated whenever $|S(\mathbf{k}^{(i+1)}) - S(\mathbf{k}^{(i)})|/|S(\mathbf{k}^{(i)})|$ is less than some prescribed tolerance.

2. Maximum Likelihood Estimation

Measurements have errors; the errors in the observations are as much a part of physical reality as is the process being observed. Meaningful parameter estimation can proceed only if one provides a mathematical model not only of the system, but also of the errors. For the errors, a probability distribution[1,2,8] is an appropriate mathematical model.

Let $P(\boldsymbol{\epsilon}_O, \boldsymbol{\psi})d\boldsymbol{\epsilon}$ be the probability that the values of all the errors in the observed variables lie within a hypercube of volume $d\boldsymbol{\epsilon}$ centered at $\boldsymbol{\epsilon}_O$. The function $P(\boldsymbol{\epsilon}, \boldsymbol{\psi})$ is called a joint probability density function. The vector $\boldsymbol{\psi}$ represents a set of unknown parameters (e.g., means, variances) which appear in the joint probability density function. The maximum likelihood method simply consists of finding those values of $\mathbf{k}$ and $\boldsymbol{\psi}$ which maximize the likelihood function, $L(\mathbf{k}, \boldsymbol{\psi}) \equiv P[\boldsymbol{\eta} - \mathbf{h}(\mathbf{x}, \mathbf{k}), \boldsymbol{\psi}]$, or, equivalently, the logarithm of the likelihood function, $S(\mathbf{k}, \boldsymbol{\psi}) = \log L(\mathbf{k}, \boldsymbol{\psi})$.

If, as is frequently assumed, the observation errors in different experiments are uncorrelated with each other, then the joint probability density $P(\boldsymbol{\epsilon}, \boldsymbol{\psi})$ is a product of the individual experiment probability densities, and

$$S(\mathbf{k}, \boldsymbol{\psi}) = \sum_{j=1}^{M} \log P[\boldsymbol{\eta}_j - \mathbf{h}(\mathbf{x}_j, \mathbf{k}), \boldsymbol{\psi}] \qquad (8)$$

If one further assumes that the errors of each experiment, $\boldsymbol{\epsilon}_j = \boldsymbol{\eta}_j - \mathbf{h}(\mathbf{x}_j,\mathbf{k})$, are independent and normally distributed with zero means and covariance matrix $\mathbf{V}_j$, then:[1,2,8]

$$P(\boldsymbol{\epsilon}_j, \mathbf{V}_j) = [(2\pi)^n \det(\mathbf{V}_j)]^{-1/2} \exp\left(-\frac{1}{2} \boldsymbol{\epsilon}_j^T \mathbf{V}_j^{-1} \boldsymbol{\epsilon}_j\right), \quad j = 1, 2, \ldots, M \tag{9}$$

where $\det(\mathbf{V}_j)$ is the determinant of $\mathbf{V}_j$. Combining Equations 8 and 9 yields:

$$S(\mathbf{k}, \mathbf{V}_1, \mathbf{V}_2, \ldots, \mathbf{V}_M) = -\frac{nM}{2}\log(2\pi) - \frac{1}{2}\sum_{j=1}^{M} \log[\det(\mathbf{V}_j)] - \frac{1}{2}\sum_{j=1}^{M} \boldsymbol{\epsilon}_j^T \mathbf{V}_j^{-1} \boldsymbol{\epsilon}_j \tag{10}$$

One interesting observation one can make concerning Equation 10 is that if all the covariance matrixes $\mathbf{V}_j$ are known, then maximizing S is equivalent to minimizing the last term on the right-hand side of Equation 10; this means that the least-squares estimates are also the maximum likelihood estimates if $q_{rsj} = [\mathbf{V}_j^{-1}]_{rs}$. (See Equation 6.)

3. Bayesian Estimation

The chemist usually has some ideas concerning the values of his parameters even before any data have been gathered. For instance, the rate constants or the diffusion constants must be positive. An estimation procedure that came up with negative values for such parameters should be entirely unacceptable. Even among the admissible values, the chemist may regard some as more plausible than others because of the information available from previous experiments.

This sort of prior information is frequently summarized in the form of a relative probability density function $P_O(\mathbf{K})$, called the prior distribution. If, e.g., $P_O(\mathbf{k}_1)/P_O(\mathbf{k}_2) = 10$, then $\mathbf{k} = \mathbf{k}_1$ is 10 times more likely than $\mathbf{k} = \mathbf{k}_2$; if the equilibrium constant of a certain chemical reaction has been estimated to be $K_O \pm \sigma$, then the prior distribution is usually set to

$$P_0(K) = \frac{1}{\sigma\sqrt{2\pi}} \exp\left[-\frac{(K - K_0)^2}{2\sigma^2}\right]$$

When one comes to estimate the parameters on the basis of the new data, the prior information can be taken into account by multiplying the likelihood function by the prior distribution to yield the posterior distribution (Bayes' theorem[1,8]). The estimation then consists of maximizing Equation 11:

$$S(\mathbf{k}, \boldsymbol{\psi}) = \log P[\boldsymbol{\eta} - \mathbf{h}(\mathbf{x}, \mathbf{k}), \boldsymbol{\psi}] + \log P_0(\mathbf{k}) \tag{11}$$

One cannot overstate the importance of using all available prior information in the form of either constraints on the parameters or prior distributions. Use of such information frequently spells the difference between convergence and nonconvergence of the estimation procedure.

B. Dynamic Models

Dynamic models are mathematical relationships between state variables and parameters which are formulated in terms of either ordinary or partial differential equations. For example, the mathematical model for a well-stirred homogeneous reaction system $A \xrightarrow{k_1} B \xrightarrow{k_2} C$ is a set of coupled ordinary differential equations (rate equations):

$$\frac{dC_A}{dt} = -k_1C_A \equiv f_A$$

$$\frac{dC_B}{dt} = k_1C_A - k_2C_B \equiv f_B$$

$$\frac{dC_C}{dt} = k_2C_B \equiv f_C \tag{12}$$

If the spatial distribution of chemical species in the above reaction system is nonuniform, then the corresponding mathematical model becomes a set of coupled partial differential equations (reaction-diffusion equations):

$$\frac{\partial C_i}{\partial t} = D_i\nabla^2 C_i + f_i, \quad i = A, B, C \tag{13}$$

where D_i is the diffusion constant of the *i*th chemical species and ∇^2 is the Laplacian operator.

Consider first a general set of *m* coupled ordinary differential equations:

$$\frac{d\mathbf{x}}{dt} = \mathbf{f}(t, \mathbf{x}(t), \mathbf{k}), \quad \mathbf{x}(t_0) = \mathbf{a} \tag{14}$$

where $\mathbf{x}$ = *m*-dimensional column vector of state variables, $[x_1,x_2,\ldots,x_m]^T$ (if Equation 14 denotes a set of coupled rate equations, then the x_i are the *m* species concentrations of the reaction system); $\mathbf{k}$ = *p*-dimensional column vector of parameters, $[k_1,k_2,\ldots,k_p]^T$ (for rate equations, the k_i are the rate constants of the elementary steps; in this chapter, we assume **k** to be time invariant for simplicity); and $\mathbf{f}$ = *m*-dimensional vector function of known form, $[f_1,f_2,\ldots,f_m]^T$ (for rate equations, the f_i are the rate laws for each of the chemical species).

The observable variables, assumed to be *n* of them, are *n* given functions of **x, k, t,** and a vector $\boldsymbol{\theta}$ of *additional* independent variables and parameters:

$$\mathbf{y} = \mathbf{h}(\mathbf{x}, t, \mathbf{k}, \boldsymbol{\theta}) \tag{15}$$

where **y** and **h** are *n*-dimensional column vectors. A common special case is that in which the state variables are observed directly, i.e., $\mathbf{y} = \mathbf{x}(t,\mathbf{k},\mathbf{a})$.

All standard parameter estimation methods previously introduced for algebraic models may be adapted to the present situation. The estimation procedure starts by solving Equation 14 (usually numerically) for **x**, substituting the solutions into Equation 15, and then evaluating any of the standard objective functions. There are, however, aspects specific to dynamic models. In addition, the incentive to use an efficient method here is particularly great because each function evaluation is itself a complex procedure requiring the solution of a set of differential equations.

Consider, e.g., least-squares estimation. For simplicity, let us assume that $\mathbf{y} = \mathbf{h}(\mathbf{x},t)$. The least-squares criteria for the cases of discrete and continuous measurements are, respectively,

$$S = \sum_{j=1}^{M} [\boldsymbol{\eta}(t_j) - \mathbf{h}(\mathbf{x}(t_j), t_j)]^T \mathbf{Q}_j[\boldsymbol{\eta}(t_j) - \mathbf{h}(\mathbf{x}(t_j), t_j)] \tag{16}$$

and

$$S = \int_0^t d\tau[\boldsymbol{\eta}(\tau) - \mathbf{h}(\mathbf{x}(\tau), \tau)]^T \mathbf{Q}(\tau)[\boldsymbol{\eta}(\tau) - \mathbf{h}(\mathbf{x}(\tau), \tau)] \tag{17}$$

where $\boldsymbol{\eta}(t)$ represents the actual experimental data at time t and where $\mathbf{Q}_j$ and $\mathbf{Q}(\tau)$ are weighting matrixes which can be chosen arbitrarily or related to the inverse of the error covariances.

To locate the minimum of the objective function S in the **k** space by means of gradient methods, partials of S, $(\partial S/\partial k_l)$, $l = 1,2,\ldots,p$, are required. Let us consider for illustration the case of discrete measurements. Differentiating Equation 16 with respect to k_l yields, for a symmetric $\mathbf{Q}_j$:

$$\begin{aligned}\frac{\partial S}{\partial k_l} &= \sum_{j=1}^{M}\sum_{r=1}^{n}\left[\frac{\partial S}{\partial \epsilon_r(t_j)}\right]\left[\frac{\partial \epsilon_r(t_j)}{\partial k_l}\right] \\ &= 2\sum_{j=1}^{M} \boldsymbol{\epsilon}(t_j)^T \mathbf{Q}_j\left[\frac{\partial \mathbf{h}(\mathbf{x}(t_j), t_j)}{\partial \mathbf{x}(t_j)}\right]\left[\frac{\partial \mathbf{x}(t_j)}{\partial k_l}\right], \quad l = 1,2,\ldots, p\end{aligned} \tag{18}$$

where

$$\boldsymbol{\epsilon}(t_j) = \boldsymbol{\eta}(t_j) - \mathbf{h}(\mathbf{x}(t_j), t_j) \tag{19}$$

and where $(\partial \mathbf{h}/\partial \mathbf{x})$ is an $n \times m$ matrix with elements $(\partial h_i/\partial x_j)$, $i = 1,2,\ldots,n$, $j = 1,2,\ldots,m$. Equation 18 also requires computation of the m-dimensional vectors of $\partial \mathbf{x}/\partial k_l)$, the elements of which, $(\partial x_i/\partial_l)$, are known as the first-order sensitivity coefficients of the x_i with respect to the k_l. Sensitivity coefficients are measures of the sensitivity of the dependent variables x_i to changes in the parameters k_l. They carry information useful to experimentalists when they try to decide which x_i to monitor in order to estimate the parameters of interest; they also carry information useful to analysts when they try to probe the detailed dynamics of their models. Sensitivity coefficients can also be used to simplify a dynamic model by determining those parameters which exert little or no effects on the model behavior. All these applications are subjects within the range of sensitivity analysis. We discuss fully the sensitivity analysis as applied to chemical kinetics in Section IV.

Sensitivity coefficients can be computed either by means of the finite difference methods or by solving sensitivity equations. The finite difference methods replace $(\partial \mathbf{x}/\partial k_l)$ by finite differences, a procedure which is error prone and unreliable. Sensitivity equations are differential equations which govern the temporal behavior of the sensitivity coefficients; they can be derived from the original dynamic model, Equation 14, by differentiating both sides of the equation with respect to the parameters k_l and then interchanging the order of differentiations with respect to t and k_l [this is permissible when $\frac{d}{dt}\left(\frac{\partial \mathbf{x}}{\partial k_l}\right)$ and $\frac{\partial}{\partial k_l}\left(\frac{d\mathbf{x}}{dt}\right)$ are continuous]. The sensitivity equations thus obtained are

$$\frac{d}{dt}\left(\frac{\partial \mathbf{x}}{\partial k_l}\right) = \mathbf{J}\left(\frac{\partial \mathbf{x}}{\partial k_l}\right) + \left(\frac{\partial \mathbf{f}}{\partial k_l}\right), \quad l = 1, 2,\ldots, p \tag{20}$$

where $\mathbf{J}$ is the Jacobian matrix $(\partial \mathbf{f}/\partial \mathbf{x})$ with elements.

$$J_{ij} = (\partial f_i/\partial x_j) \tag{21}$$

The initial conditions of the sensitivity equations in Equation 20 are

$$\left(\frac{\partial \mathbf{x}}{\partial k_l}\right)(t_0) = 0, \quad l = 1, 2, \ldots, p \tag{22}$$

because $\mathbf{x}(t_0)$ is independent of $\mathbf{k}$. Methods for solving sensitivity equations of both ordinary and partial differential systems are detailed in Section IV.

The least-squares estimation of the dynamic model (Equation 14) proceeds as follows.

1. Guess the initial values $\mathbf{k}^{(1)}$.
2. Integrate Equation 14 from zero to t_M and evaluate S by means of Equation 16 (or Equation 17 in the case of continuous measurements).
3. Integrate the sensitivity equations in Equation 20 from zero to t_M.
4. Evaluate $(\partial S/\partial k_l)$ from Equation 18, $l = 1,2,\ldots,p$.
5. Form updated estimates of k_l by means of any of the gradient methods.
6. Repeat steps 2 to 5 until $|S(\mathbf{k}^{i+1}) - S(\mathbf{k}^{(i)})|/|S(\mathbf{k}^{(i)})|$ is less than some prescribed tolerance.

It is a straightforward matter to extend the previous discussion to dynamic models for mulated in terms of partial differential equations. The only difference is that, in the partial differential systems, state variables are functions of both space and time and model parameters may enter into the boundary conditions (e.g., heat and mass transfer coefficients on the boundary) as well as the differential equations themselves. Thus, objective functions like the least-squares criterion may look like:[1,2]

$$S = \sum_{j=1}^{M} \sum_{r=1}^{M_1} \sum_{s=1}^{M_2} \boldsymbol{\epsilon}_j(\mathbf{r}_r, t_s)^T \mathbf{Q}_{rsj} \boldsymbol{\epsilon}_j(\mathbf{r}_r, t_s) \tag{23}$$

where $\mathbf{r}$ denotes spatial location. The parameter estimation procedure nonetheless follows almost line by line the procedure presented earlier for the models of ordinary differential equations.[1,2]

C. Confidence Limits

It is not enough just to compute the estimates of parameters. What is equally important is that one must also know how reliable the estimates are. Due to the existence of random experimental errors, one cannot expect to obtain exactly the same estimates from different data samples, even if the samples were gathered under similar conditions. Meaningful estimates must therefore be augmented with information on their variability. For example, a statement such as $k = 3 \pm 0.2$ is much more meaningful than the statement $k = 3$.

To gather information on the variability of the estimates, one must estimate their probability distribution. (Notice that estimates are themselves random variables because they are computed from observations which contain random errors.) The means of the estimated distribution are, then, the estimated values of the parameters, and the covariance matrix of the distribution provides a measure of the reliability of the estimates. Let the mean of a random variable Z be denoted $\langle Z \rangle$. The covariance matrix $\mathbf{P}$ of p random variables Z_i, $i = 1,2,\ldots,p$, is, then, a pxp matrix defined by

$$\mathbf{P}_{ij} = \langle (Z_i - \langle Z_i \rangle)(Z_j - \langle Z_j \rangle) \rangle \quad (24)$$

The diagonal elements of **P** are the individual variances, the square roots of which are the *standard deviations.*[1,2,8] The off-diagonal elements of **P** are the covariances, which measure the interdependence of the estimates of various parameters.[1,2,8]

It can be shown[1] that, for a wide class of maximum likelihood estimates with normal distributions, the covariance matrix **P** of the estimates is related to the likelihood function L by

$$[\mathbf{P}^{-1}]_{ij} = (\partial^2 \log L/\partial k_i \partial k_j)_{\mathbf{k}=\hat{\mathbf{k}}} \quad (25)$$

where $\hat{\mathbf{k}}$ is the optimal estimate of **k**. This is quite reasonable because the parameter estimation problem seeks the global minimum of the objective function S(**k**); if S is little affected by changes in a certain parameter, one would have doubts concerning the reliability of its estimate. Therefore, estimation reliability is connected with the dependence of S on the parameters. At the minimum of S, ($\partial S/\partial \mathbf{k}$) vanishes; the effects of the parameters on S, then, are summarized in the second-order derivatives.

After computing **P**, confidence limits may be calculated. In the case of a single parameter k, confidence limits are the limits on either side of the estimate which insure that, if a large number of replicate experiments is performed, the true value of k will lie within these limits a specified percentage of times. These limits are simply a product of a fixed number β (depending on the percentage) and the standard deviation.[1,2] For example, if σ is the standard deviation for an estimate of k, the 90% confidence limits on k are $\hat{k} \pm 1.6449\ \sigma$, where $\hat{k}$ is the estimate. When there are *p* parameters, the above confidence interval must be replaced by a *p*-dimensional confidence region hyperellipse. The confidence region hyperellipse is calculated as follows. Let **b** be a unit vector in the *p*-dimensional parameter space. If $\hat{\mathbf{k}}$ is an estimate of **k** and **P** the corresponding covariance matrix, then the confidence interval along the direction of **b** within which the projection of the true value of **k** on **b** will lie a specified percentage of times (in a large series of similar experiments) is $\mathbf{b}^T\hat{\mathbf{k}} \pm \beta (\mathbf{b}^T\mathbf{P}\mathbf{b})^{1/2}$, where β depends on the confidence level. The hyperellipse is the region containing the confidence intervals for all directions of **b** on the unit hypersphere centered at $\hat{\mathbf{k}}$.

Usually, the individual confidence intervals provide a rough guide to the joint confidence region if the individual estimates are uncorrelated. However, if the estimates are highly correlated, individual confidence distributions will have a very wide spread with the joint confidence region contained in a narrow oblique band. This situation can cause rather misleading inferences to be drawn if more than one parameter is estimated by information of individual limits alone.[2] Figure 1 provides one such example. As one can see from the figure, an estimate of $k_1 = 0.75$ is not at all unusual judging from the individual confidence distribution for k_1. Similarly, an estimate of $k_2 = 0.75$ individually is not unusual. However, the pair of estimates (0.75, 0.75) corresponds to a point well outside the 95% confidence region. The explanation is simply that, while for *some* value of k_2 the estimate of $k_1 = 0.75$ is not contradicted by the data and for *some* value of k_1 the estimate of $k_2 = 0.75$ is not contradicted by the data, the joint occurrence of these two estimates is contracted at a fairly high significance level.

D. Design of Experiments

The objective of an experiment is to gain relevant information. If the experiments are carried out following a well-planned program, the data gathered can be most informative. On the other hand, for poorly planned experiments, the data can display a highly disperse distribution, which renders the estimates rather uncertain.

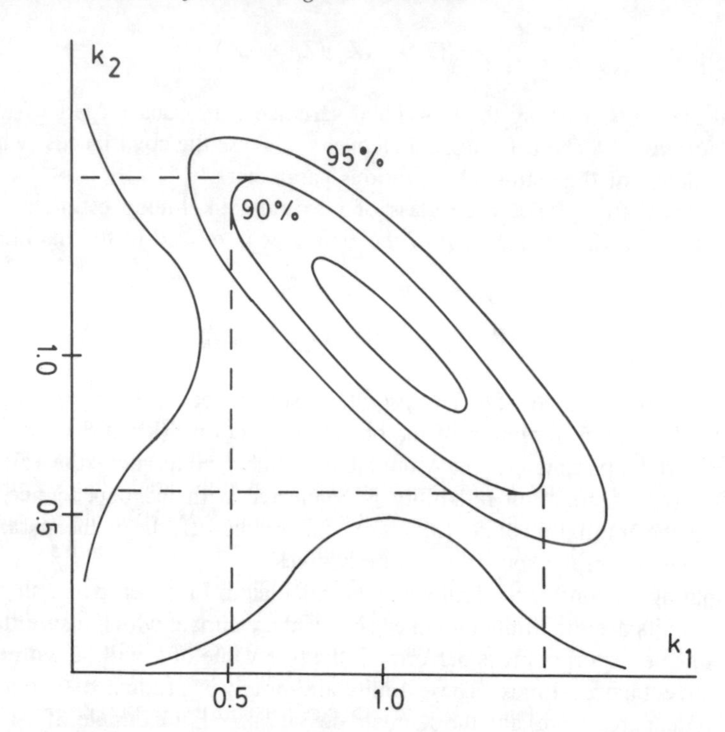

FIGURE 1. Confidence region with individual distributions for the estimation of two parameters, k_1 and k_2.

As an example for the improperness of experimental design, suppose that the model equation is $y = k_1x_1 + k_2x_2$. If all experiments were performed under the condition $x_1 \approx x_2$, then $y \approx (k_1 + k_2)x_1$ and it is impossible to estimate k_1 and k_2 individually — only their sum.

In 1948, Shannon[9] showed that the unique (aside from a positive multiplicative factor) suitable measure of uncertainty associated with a probability density function $P(\mathbf{k})$, where $\mathbf{k}$ is a random vector, is

$$I(p) = -\langle \log P(\mathbf{k}) \rangle = -\int P(\mathbf{k}) \log P(\mathbf{k})\, d\mathbf{k} \tag{26}$$

Information is gained by reducing uncertainty.

Suppose one now intends to measure certain dependent variables $\mathbf{y}$ of a system at given values of the independent variables (or experimental conditions) $\mathbf{x}$, such that estimates of the parameters $\mathbf{k}$ can be obtained (or improved, if some prior experiments have been done). Designing the experiment means choosing in some rational way the values of $\mathbf{x}$ at which $\mathbf{y}$ is to be measured. Let $P_O(\mathbf{k})$ and $P^*(\mathbf{k})$ be, respectively, the prior and posterior distributions in the usual Bayesian sense. The amount of information H that can be gained by performing the present experiment equals the reduction in uncertainty from the prior to the posterior distributions:[10]

$$H = I(P_0) - I(P^*) \tag{27}$$

One aims to design the experiment in such a way that H is maximized, or, equivalently, in such a way that $I(P^*)$ is minimized. [Notice that $I(P_0)$ is unaffected by the present experiment.]

Let P^* be a multivariate normal distribution with a covariance matrix $\mathbf{V}$. Then:

$$I(p^*) = -\langle \log p^* \rangle$$
$$= \frac{n}{2}(1 + \log 2\pi) + \frac{1}{2}\log[\det(\mathbf{V})] \tag{28}$$

This shows that the experimental design criterion is to choose the experimental conditions **x** such that det(**V**) is minimized;[11] thus, a distribution with a small variance contains more information than one with a large variance.

III. SIMULATION OF REACTION PROCESSES

The course of a chemical reaction is governed by a set of molecular-level chemical transformations known as the elementary steps. The whole set of elementary steps is called the reaction mechanism. Before the advent of powerful computational tools, reaction mechanisms developed were more a function of solvable mathematics than of the real-world chemistry. With the development of large digital computers and efficient programs, chemists are now capable of dealing with mechanisms comprised of tens to hundreds of elementary steps. Massive kinetic models are now being used, especially in the field of combustion, to study the underlying molecular processes of an observed chemical phenomenon, and to unify the manifold experimental data that have accumulated over the years.

In this section, the equations for the dynamic models of both (well-stirred) *homogeneous* and *inhomogeneous* (flow) systems are discussed. Basics of the numerical techniques for solving these equations are also given.

A. Homogeneous Systems

In a well-stirred reactor, the composition and physical properties of the reaction system are the same everywhere. Hence, there exists just one independent variable, the time t. Systems without dependence on spatial coordinates are called homogeneous.

Suppose that a mechanism with p elementary steps is proposed for some chemical reaction. By law of mass action, the corresponding rate equations are

$$\frac{dx_i}{dt} = \sum_{l=1}^{p/2} (\bar{\nu}_{il} - \nu_{il})\left[k_l \prod_{j=1}^{m} x_j^{\nu_{jl}} - \bar{k}_l \prod_{j=1}^{m} x_j^{\bar{\nu}_{jl}}\right] \equiv f_i(\mathbf{x}, \mathbf{k}),$$
$$i = 1, 2, \ldots, m \tag{29}$$

Here x_i is the concentration of species i; $\nu_{il}(\bar{\nu}_{il})$ is the stoichiometric coefficient of the species i on the reactant (product) side of the elementary step l; k_l ($\bar{k}_l$) is the forward (backward) rate constant for the elementary step ℓ; and **k** is a p-dimensional column vector of all the rate constants in the mechanism, $[k_1, k_2, \ldots, k_{p/2,} \bar{k}_1, \bar{k}_2, \ldots, \bar{k}_{p/2},]^T$.

Equation 29 describes just the chemistry of a homogeneous system. To specify the state of the system completely, other relationships, depending on the reaction conditions, are needed. For example, a homogeneous gas-phase reaction under the constant-temperature, constant-volume condition can have pressure changes. In order to calculate them, the ideal gas law is required:

$$P = RT \sum_{i=1}^{m} x_i \tag{30}$$

where P is the pressure, T is the temperature, and R is the gas constant. On the other hand, if the gas-phase reaction occurs under the constant-volume adiabatic condition, then both P and T can vary. One therefore requires, in addition to Equation 30, the relation of internal energy U conservation:

$$\frac{d}{dt}U = 0 \tag{31}$$

which allows temperature changes to be calculated:[12]

$$\frac{dT}{dt} = \sum_i \frac{dx_i}{dt}\left(H_i - \frac{P}{\sum_j x_j}\right) \Big/ \left(\frac{P}{T} - \sum_j x_j C_{pj}\right) \tag{32}$$

Here, H_i and C_{pj} are, respectively, molar enthalpy and molar constant-pressure heat capacity of the *i*th species. Notice that the values of rate constants in Equation 29 depend on the temperature and must, in the present case, be so evaluated as to conform to the current temperature. Further development of the equations which describe the evolution of homogeneous reaction systems can be found in, e.g., Reference 12.

The construction of a reaction mechanism usually starts with the selection of chemical species. This step is, of course, critical. If too many species are selected, the mechanism becomes unnecessarily complicated; if too few species are selected, important elementary reactions may be left out. After the selection of species, elementary steps are developed. Here, one may use an intuitive approach in which paths from the reactants to the products are chosen because of their likely chemical importance, or one may use an extensive approach in which all conceivable reactions among the selected species are chosen.[3,13-15]

The intuitive approach works best in simple systems, but may err on the side of having too few reactions in complex systems. The extensive approach, on the other hand, produces massive mechanisms, the complexity of which grows exponentially with the number of species. For example, when a given 25 species are selected for methane-air combustion,[13] 322 reactions (and their reverse reactions) can be constructed; when 14 more species are added, 1078 reactions can be constructed. It is important, therefore, to prescreen by means of sensitivity analysis the mechanism developed by the extensive approach so that kinetically insignificant steps can be weeded out. Sensitivity analysis can also be useful in providing information concerning kinetic details of the reaction mechanism (both simple and complex), and helping to identify those significant but uncertain rate constants so that further measurements can be planned.

We conclude this section by noting that, aside from the above methodological considerations, success of a mechanism-building process also depends essentially on the common sense, chemical knowledge, past experience, and imagination of the chemist.

1. Phenomenon of Stiffness

Integration of most kinetic models can only be done with the help of a computer. Closed-form solutions are rarely obtainable for systems of practical interest.

Consider Equation 29, for example. To solve these *m* coupled ordinary differential equations, one usually proceeds stepwise by calculating a sequence of approximations $\mathbf{u}_j \approx \mathbf{x}(t_j)$ on a set of discrete time points, $t_{j+1} = t_j + \Delta t_j$, $j = 1,2,\ldots, M$, where t_1 = initial time, t_{M+1} = final time, and Δt_j = step size of the *j*th integration step. A simple example of such discrete variable methods is provided by the first-order Taylor series approximation to the solution $\mathbf{x}$ in the neighborhood of some t_j:

$$\mathbf{x}(t_j + \Delta t_j) \approx \mathbf{x}(t_j) + \Delta t_j \frac{d\mathbf{x}(t_j)}{dt} = \mathbf{x}(t_j) + \Delta t_j\, \mathbf{f}(\mathbf{x}(t_j), \mathbf{k}) \tag{33}$$

The amount by which the solution **x** fails to satisfy Equation 33 is known as the local truncation error (made in the step from t_j to $t_j + \Delta t_j$). For the formulae of Equation 33, the local truncation error is of second order in Δt_j, i.e., $O[(\Delta t_j)^2]$; the error therefore decreases as Δt_j gets smaller. The global truncation error is the accumulation of local truncation errors made at each step. Replacing $\mathbf{x}(t_j)$ in Equation 33 by the numerical approximation $\mathbf{u}_j$ yields:

$$\mathbf{u}_{j+1} = \mathbf{u}_j + \Delta t_j\, \mathbf{f}(\mathbf{u}_j, \mathbf{k}) \tag{34}$$

Equation 34 is known as the Euler's method and can be used to advance the approximation $\mathbf{u}_j$ from initial time to any time point of interest by proper choice of step sizes.

Now, mechanistic models of reasonable complexity in chemical kinetics usually involve rate constants of vastly different magnitudes, a phenomenon well known to numerical analysts as stiffness,[16-20] which can cause serious numerical problems when inappropriate numerical methods are used. To illustrate this point, let us consider a hypothetical complex reaction mechanism:

$$A \xrightarrow{k_1 = 1999.9} B \xrightarrow{k_2 = 0.1} \text{other species and elementary steps}$$

$$A \xrightarrow{k_3 = 0.1} \text{other species and elementary steps} \tag{35}$$

The rate equations for species A and B in the above mechanism are

$$\frac{dx_A}{dt} = -2000\, x_A$$

$$\frac{dx_B}{dt} = 1999.9\, x_A - 0.1\, x_B \tag{36}$$

where x_A and x_B are the concentrations of A and B, respectively. If the initial conditions are $x_A(0) = 1$ and $x_B(0) = 0$, then $x_A(t) = e^{-2000t}$ and $x_B(t) = e^{-0.1t} - e^{-2000t}$. The solutions x_A and x_B are seen to consist of a fast transient, e^{-2000t}, and a slow transient, $e^{-0.1t}$. The fast transient is dead at around $t = 0.01$, and the slow transient at around $t = 200$.

Suppose now that the experiment measures both A and B until $t = 100$. Let us apply the Euler's method to compute model predictions of A and B in the time range 0 to 100. One expects that once the fast transient in the solutions has died away, the system is governed by a slowly varying $e^{-0.1t}$ and, presumably, a large step size could be used in integrating Equation 36. As we shall see, such is not the case.

The application of Equation 34 to Equation 36 yields:

$$v_{j+1} = (1 - 2000\, \Delta t_j)\, v_j$$

$$w_{j+1} = 1999.9\, \Delta t_j v_j + (1 - 0.1\, \Delta t_j)\, w_j \tag{37}$$

where v_j and w_j are the numerical approximations of $x_A(t_j)$ and $x_B(t_j)$, respectively. Suppose that the deviations of v_j and w_j from the true solutions are ϵ_A and ϵ_B, respectively, i.e., $v_j = x_A(t_j) + \epsilon_A$ and $w_j = x_B(t_j) + \epsilon_B$. Equation 37 then shows that v_{j+1} and w_{j+1} are in error at least by $(1 - 2000\, \Delta t_j)\epsilon_A$ and $1999.9\, \Delta t_j \epsilon_A + (1 - 0.1\, \Delta t_j)\epsilon_B$, respectively. (Recall that there are also local truncation errors associated with application of Equation 33 and, hence, Equation 34.) Therefore, in order that the errors committed at the step j do not get amplified as one proceeds to the step j + 1 (or, in order that *stable* numerical solutions can be maintained), the step size Δt_j must be smaller than ca. 1/2000. This restriction on step size persists even after t = 0.01 where fast transient has died away completely. The problem is now evident: to reach t = 100, using step sizes smaller than 1/2000 requires at least 2×10^5 integration steps; this number of steps not only necessitates an enormous amount of computations, but also causes serious accumulation of round-off errors. Round-off errors are those which arise from inexact representation of numbers by means of a finite number of digits in the calculation (of each step), e.g., representing $^1/_3$ inexactly by 0.3333. Round-off error acts in the same way as an additional local truncation error.

Stiffness problem involved in solving the rate equations numerically can now be summarized as follows. For a mechanistic model with rates of vastly different magnitudes, the time domain of interest is determined by the slowest rate, while the integration step size is determined by the fastest rates if an inappropriate numerical method is used; this conflict leads to impossibly long computing time which may even cause a total failure of the computed results.

A more formal definition of stiffness is given as follows.[18] Consider a general linear constant coefficient ordinary differential system:

$$\frac{d\mathbf{x}(t)}{dt} = \mathbf{A}\,\mathbf{x} + \mathbf{b}(t) \tag{38}$$

where **A** is an $m \times m$ matrix whose eigenvalues λ_i, i = 1,2,..., m, are assumed distinct. The Equation 38 system is said to be stiff if

1. $\mathrm{Re}\,\lambda_i < 0$, i = 1,2..., m, where $\mathrm{Re}\,\lambda_i$ denotes the real part of λ_i.
2. $\max|\mathrm{Re}\lambda_i|/\min|\mathrm{Re}\lambda_i| = \lambda >> 1$, where $\max|\mathrm{Re}\lambda_i|$ and $\min|\mathrm{Re}\lambda_i|$ denote, respectively, the maximum and minimum of all $|\mathrm{Re}\lambda_i|$, i = 1,2,..., m; the ratio λ is called the stiffness ratio, which is a measure of the stiffness of the system. For the simple example Equation 36, $\lambda = 2 \times 10^4$; this is considered a modest ratio. In chemical kinetics, $\lambda = 10^6$ is not uncommon.

A variable coefficient linear system $d\mathbf{x}/dt = \mathbf{A}(t)\mathbf{x} + \mathbf{b}(t)$ or a nonlinear system $d\mathbf{x}/dt = \mathbf{f}(t,\mathbf{x})$ is said to be stiff in an interval if for t in that given interval the eigenvalues $\lambda(t)$ of $\mathbf{A}(t)$, or of the Jacobian matrix $[\partial\mathbf{f}(t)/\partial\mathbf{x}(t)]$, respectively, satisfy points 1 and 2 above.

Finally, we note that it is possible by a simple modification of the Euler's method to remove the above severe step size limitation. Thus, replacing $\mathbf{f}(\mathbf{u}_j,\mathbf{k})$ on the right-hand side of Equation 34 by $\mathbf{f}(\mathbf{u}_{j+1},\mathbf{k})$, we obtain:

$$\mathbf{u}_{j+1} = \mathbf{u}_j + \Delta t_j\, \mathbf{f}(\mathbf{u}_{j+1}, \mathbf{k}) \tag{39}$$

Equation 39 is called the "implicit" (or backward) Euler's method because the unknown $\mathbf{u}_{j+1}$ is also contained on the right-hand side of the equation in a (possibly nonlinear) function **f**. Application of Equation 39 to Equation 36 gives:

$$v_{j+1} = \frac{v_j}{1 + 2000\,\Delta t_j}$$

$$w_{j+1} = \frac{1}{1 + 0.1\,\Delta t_j}\left(w_j + \frac{1999.9\,\Delta t_j}{1 + 2000\,\Delta t_j}\,v_j\right) \tag{40}$$

Clearly, errors associated with v_{j+1} and w_{j+1} are now $\epsilon_A/(1 + 2000\,\Delta t_j)$ and $[\epsilon_B + 1999.9\,\Delta t_j\epsilon_A/(1 + 2000\,\Delta t_j)]/(1 + 0.1\,\Delta t_j)$, respectively. The method is therefore stable for *all* choice of $\Delta t_j > 0$. Numerical methods enjoying this kind of property are known to be absolutely stable (or, A stable).

Although being absolutely stable, the implicit Euler's method is still not an ideal candidate for general-purpose rate equation solvers. This is because the method has a local truncation error of only second-order in step size (see Equation 33); therefore, the accuracy of the numerical results is generally low unless step sizes are again restricted to small values.

Presently, there exists an extensive literature[16,20-31] on numerical methods developed for stiff systems. As one expects from the above discussion, the answer to the stiffness problem appears to lie in the use of implicit numerical procedures. In the following section, one such algorithm, the use of backward differentiation formulae, is presented for illustrative purpose.

2. Backward Differentiation Algorithm

By far the most common method for solving stiff differential systems is the use of backward differentiation formulae.[16,20,21] When applied to Equation 29, the formulae are of the form:

$$\boldsymbol{\mu}_{l+q} = \sum_{j=0}^{q-1} \alpha_j \, \mathbf{u}_{l+j} + \Delta t \beta_q \, \mathbf{f}_{l+q}, \quad l = 0, 1, 2, ... \tag{41}$$

where $\mathbf{u}_l$ is the numerical approximation to $\mathbf{x}(t_l)$, q is a positive integer, $\mathbf{f}_{l+q} = \mathbf{f}(u_{l+q},\mathbf{k})$, and Δt is a constant step size; the α_j and β_q are constants with $\alpha_0 \neq 0$ and $\beta_q \neq 0$. Equation 41 is also called a linear q-step method because q past values of the numerical approximation, $\mathbf{u}_{l+j}$, $j = 0,1,2,...,q - 1$, are needed in order to compute the unknown $\mathbf{u}_{l+q}$. The scheme is clearly implicit. The implicit Euler's method, Equation 39, is a special case of Equation 41 with $q = 1$ and $\alpha_0 = \beta_1 = 1$.

The order of a numerical method is said to be *r* if the local truncation error of the method is of order $r + 1$ in step size Δt, $O[(\Delta t)^{r+1}]$. For Equation 41, the order is equal to the step number q. This is understandable if one recalls that a *q*th-degree polynomial can be constructed from $q + 1$ distinct points.

The backward differentiation formulae (Equation 41) of order one to six constitute a useful class of stiffly stable methods[16] and have been used successfully by Gear[16,22,23] in solving general stiff differential systems. With Gear's method, Δt is restricted to small values, as it should be by the requirement of accuracy, only where the solution is relatively active; then in regions where the solution is inactive, also called regions of stiffness, the backward differentiation formulae assure that Δt is no longer restricted by the time constants of the fast transients, unless or until the fast transients become active again. Currently, there are available general-purpose Gear's method computer codes[24-26] which are finely tuned for user use.

We now briefly describe how the backward differentiation formulae (Equation 41), can be implemented. To begin with, q starting values $\mathbf{u}_0,\mathbf{u}_1,...,\mathbf{u}_{q-1}$ are needed. These, except $\mathbf{u}_0$ which is given by the initial condition, can be generated by means of, say, the implicit Euler's method. Equation 41 is then solved step by step for $\mathbf{u}_{l+q}$, $l = 0,1,2,....$ If the function $\mathbf{f}$ is nonlinear in $\mathbf{u}_{l+q}$, we must solve Equation 41 iteratively, and it is to our advantage to choose the initial guess $\mathbf{u}_{l+q}^{(0)}$ as accurately as possible. A convenient choice is provided by the use of an "explicit" multistep method (Equation 42):

$$\boldsymbol{\mu}_{l+q}^{(0)} = \sum_{j=0}^{q-1} \tilde{\alpha}_j \, \mathbf{u}_{l+j} + \Delta t \tilde{\beta}_q \, \mathbf{f}_{l+q-1} \tag{42}$$

Equation 42 is called the predictor. The original implicit method (Equation 41) is called the corrector, and the combined process of Equations 41 and 42 is called a predictor-corrector method.

Much of the computing effort in a predictor-corrector method is associated with the iterative solution of corrector equation. For stiff systems, the Newton-Raphson method is usually applied. To solve the vector equation $\mathbf{F}(\mathbf{u}) = \mathbf{0}$, the Newton-Raphson method (*cf.* discussions after Equation 7) is

$$\mathbf{u}^{(i+1)} = \mathbf{u}^{(i)} - \mathbf{J}(\mathbf{u}^{(i)})^{-1}\,\mathbf{F}(\mathbf{u}^{(i)}), \quad i = 0, 1, 2, \ldots \tag{43}$$

where $\mathbf{J}$ is the Jacobian matrix $(\partial\mathbf{F}/\partial\mathbf{u})$ with elements $(\partial F_i/\partial u_j)$. Applying Equation 43 to Equation 41 then gives:

$$\mathbf{u}_{l+q}^{(i+1)} = \mathbf{u}_{l+q}^{(i)} - [\mathbf{I} - \Delta t\beta_q(\partial\,\mathbf{f}(\mathbf{u}_{l+q}^{(i)}, \mathbf{k})/\partial\mathbf{u}_{l+q}^{(i)})]^{-1} \cdot \left[\mathbf{u}_{l+q}^{(i)} - \sum_{j=0}^{q-1} \alpha_j\,\mathbf{u}_{l+j}^{(i)} - \Delta t\beta_q\,\mathbf{f}(\mathbf{u}_{l+q}^{(i)}, \mathbf{k})\right], \quad i = 0, 1, 2, \ldots \tag{44}$$

where $\mathbf{I}$ is the $m \times m$ unit matrix. In practice, the iteration is sometimes terminated after a fixed number of operation cycles in order to cut down the computation cost. Further details concerning the implementation of backward differentiation formulae (e.g., variable-order-variable-step size procedure) can be found in References 16 and 20 to 22. Discussions on other stiff schemes for solving the coupled rate Equation 29 (e.g., implicit Runge-Kutta methods, Taylor series methods, and singular perturbation methods) can be found in References 16, 20 to 22, and 27 to 31.

3. Stochastic Approach

Being a set of coupled ordinary differential equations, Equation 29 yields a unique solution $\mathbf{x}$ which is a continuous function of time for a given set of initial concentrations. Therefore, the course of a reaction is described by Equation 29 as being deterministic and continuous.

However, a reaction process cannot be deterministic for a given set of initial reactant concentrations because the microscopic state of a system is determined not by the crude information of species populations, but by precise positions and momenta of *all* the molecules in the system, if classical mechanics holds. Concentrations therefore fluctuate. For large systems, these fluctuations are usually small compared to the macroscopic average values (this can be shown explicitly for macroscopic systems in thermodynamic equilibrium, except at points of phase transitions). However, for systems containing a small number of molecules,[32,33] such as many reactions occurring in biological cells, and for systems undergoing nonlinear kinetics far away from equilibrium and being in the vicinity of an instability of the macroscopic rate equations,[34-36] fluctuations become an essential aspect of the evolution, and it is necessary to construct a theory of fluctuations to describe the process.

A chemical process also cannot be continuous in time because molecular populations change by only discrete integer amounts. For macroscopic systems, this deviation from continuity, like fluctuations, is insignificant; however, for small systems, discreteness of the concentration-time profiles should become detectable.

In recent years, developments in theory and experiment have enhanced enormously the need for a detailed understanding of fluctuations in nonequilibrium systems near points of instability. Due to the mechanistic complexity of such systems, however, application of even approximate theoretical methods is a formidable task.

The reactive molecular dynamics[37,38] is a computer simulation algorithm at microscopic level in which the exact classical dynamics, including elastic, inelastic, and reactive scattering processes, of a collection of molecules are followed as they interact according to classical mechanics. The microscopic data generated over different initial dynamic conditions then yield the necessary statistical information concerning, e.g., the microscopic average values and the sizes of fluctuations. Practical application to date of the reactive molecular dynamics, however, is limited to systems containing just hundreds to thousands of particles involved in chemical reactions for which reactive probabilities do *not* vary too widely.[38]

In order to study laboratory systems with widely different rate constants, a higher level method is invoked. This so-called stochastic approach treats only the numbers of molecules in a small cell or volume element, and considers the evolution, and hence the fluctuations, as random processes. (Spatial inhomogeneity can be handled in a similar fashion by taking into account diffusion between adjacent cells.[39]) The stochastic approach can be considered intermediate between the macroscopic description, like Equation 29, which is also known as the deterministic approach, and the complete microscopic treatment. The first stochastic treatment of chemical reactions dates back to 1940 when Kramers[32,40] considered a chemical reaction a Brownian motion of particles, whose passage over a potential barrier represents the rate of decomposition.

Traditionally, the stochastic time evolution of a reaction system is calculated by means of a master equation[32-34] of the form:

$$\frac{\partial P(\mathbf{x}, t)}{\partial t} = \sum_{j=1}^{p} [B_j(\mathbf{x}, t) - a_j(\mathbf{x}, t)\, P(\mathbf{x}, t)] \tag{45}$$

where $P(\mathbf{x},t)$ = the probability that the system is in the state $\mathbf{x}$ (i.e., the system has x_i molecules of species i with $i = 1,2, \ldots, m$) at time t; $B_j(\mathbf{x},t)dt$ = the probability at time t that the system arrives at the state $\mathbf{x}$ in a time increment dt by undergoing a single *j*th elementary reaction; and $a_j(\mathbf{x},t)dt$ = the probability at time t that the system leaves the state $\mathbf{x}$ in a time increment dt by undergoing a single *j*th elementary reaction. The master Equation 45 expresses the evolution of the probability distribution function P as a competition between the gain terms B_j and the loss terms a_jP. In view of its multivariate nature, Equation 45 is very difficult to solve, even on a computer.

Recently, an interesting stochastic simulation algorithm was proposed by Gillespie.[41-43] With Gillespie's method, the attention is focused on determining for the system the elapsed time τ needed for an elementary reaction to occur and the identity of the reaction thus occurred. To calculate the time increment τ, one uses Equation 46:

$$\tau = \ln\left(\frac{1}{r_1}\right) \Big/ \sum_{j=1}^{p} a_j \tag{46}$$

where r_1 is a computer-generated random number between zero and one and a_j, defined in the preceding paragraph, is related to the current numbers of the molecules in the system, x_i, $i = 1,2, \ldots, m$, and the macroscopic elementary rate constant k_j of Equation 29 in a simple fashion.[41] Notice that the sum $\sum_{j=1}^{p} a_j$ is a measure of the total probability per unit time that any one of the *p* elementary reactions occurs.

To determine the identity of the elementary reaction occurring after an elapsed time τ, one generates a second random number r_2 between zero and one, and then calculates the desired reaction number l according to

$$\sum_{j=1}^{l-1} a_j \Big/ \sum_{j=1}^{p} a_j < r_2 \leq \sum_{j=1}^{l} a_j \Big/ \sum_{j=1}^{p} a_j \tag{47}$$

The meaning of Equation 47 becomes clear by noting that $a_j/\sum_{j=1}^{p} a_j$ is simply the probability that the next molecular event will be j.

Having determined τ and l, the state **x** of the system is adjusted by subtracting the stoichiometric coefficients for reactants and adding the coefficients for products of the *l*th elementary step. This procedure continues until the final time point has been reached. A statistical ensemble is generated by repeated simulation of the chemical evolution using different sequences of random numbers. Within limits imposed by computer time restrictions, ensemble averages and relevant statistical information can then be evaluated to any desired degree of accuracy.

Let us use $P(\tau,l;\mathbf{x},t)$ to denote the probability per unit time that the system at state **x** and time t will undergo a single *l*th elementary reaction in a time increment τ. Then, Gillespie[42] showed that Equations 46 and 47 are equivalent to a probability density function of the form:

$$P(\tau, l; \mathbf{x}, t) = \begin{cases} a_l(\mathbf{x}, t) \exp\left[-\tau \sum_{j=1}^{p} a_j(\mathbf{x}, t)\right], & \text{if } 0 \leq \tau < \infty \text{ and } l = 1, 2, \ldots, p \\ 0, & \text{otherwise} \end{cases} \quad (48)$$

Equation 48 is also a rigorous result within the framework of stochastic formulation and, consequently, is equivalent to the master equation approach.

The algorithm (Equations 46 and 47) provides a discrete and stochastic description for the chemical evolution, and is attractive in that (1) the algorithm is very easy to code in computer language, (2) the algorithm requires very little computer memory space, and (3) no differential system is solved and hence no troublesome stiffness problem ensues.

On the minus side, however, the algorithm requires lengthy computation time.[41,45] This can limit the total number of elementary reactions, the total numbers of molecules, and the total system time that can be simulated. The algorithm is still not to be compared with the deterministic algorithm (Equation 29) in situations where the deterministic approach offers a valid description.[45]

Some interesting recent developments of stochastic theory can be found in, e.g., References 39 and 46.

B. Inhomogeneous Systems

So far, we have ignored the spatial coordinates of a chemical reaction system, except for a brief mention in the preceding section of stochastic solution of reaction-diffusion problems. However, for many systems like combustion and electrochemical processes, the system composition and physical properties change with position, and sometimes even the system itself is in motion. This implies that transport properties, including heat conduction and diffusion, contribute to the evolution of the system, and these spatial effects coupled with the forces generated by the chemical reaction may cause the fluid dynamics of the system to become important.

The temporal behavior of an inhomogeneous system is governed by four fundamental conservation laws which are the basis of all physical and chemical problems: conservation of mass, momentum, energy, and atoms. These four conservation laws are represented by four fundamental continuity equations in the form[47,48]

$$\partial\rho(\mathbf{r}, t)/\partial t = -\nabla \cdot \mathbf{J}(\mathbf{r}, t) + q(\mathbf{r}, t) \quad (49)$$

where **r** denotes spatial location. The function $\rho(\mathbf{r},t)$ denotes the local density of the quantity under consideration, i.e., mass density, components of momentum density, energy density, or chemical species concentrations, in the neighborhood of **r** and at the time t. The vector $\mathbf{J}(\mathbf{r},t)$ denotes the flux vector of ρ, which is defined as the amount of the quantity transported per unit area per unit time at **r** and t. The term $q(\mathbf{r},t)$ denotes the rate of formation of the quantity per unit volume at **r** and t due to the presence of sources.

Now, according to the divergence theorem, the divergence of the flux, $\nabla \cdot \mathbf{J}(\mathbf{r},t)$, represents the rate of loss due to the flow of the quantity per unit volume at **r** and t. Therefore, the meaning of Equation 49 is clear: it is simply that, in the neighborhood of any given point **r** in space and at any given instant t, the rate of change of ρ equals the rate of loss due to the flow plus the rate of gain due to the source.

Among the four continuity equations, the simplest is that of mass, in which ρ is the mass density, **J** is ρ times **v**, with **v** being the mass average velocity, and q is the rate of mass formation per unit volume, which is zero when the system is closed. Below, we discuss the species continuity equations for illustrative purpose. Details on other equations can be found in References 47 and 48. A brief introduction to the numerical methods for solving partial differential systems is also given using species continuity equations as examples.

Let x_i be the molar concentration of chemical species i. Then, the continuity equation for the *i*th chemical species reads:

$$\partial x_i/\partial t = -\nabla \cdot \mathbf{J}_i + f_i, \quad i = 1, 2, \ldots, m \tag{50}$$

where $\mathbf{J}_i$ is the total molar flux of species i and f_i is the rate of chemical change of species i given by the mass action law (Equation 29). The flux $\mathbf{J}_i$ is usually separated into two parts: one due to the overall system flow (convection) and one due to the molecular agitations (diffusion). For the purpose of discussion, we assume that convection is absent from the system. Thus, $\mathbf{J}_i$ becomes the diffusive flux, which according to Fick's first law of diffusion,[48,49] is

$$\mathbf{J}_i = -D_i \nabla x_i, \quad i = 1, 2, \ldots, m \tag{51}$$

51 into Equation 50 gives:

$$\partial x_i/\partial t = \nabla \cdot (D_i \nabla x_i) + f_i, \quad i = 1, 2, \ldots, m \tag{52}$$

phenomena described in Equation 52 include air pollutions, electrochemical processes,[50] and biochemical processes involving enzymatically active membranes.[51] To solve Equation 52, initial and boundary conditions of the x_i must be supplied.

As one would have expected, many outstanding questions described by Equation 52 are beyond even approximate resolution with available theoretical methods. One must therefore resort to computers. However, unlike the situation with ordinary differential systems, numerical methods for solving general partial differential equations are not so well developed. This is especially true for problems in more than one space variable.

To simplify matters, let us consider Equation 52 in one space variable r. We assume that the diffusion constants D_i are independent of the species concentrations x_i and r. Hence,

$$\partial x_i/\partial t = D_i(\partial^2 x_i/\partial r^2) + f_i, \quad i = 1, 2, \ldots, m \tag{53}$$

There are two classes of numerical methods which are commonly applied to solve partial

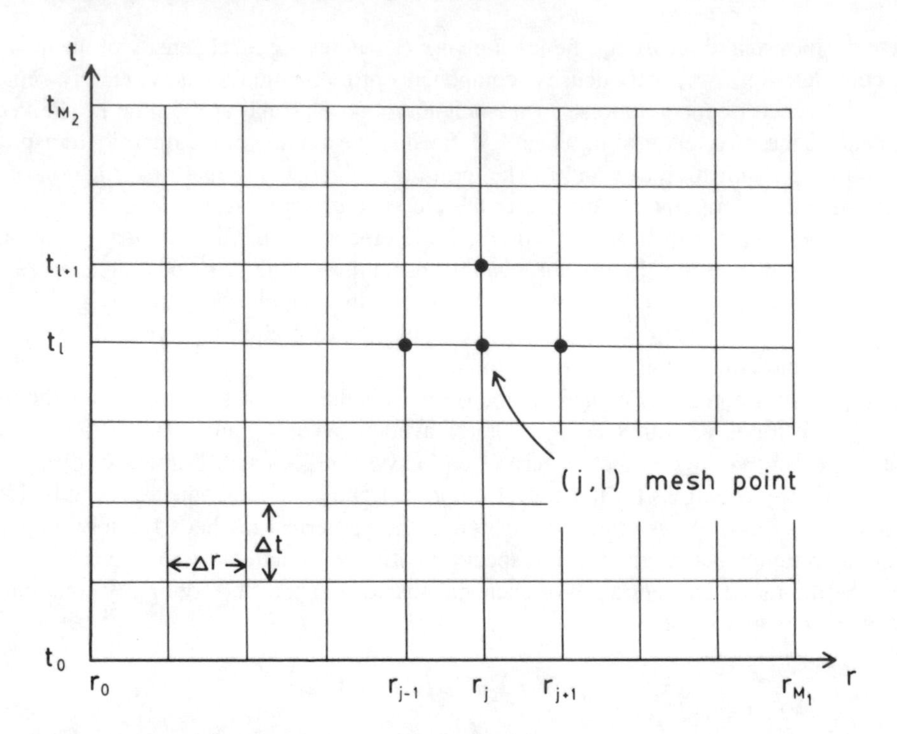

FIGURE 2. The finite-difference mesh in the solution of one-dimensional reaction-diffusion equation.

differential equations like Equation 53: the finite difference and the finite element techniques. In the finite difference method, the space-time domain of the problem is divided into a mesh of smaller regions, and derivatives of the solutions are approximated by finite differences. For example, a simple explicit finite-difference approximation to Equation 53 is[53,54]

$$u_j^{l+1} = u_j^l + \Delta t[D_i(u_{j+1}^l - 2u_j^l + u_{j-1}^l)/(\Delta r)^2 + f_i(\mathbf{u}_j^l, k)] \tag{54}$$

where u_j^l and $\mathbf{u}_j^l$ are numerical approximations to $x_i(r_j,t_l)$ and $\mathbf{x}(r_j,t_l)$, respectively; $r_j = r_0 + j\Delta r$, $t_l = t_0 + l\Delta t$, $j = 0,1,2, \ldots, M_1$, $l = 0,1,2, \ldots, M_2$; Δr and Δt are constant mesh lengths along the r and t axes, respectively; and the space-time domain of the problem is defined by $r_0 \leq r \leq r_{M_1}$, $t_0 < t < t_{M_2}$. (See Figure 2.) Equation 54 is seen to consist of a forward difference approximation in t with an error of the order of Δt and a central difference approximation in r with an error of the order of $(\Delta r)^2$. Equation 54 is very simple to use, but suffers from the disadvantage that the ratio $D_i\Delta t/(\Delta r)^2$ is strictly limited in order to maintain numerical stability.

To remove the stability limitation, an implicit difference scheme of the Crank-Nicolson type may be used. With reference to Equation 53, this is[54]

$$\frac{u_j^{l+1} - u_j^l}{\Delta t} = \frac{D_i}{2(\Delta r)^2}[(u_{j+1}^{l+1} - 2u_j^{l+1} + u_{j-1}^{l+1}) + (u_{j+1}^l - 2u_j^l + u_{j-1}^l)]\, f_i\left[\frac{1}{2}(\mathbf{u}_j^{l+1} + \mathbf{u}_j^l), \mathbf{k}\right] \tag{55}$$

Equation 55 can be obtained from Equation 53 by replacing $\mathbf{x}$ and its derivatives by the means of their numerical approximations at t_l and t_{l+1}. Equation 55 has the advantage of remaining stable for all values of $D_i\Delta t/(\Delta r)^2$, the limit being set by the accuracy required. Unfortunately, the possible nonlinearity of f_i complicates the solution of Equation 55 and thereby increases the computation effort at each operation cycle. Some three-time-level schemes,[54] e.g., have been proposed to deal with such nonlinearity.

The finite difference method may also be implemented by discretizing only the spatial domain. This is also known as the method of lines. Thus, replacing spatial derivatives in Equation 53 by their finite-difference approximations and taking proper account of the boundary conditions, the partial differential system (Equation 53) is reduced to a set of $m(M_1 + 1)$ coupled ordinary ones, which can then be solved by the available efficient ordinary differential equation solvers.[23-26] The recent development in the numerical solution of ordinary differential systems has made this method of lines attractive for a number of problems like Equation 53.

In applying the alternative finite element technique[57-59] to Equation 53, the spatial domain of the problem is usually discretized first. The solutions x_i are then represented in terms of a set of basis or shape functions, $\phi_j(r)$, $j = 1,2, \ldots, L$:

$$x_i(r, t) \approx u(r, t) = \sum_{j=1}^{L} a_j(t)\, \phi_j(r) \tag{56}$$

These basis functions are defined in a piecewise manner and are continuous across mesh boundaries and nonzero over only a few, usually one to two (adjacent) mesh elements. Therefore, in each subinterval, the solutions are related in fact to just a few basis functions which are nonzero in that interval. Figure 3 shows four such piecewise cubic Lagrangian basis functions,[57] each being nonzero over at most two adjacent elements.

Substitution of Equation 56 into Equation 53 produces a nonzero residual R(r,t) given by

$$R(t, t) = \partial u(r, t)/\partial t - [D_i(\partial^2 u(r, t)/\partial r^2) + f_i(\mathbf{u}(r, t), \mathbf{k})] \tag{57}$$

The expansion coefficients $a_j(t)$ are then determined by requiring that an integration of the weighted residual over the entire spatial domain be zero:

$$\int dr\, w_j(r)\, R(r, t) = 0, \quad j = 1, 2, \ldots, L \tag{58}$$

where the w_j are weighting functions. One obvious choice of the weighting functions is $w_j(r) = \phi_j(r)$, $j = 1,2, \ldots, L$, which constitutes the Galerkin method.[57-59] In the Galerkin method, the residual R is thus orthogonal to all basis functions.

Equation 58 is clearly a set of coupled ordinary differential equations and can be solved as before by any efficient ordinary differential equation solvers.[23-26] If one desires, Equation 58 can also be solved in the finite element manner by discretizing the time domain, representing the coefficients $a_j(t)$ in terms of basis functions, and then applying the weighted residual technique. The stability characteristics of the formulae obtained this way, however, should be examined carefully.

There are some recent developments[60,61] of the finite element technique which attempt to minimize the error in the polynomial representation of the solution, Equation 56, by adjusting the location of mesh boundaries. Encouraging results are reported for problems involving moving steep fronts[60] or discontinuities.[61]

Finally, we note that most of the computer time in solving partial differential problems like Equation 53 is spent in solving the huge linear algebraic systems required by the implicit numerical technique. Advances in the simulation of inhomogeneous systems are, therefore, closely linked with the development of efficient ways for solving matrix equations.

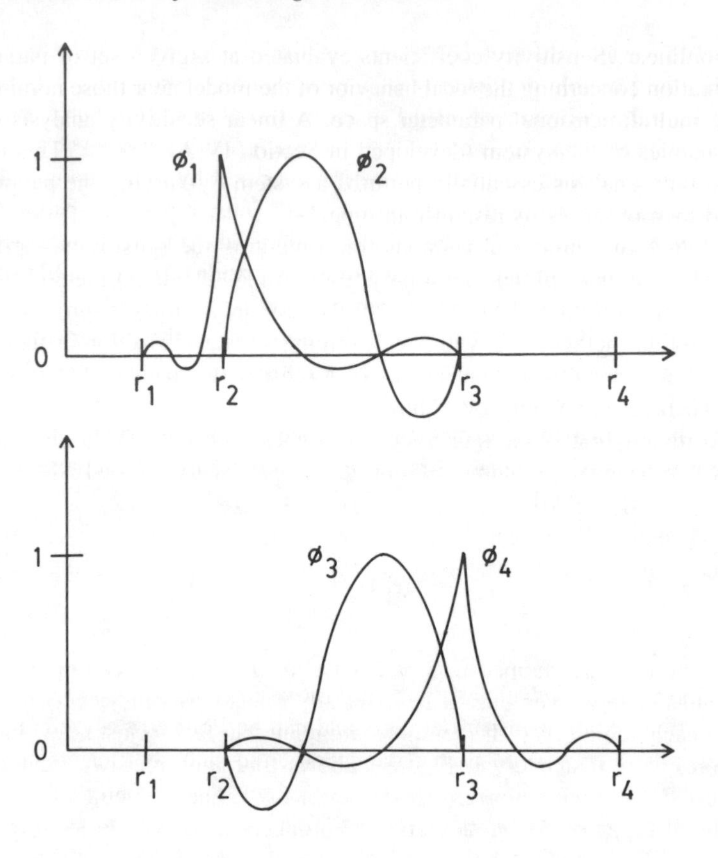

FIGURE 3. Piecewise cubic Lagrangian basis functions ϕ_l, l = 1,2,3,4, over three mesh elements.

IV. SENSITIVITY ANALYSIS

The recent interest in the computer simulation of complex physical and chemical phenomena has brought about considerable research in the area of sensitivity analysis,[4,14,15,62-88] a study which probes the response of a dynamic system to parameter variations. When a mechanistic model involves inaccurately determined rate constants, the information supplied by a sensitivity analysis is especially valuable. Such information can contribute directly to parameter estimation, experimental plannings, or the assessment of a modeling result. It can also be used to simplify a reaction model by determining kinetically important and unimportant elementary steps. Even for design studies, information about the sensitivity of a reactor model to initial reactant concentrations, temperature, and so on can be useful in optimizing certain species profiles. Other areas in which sensitivity analysis finds applications are quantum dynamics,[89,90] reactor physics,[91,92] dynamical engineering systems,[93,94] and socioeconomic problems.[95]

When the dynamic model consists of just a few parameters, the desired sensitivity information can be obtained by directly varying the parameters and computing the model responses. However, as the number of parameters gets larger, the many different ways of choosing the parameters to be varied simultaneously complicates the situation and renders the approach impractical.

An alternative method is to compute the partial derivatives of the model variables with respect to model parameters. These partials are called sensitivity coefficients. If only first-order sensitivity coefficients are considered, the analysis is called linear; otherwise, the analysis is nonlinear. Sensitivity coefficients evaluated at a given set of parameter values supply information concerning the local behavior of the model near those nominal parameter values in the multidimensional parameter space. A linear sensitivity analysis can probe in detail the dynamics of the system (developed in Section IV.A).[71,73,96-98] This is so because a linear sensitivity analysis essentially perturbs a system by varying the parameters infinitesimally and then observes its response in time.

Thus far, four basic numerical methods for computing the sensitivity coefficients have been proposed: Fourier amplitude sensitivity test (FAST),[62,63] direct method (DM),[64,65,82,83] Green's function method (GFM),[66-70,76,80,84,85] and polynomial approximation method (PAM).[73-75] All four methods deal with the deterministic rate or reaction-diffusion equations, on which our attention in this section is focused. A sensitivity procedure for treating stochastic systems has been given recently by Rabitz.[77]

Let us next review briefly the four basic approaches — FAST, DM, GFM, and PAM — starting with homogeneous systems. A more thorough discussion of PAM is presented in Section IV.B.

In the FAST approach,[62,63] each parameter k_l of the system, Equation 29, is varied as a periodic function of a search variable s:

$$k_l = k_l^{(0)} \exp[f_l(\sin \omega_l s)], \quad l = 1, 2, \ldots, p \tag{59}$$

where $k_l^{(0)}$ is the nominal value of k_l, ω_l is a positive integer, and f_l is a function of given form. The solutions x_i are then periodic in s and can be Fourier analyzed. By assigning a different integer frequency ω_l to each parameter k_l, the sine coefficients of the ω_l in the resulting Fourier series are related to the first-order sensitivity coefficients of the k_l. The distinct feature of FAST is that the first-order sensitivity coefficients are obtained as averages over the uncertainties of all the parameters in the system. Unfortunately, the method demands relatively long computation time[66,99] and becomes rather costly for complex systems unless correlated parameters can be identified to reduce the required number of independent sample points.[99]

In DM,[64,65] the emphasis is on a set of sensitivity differential equations which govern the temporal behavior of the sensitivity coefficients. The sensitivity equations are derived by differentiating the dynamical model (Equation 29) with respect to the parameters. There is one such equation for each sensitivity coefficient. For example, suppose that a first-order sensitivity coefficient vector $(\partial \mathbf{x}/\partial k_l)$ is desired. Then the corresponding sensitivity equation is given by Equation 20 with $\mathbf{f}$ defined by Equation 29. Equations 20 and 29 may be treated as a single set of coupled ordinary differential equations and solved simultaneously to yield $\mathbf{x}$ and $(\partial \mathbf{x}/\partial k_l)$;[65,100] or they may be decoupled by preparing and storing the solution $\mathbf{x}$ to Equation 29 in advance, and subsequently using interpolated $\mathbf{x}$ to provide $\mathbf{J}$ and $(\partial \mathbf{f}/\partial k_l)$ such that Equation 20 can be solved.[76,77,101] Recently, a more efficient decoupled algorithm is proposed in which the sensitivity coefficients $(\partial \mathbf{x}/\partial k_l)$ are calculated by alternating the solution of Equation 20 with the solution of Equation 29.[83] The DM approach is straightforward to apply. However, since realistic mechanistic models in chemical kinetics usually involve a large number of parameters and integration of these stiff systems usually requires a large number of steps, the method could also be expensive.

The GFM[66] approach works with the Green's matrix equation which is derived from sensitivity differential equations. The number of vector differential equations to be solved equals the number of components in the system. Therefore, when applied to a system with more parameters than components, significant computational savings could be achieved. It has been pointed out,[67] however, that for practical reasons, GFM is limited to a linear analysis on a small set of discrete time points only. In addition, the computation cost of GFM is still high. Recently, two variants of GFM, the scaled Green's function method[69,70,72] and the analytically integrated Magnus modification,[76,77] have been proposed which attempt to alleviate these minus points.

In the last approach, PAM,[73] the very simple idea of polynominal approximation is invoked. PAM uses prior knowledge of the system temporal behavior to divide the entire time domain into subintervals; the division is such that, within each subinterval, the system can be mimicked by low-degree interpolation polynomials. This forces all parametric dependence of the system to reside in the expansion coefficients and transforms the sensitivity differential equations into a set of algebraic equations. The prior knowledge of the system temporal behavior comes from a preliminary simulation study of the system, which is always available in any kind of modeling studies. A basic reason for adopting such a procedure is that knowledge of the regions where the system varies rapidly with time and where the system stays essentially unchanged enables one to describe most efficiently the system behavior by polynomials. For example, no effort will now be wasted in a large slowly varying region because we know exactly where it is and can take a single giant interval to cover it. On the other hand, the usual ordindary differential system solvers will generally take quite a number of steps to cover the same region because of the difficulty in step-changing mechanism.

PAM offers several advantages. First, the major computational effort of PAM is proportional to the number of species in the system, not to the number of parameters. (GFM also shares this advantage.) Second, the number of time subintervals is kept at a minimum in PAM, which means that the number of evaluations of **J**, $(\partial \mathbf{f}/\partial k_l)$, and other quantities needed in the sensitivity differential equations is also at a minimum (for sufficiently complex systems, these evaluations may be very costly). Third, the computer coding of PAM is comfortably simple because essentially all that one requires is an efficient matrix-inversion program (see Section IV.B), which is widely available. Fourth, if one desires, the second- and higher-order sensitivity cofficients can be generated quite readily by PAM once first-order ones are known.

There is no doubt that the high degree of interest in sensitivity analysis will continue to motivate further technique developments and applications of sensitivity algorithms.

The four basic approaches we have just described for homogeneous systems can be extended to inhomogeneous systems without much conceptual difficulty. In fact, extensions of DM,[82,84,85] GFM,[80] and PAM[75] have already been reported. As for FAST, although its authors based their formulation on homogeneous systems,[62,63] the mathematics appear equally applicable to inhomogeneous systems as well. In spite of all these developments, however, little computational work has been done to date, because of the complexity associated with partial differential systems. Assessment of the efficiency of each sensitivity method must await further applications to more complex reaction models in the future. Again, as with homogeneous systems, we expect such applications and certainly improvements in algorithms to be in the offing.

A. Sensitivities and Kinetic Information

In this section, we will discuss the significance of sensitivity information and the way such information can be used in extracting kinetic details from a reaction system.

Consider the first-order sensitivity coefficients, $(\partial x_i/\partial k_l)$, $i = 1,2, \ldots, m$, $l = 1,2, \ldots, p$. Here x_i denotes the concentration of species i and k_l denotes the rate constant of the lth elementary step. The coefficient $(\partial x_i/\partial k_l)$ measures the first-order (linear) response of x_i to the variation of k_l, as is evident from a first-order Taylor series expansion of x_i in k_l. To compare different sensitivity coefficients sensibly, normalization is required:

$$S_{il} = \frac{k_l}{x_i}\left(\frac{\partial x_i}{\partial k_l}\right) \tag{60}$$

The normalized sensitivity coefficients S_{il} eliminate artificial variation due to the magnitudes of x_i and k_l such that only percentage changes in the x_i caused by percentage changes of the k_l are compared with each other.

Statement 1: If $|S_{il}|>|S_{ir}|$, then the concentration of species i is influenced more by the *l*th reaction than by the *r*th reaction. A positive S_{il} means that the step l has a constructive effect on species i; a negative S_{il} means that the step l has a destructive effect on species i. If *all* the species exhibit very low sensitivity to a certain number of steps throughout the entire time domain of interest, then these steps usually may be eliminated from the mechanism without altering much the kinetic behavior of the model.[71,73,96]

Some clarifications are needed for the above statement. First, it is important to realize that, in some cases, the distinction between low and high sensitivities may be ambiguous. For example, although a step with a normalized sensitivity coefficient of value 1.0 is clearly kinetically much more important than a step with coefficient of value 0.01, there is really not much difference in kinetic importance between a coefficient of value 0.4 and a coefficient of value 0.5.

Second, it is important to realize that the sensitivity coefficients are solutions to the sensitivity differential equations with given initial conditions at given initial time t_0. (This is true for both homogeneous and inhomogeneous systems.) If the sensitivity coefficients are computed at a time t_1 much later than t_0, then the meaning of these coefficients may be obscured due to an accumulation of kinetic effects over the interval from t_0 to t_1. Let us use a simple system:

$$A \xrightarrow{k_1} B \xrightarrow{k_2} C \tag{61}$$

to illustrate this point.[45] The concentrations of A and B of this system are given by Equation 3; the rate equations are given by Equation 12. If one perturbs the system (Equation 61) by increasing at the time t_1 the rate constant k_1 by a small amount, then the concentration of B will increase compared with that of the unperturbed system, (at least) right after t_1. One can prove this easily by means of Equation 3, or by performing a numerical experiment in which the rate constant k_1 is increased instantaneously at t_1 and then the rate equations solved numerically for the perturbed solutions. The sensitivity coefficient $(\partial C_B/\partial k_1)$ must, therefore, be positive at all times as is evident from a first-order Taylor series expansion of C_B in k_1.

Now let us calculate the coefficient $(\partial C_B/\partial k_1)(t_1)$ by integrating the sensitivity differential Equation 20 from t_0 with initial conditions Equation 22. The result is

$$(\partial C_B/\partial k_1)(t_1) = \frac{C_A^{(0)}}{k_2 - k_1}\left[\frac{k_2}{k_2 - k_1}(e^{-k_1 t} - e^{-k_2 t}) - k_1 t e^{-k_1 t}\right] \tag{62}$$

where $t = t_1 - t_0$ denotes the duration of integration, t_0 denotes the initial time, and $C_A^{(0)}$ denotes the initial concentration of A. Equation 62 may also be obtained by differentiating Equation 3 with respect to k_1. For small t:

$$(\partial C_B/\partial k_1) \approx t C_A^{(0)} \tag{63}$$

which is positive. For sufficiently large t, however, $(\partial C_B/\partial k_1)$ becomes negative, a result proved easily by considering separately the cases $k_1 > k_2$ and $k_2 > k_1$. At the time t_* given by

$$t_* = \left(\frac{k_2}{k_1}\right) \frac{1 - e^{-(k_2 - k_1)(t_* - t_0)}}{k_2 - k_1} + t_0 \tag{64}$$

the coefficient $(\partial C_B/\partial k_1)$ is zero. The time t_* is independent of the initial concentrations, $C_A^{(0)}$ and $C_B^{(0)}$. Figure 4 represents pictorially the cause for the sign change of $(\partial C_B/\partial k_1)$.

Clearly, the coefficient $(\partial C_B/\partial k_1)(t_1)$ computed by integrating the sensitivity differential equation from an initial time t_0 so early that $t_1 > t_*$ is incapable of reflecting the real kinetic situation of the process $A \xrightarrow{k_1} B \xrightarrow{k_2} C$ at t_1. A close analog in actual laboratory practice is provided by the act of studying, improperly, the dynamics of a reaction system by perturbing the system at a time t_0 and then measuring its response at a much later time t_1.

The above analysis suggests[102] that, to study the dynamics of a general reaction system around a given instant t_1 by means of normalized sensitivity coefficients, one should always apply the initial condition, say Equation 22, of the sensitivity differential equations, say Equation 20, at a time t_0 reasonably close to t_1. Thus, if we calculate $(\partial C_B/\partial k_1)(t_1)$ for Equation 61 always for small duration by using a variable initial time t_0 chosen close to t_1, we will always get a positive value in conformity with the real kinetic situation.

A more complex example is supplied by the Oregonator:[87,103]

$$\begin{aligned}
A + Y &\xrightarrow{k_1} X, & k_1 &= 1.34 \\
X + Y &\xrightarrow{k_2} P, & k_2 &= 1.60 \times 10^9 \\
A + X &\xrightarrow{k_3} 2X + Z, & k_3 &= 8.00 \times 10^3 \\
2X &\xrightarrow{k_4} Q, & k_4 &= 4.00 \times 10^7 \\
Z &\xrightarrow{k_5} Y, & k_5 &= 1.00
\end{aligned} \tag{65}$$

with initial concentrations $A^{(0)} = 6.00$, $X^{(0)} = 5.025 \times 10^{-11}$, $Y^{(0)} = 3.00 \times 10^{-7}$, and $Z^{(0)} = 2.41 \times 10^{-8}$, in arbitrary units.[87] Figure 5 shows the concentration-time profiles of X, Y, and Z. At the time $t = 3.6$ where X is in a steady state (Figure 5), the modified[88] normalized sensitivity coefficients $\frac{k_l}{X}\left(\frac{\partial X}{\partial k_l}\right)_\tau$, $l = 1,2, \ldots ,5$, where τ indicates the period of oscillation, computed by integrating the sensitivity differential equations with zero initial conditions at $t = 0$, are[102] of the order of 10 or more with $\frac{k_1}{X}\left(\frac{\partial X}{\partial k_1}\right)_\tau \approx -27$ and $\frac{k_2}{X}\left(\frac{\partial X}{\partial k_2}\right)_\tau \approx 25$. This result is rather difficult, if not impossible, to interpret. On the other hand, the regular normalized sensitivity coefficients of X at $t = 3.6$, computed by applying the zero initial conditions at $t = 3.1$, are[102] of the order of 10^{-3} or less except that $\frac{k_1}{X}\left(\frac{\partial X}{\partial k_1}\right)_\tau \approx -\frac{k_2}{X}\left(\frac{\partial X}{\partial k_2}\right)_\tau \approx 1$, which indicates that in the steady state of X, the two most influential reactions on X are steps 1 and 2 with step 1 constructive in nature and step 2 destructive in nature, all in agreement with an analysis given by Field and Noyes.[103]

In applying the variable-initial-time procedure to study the dynamics of a system, one should also note that the initial time t_0 must not be located too close to the observation time

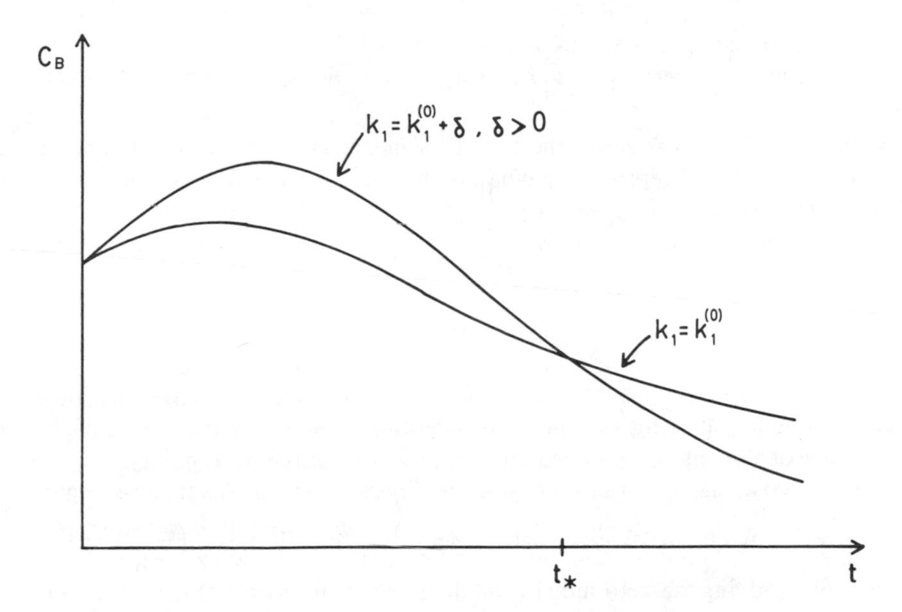

FIGURE 4. Concentration-time profiles of species B for the system $A \xrightarrow{k_1} B \xrightarrow{k_2} C$ with two different values of k_1.

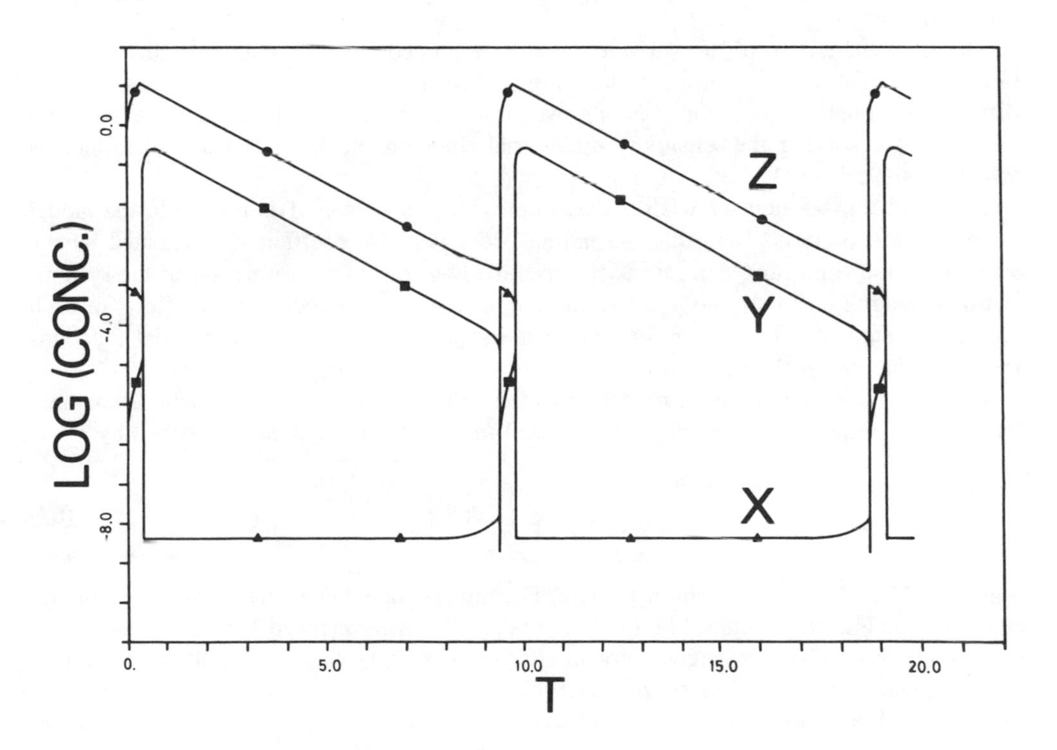

FIGURE 5. Concentration-time profiles of X, Y, and Z for the Oregonator.

t_l; otherwise, the computed sensitivity coefficients will merely reflect the rates of those elementary reactions *directly* producing or consuming the species under consideration, as shown by Equation 66:

$$\frac{S_{il}(t_1)}{t_1 - t_0} \approx \left(\frac{dS_i}{dt}\right)(t_0)$$
$$= \sum_j J_{ij}(t_0)\, S_{jl}(t_0) + (\partial f_i/\partial k_l)(t_0)$$
$$= (\partial f_i/\partial k_l)(t_0) \tag{66}$$

with f_i given by Equation 29; the partial $(\partial f_i/\partial k_l)$ is nonzero only if f_i directly involves the elementary step l. To bring out the more interesting indirect secondary or higher order effects, a reasonable distance between t_0 and t_l must therefore be kept.

We may use data again from the Oregonator (Equation 65) to illustrate the importance of indirect kinetic effects: the five normalized sensitivity coefficients $\frac{X}{k_l}\left(\frac{\partial X}{\partial k_l}\right)$ at $t = 0.390$ computed by applying the zero initial conditions at $t = 0.389$ are, for l from 1 to 5, 0.26×10^{-2}, -3.4, 1.3, -0.46×10^{-1}, and -2.4, respectively.[102] Clearly, step 5 can exert a strong negative influence on X through the indirect passage via step 2. Note that, in this case, a rather short distance between t_0 and t_l is sufficient to bring out the indirect kinetic effects of step 5.

The implementation of the variable-initial-time procedure starts with selecting the time points of interest where dynamics of the system are desired and then integrating the sensitivity differential equations near those points over intervals of reasonable length. An excellent candidate for solving the sensitivity differential equations in this interval-wise fashion is PAM. (See Section IV.B.)

The third remark that we wish to make concerning statement 1 is related to the model simplification process. Note that the normalized sensitivity coefficients computed by the variable-initial-time procedure are useful in providing dynamic information of the system, but their magnitudes carry no absolute meaning. To assess the accumulated effects of each elementary step upon the system, one must still integrate the sensitivity differential equations from the true initial time $t_0 = 0$ over the entire time domain of interest.

The effort associated with a model simplification process may be reduced considerably by working with the time-averaged normalized sensitivity coefficients $\overline{S}_{il}$ defined by[96]

$$\overline{S}_{il} = \frac{1}{T}\int_0^T dt|S_{il}| \tag{67}$$

where [0,T] is the time domain of interest. Equation 67 may be evaluated by means of any standard quadrature formulae like the trapezoidal rule. A normalized sensitivity coefficient of magnitude smaller than a given bound ϵ throughout [0,T] will necessarily give rise to a time-averaged normalized sensitivity coefficient smaller than ϵ. Therefore, the $\overline{S}_{il}$ may be used to rank the elementary reactions and serve as a very convenient reference for simplifying the reaction mechanisms.

It should be noted, however, that a normalized sensitivity coefficient of magnitude considerably exceeding the bound ϵ for some time while otherwise remaining smaller than ϵ may still give rise to a time-averaged normalized sensitivity coefficient smaller than ϵ. In practice, this situation can be detected by inserting a check post into the quadrature program or elsewhere such that whenever the magnitude of a normalized sensitivity coefficient exceeds the given bound ϵ, a warning or related information is printed out.[96]

Recently, an interesting model reduction method based on the principal component analysis[1] of nonlinear parameter estimation was proposed.[79] The method uses the eigenvalues of the matrix of sum of squares of normalized sensitivity coefficients over a set of discrete time points as a measure of kinetic importance. Model reduction is effected by eliminating steps associated with those eigenvectors of small enough eigenvalues.

Finally, we note that, experimentally, one sometimes employs the technique of doping the system with chemical species. Doping summarizes the effects of a given chemical species upon the system behavior and serves as complementary evidence in establishing the dominant reaction pathways. To calculate the theoretical response of a system to doping, we simply apply the variable-initial-time procedure to compute the sensitivity coefficients with respect to "initial" species concentrations at the time point of doping.[96] Over a proper short period of time, the computed normalized sensitivity coefficients then represent the influence of doping various chemical species upon the system dynamics near t_0. Note that the initial conditions for the sensitivity differential equations of the sensitivity coefficients with respect to initial species concentrations are *not* zero ones like Equation 22. (See Section IV.B.)

Statement 2: If a species concentration x_i depends on the parameters α and β in the combination $\mu\alpha + \nu\beta$ over a certain period of time $t_1 \leqslant t \leqslant t_2$, where μ and ν are constants, i.e.,

$$x_i = x_i(t, \mu\alpha + \nu\beta), \; t_1 \leqslant t \leqslant t_2 \tag{68}$$

then:[71]

$$\frac{S_{i\alpha}(t)}{S_{i\beta}(t)} = \frac{\mu\alpha}{\nu\beta}, \; t_1 \leqslant t \leqslant t_2 \tag{69}$$

where $S_{i\alpha} = \frac{\alpha}{x_i}\left(\frac{\partial x_i}{\partial \alpha}\right)$ and $S_{i\beta} = \frac{\beta}{x_i}\left(\frac{\partial x_i}{\partial \beta}\right)$. If a species concentration x_i depends on the parameters α and β in the combination $\alpha^\mu\beta^\nu$ over a certain period of time $t_1 \leqslant t \leqslant t_2$, where μ and ν are constants, i.e.,

$$x_i = x_i(t, \alpha^\mu\beta^\nu), \; t_1 \leqslant t \leqslant t_2 \tag{70}$$

then:[71]

$$\frac{S_{i\alpha}(t)}{S_{i\beta}(t)} = \frac{\mu}{\nu}, \; t_1 \leqslant t \leqslant t_2 \tag{71}$$

Let us prove Equation 71. Equation 69 can be proven following similar lines.[71] The functional dependence (Equation 70) of x_i upon α and β implies that the simultaneous variation of α and β will not affect x_i, provided the variation leaves the combination $\alpha^\mu\beta^\nu$ unaltered. Thus,

$$dx_i = (\partial x_i/\partial\alpha)\, d\alpha + (\partial x_i/\partial\beta)\, d\beta = 0 \tag{72}$$

$$d(\alpha^\mu\beta^\nu) = 0 \tag{73}$$

Equation (73) is equivalent to

$$\mu d\alpha/\alpha + \nu d\beta/\beta = 0 \qquad (74)$$

Substitution of Equation 74 into Equation 72 then yields Equation 71. Note that the derivation of Equation 71 from Equation 70 is independent of the way (i.e., the variable-initial-time or the regular fixed-initial-time procedure) normalized sensitivity coefficients are computed.

The message carried by statement 2 is that a constant ratio of any two normalized sensitivity coefficients, $S_{i\alpha}$ and $S_{i\beta}$, over a certain period of time implies a possible functional dependence of x_i upon the combination of parameters α and β in the form $\mu\alpha + \nu\beta$ or $\alpha^{\mu}\beta^{\nu}$. In practice, the cases of $\mu\alpha + \nu\beta$ and $\alpha^{\mu}\beta^{\nu}$ dependence may be distinguished from each other by direct numerical verification. Note that, although only simple combinations of two parameters are considered, more complex ones can be constructed from them easily.[71] For example, the dependence of

$$x_i = x_i(\alpha\beta/\gamma, t) \qquad (75)$$

can give rise to three relationships:

$$S_{i\alpha}/S_{i\beta} = 1 \qquad S_{i\alpha}/S_{i\gamma} = -1 \qquad S_{i\beta}/S_{i\gamma} = -1 \qquad (76)$$

By an examination of the normalized sensitivity coefficient-time profiles followed by direct numerical verification, any reasonably complex functional dependence of species concentrations upon parameters like rate constants and initial reactant concentrations may be unraveled.

Next, we present two illustrative examples. The first example is taken from the Oregonator. For this system, the normalized sensitivity coefficients computed by the variable-initial-time procedure in the interval from $t \approx 0.39$ to 9.0 (see Figure 5) showed that[102] Equation 77 is always satisfied:

$$\frac{k_1}{X}\left(\frac{\partial X}{\partial k_1}\right) \Big/ \frac{k_2}{X}\left(\frac{\partial X}{\partial k_2}\right) \approx -1 \qquad (77)$$

and that $\frac{k_1}{X}\left(\frac{\partial X}{\partial k_1}\right)$ is always positive and $\frac{k_2}{X}\left(\frac{\partial X}{\partial k_2}\right)$ is always negative. According to statement 2, this then implies that the concentration X is a function of k_1/k_2, in agreement with the analysis given by Field and Noyes.[103]

The second example is supplied by the Chapman[104] mechanism for atmospheric ozone kinetics:

$$\begin{aligned} O + O_2 &\xrightarrow{M} O_3, & k_1 &= 1.63 \times 10^{-16} \\ O + O_3 &\longrightarrow 2O_2, & k_2 &= 4.66 \times 10^{-16} \\ O_2 &\longrightarrow 2O, & k_3 &= 5.00 \times 10^{-11} \\ O_3 &\rightarrow O + O_2, & k_4 &= 2.5 \times 10^{-4} \end{aligned} \qquad (78)$$

The initial conditions are $O^{(0)} = 1.0 \times 10^6$, $O_3^{(0)} = 1.0 \times 10^{12}$, and $O_2^{(0)} = 3.7 \times 10^{16}$, in units of number of molecules, centimeters, and seconds. Figure 6 shows the normalized sensitivity coefficients of O with respect to k_1 and k_4 computed by integrating the sensitivity differential equation continuously from $t = 0$ to 3.6×10^6 sec.[71,73] Clearly, the ratio of these two normalized sensitivity coefficients remains a constant value of -1 from $t \approx 1$ to

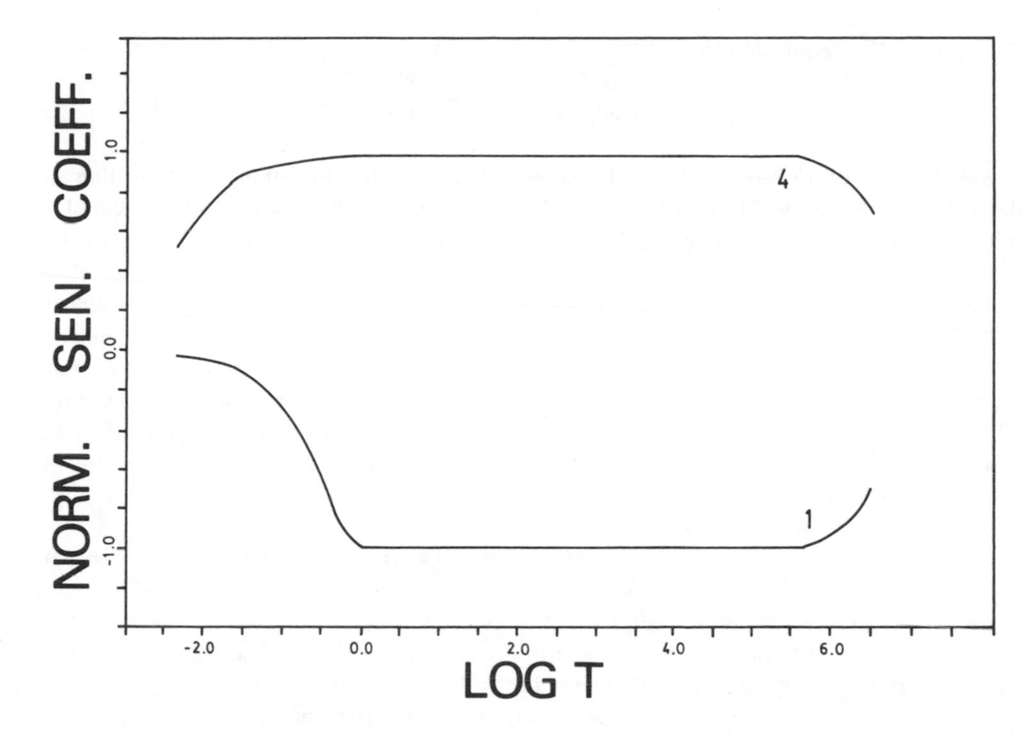

FIGURE 6. Normalized sensitivity coefficients of oxygen atom concentration with respect to k_1 and k_4, indicated in the figure by the numerals 1 and 4, respectively, for the Chapman mechanism of ozone kinetics.

3×10^5 sec, implying $O = O(k_4/k_1)$. This inference has been verified by both theoretical and numerical analysis.[71,73] Further examples may be found in References 71, 73, 96, and 98.

The functional dependence of species concentrations upon parameters may also be studied by an application of the principal component analysis in nonlinear parameter estimation[1] as suggested recently by Vajda et al.[79]

Statement 3: A wealth of kinetic information can be brought out by properly combining different sensitivity coefficients. For example, sensitivity of a species concentration x_i to the variation of the ratio of initial reactant concentrations, $u = x_j(0)/x_l(0)$, is simply:[96]

$$(\partial x_i/\partial u)_{v \text{ fixed}} = \frac{v}{(1 + u)^2}\left\{\left[\frac{\partial x_i}{\partial x_j(0)}\right] - \left[\frac{\partial x_i}{\partial x_l(0)}\right]\right\} \tag{79}$$

where $v = x_j(0) + x_l(0)$. A second example is the sensitivity of the period τ of oscillation in an oscillatory kinetic system to the variation of a parameter α:[87,88,105]

$$\left(\frac{\partial \tau}{\partial \alpha}\right) = \left[\left(\frac{\partial x_i}{\partial \alpha}\right)(t) - \left(\frac{\partial x_i}{\partial \alpha}\right)(t + \tau)\right] \Big/ \left(\frac{\partial x_i}{\partial t}\right)(t), \quad i = 1, 2, \ldots, m \tag{80}$$

Another example is given by the derived sensitivity coefficients,[67,77] which are useful measures for the effects on the determined values of a certain number of rate constants by the uncertainties of other rate constants in the system. Derived sensitivity coefficients are related to the usual sensitivity coefficients $(\partial x_i/\partial \alpha)$ by a linear algebraic matrix equation.[67,77] Further examples may be found in References 77, 88, 96, 97, 102, and 105.

The information contents of higher order sensitivity coefficients can be analyzed similarly as above;[71] however, computing those coefficients usually demands sizable extra computation time and produces a vast amount of numerical output which is difficult to handle. The systematics of sensitivity coefficient usages in chemical kinetics is presently under active development and new algorithms are still being proposed from time to time by researchers in the field.

Due to its complexity and the fact that little related computational work has been done, the sensitivity of a system with respect to temporally and spatially varying parameters was not discussed in the above passages. Such materials may be found in References 65, 77, 84, 85, and 92.

B. Polynomial Approximation Method

We present in this section the method of polynomial approximation (PAM)[73-75] for solving sensitivity differential equations. PAM enjoys a rather fast computation speed and is simple to use. The major computation effort of PAM is proportional to the number of species in the system, not to the number of parameters. PAM is also well suited to the application of variable-initial-time procedure mentioned in the previous section. These features make PAM particularly attractive towards large kinetic models wherein large numbers of parameters are involved.

1. Homogeneous Systems

Consider the general nonlinear first-order *m*-vector ordinary differential Equation 14 or 29, which we reproduce here for convenience:

$$\frac{d\mathbf{x}}{dt} = \mathbf{f}(t, \mathbf{x}, \mathbf{k}), \quad \mathbf{x}(t_0) = \mathbf{a} \tag{81}$$

The time propagation of a first-order sensitivity coefficient $\mathbf{w} = (\partial \mathbf{x}/\partial \alpha)$, where α denotes a parameter (k_l, or a_i), is governed by the following sensitivity differential equation (see Equation 20):

$$\frac{d\mathbf{w}}{dt} - \mathbf{J}\,\mathbf{w} = \mathbf{s} \tag{82}$$

where

$$\mathbf{s} = (\partial \mathbf{f}/\partial \alpha) \tag{83}$$

and $\mathbf{J}$ is the $m \times m$ Jacobian matrix with elements $J_{ij} = (\partial f_i/\partial x_j)$. The inhomogeneous term $\mathbf{s}$ is independent of $\mathbf{w}$.

The initial condition $\mathbf{w}(t_0)$ of Equation 82 depends on whether α refers to an initial condition for one of the component x_j, i.e.,

$$\mathbf{w}(t_0) = \mathbf{0} \tag{84}$$

if α is not an initial condition, or

$$\mathbf{w}(t_0) = (0, 0,\ldots, 0, 1, 0,\ldots, 0)^T \tag{85}$$

with 1 at the *j*th position if $\alpha = a_j$.

Equation 82 is, in fact, the generic form for sensitivity differential equations governing sensitivity coefficients of all orders (first and higher). We shall develop PAM based on solving Equation 82 from $t = c$ to $t = d$. The coefficient $\mathbf{w}(c)$ is assumed known from initial conditions (Equation 84 or 85) or from prior computation.

Suppose now that polynomials of degree q suffice to mimic the temporal behavior of $\mathbf{x}$ in [c,d], i.e.,

$$x_i(t, \mathbf{k}, \mathbf{a}) = \sum_{r=0}^{q} c_{ir}(\mathbf{k}, \mathbf{a})\, L_r(t), \quad i = 1, 2,\ldots, m \tag{86}$$

where L_r is an *r*th-degree polynomial. Then, all parameter dependences of x_i reside in the expansion coefficient c_{ir}, and

$$w_i = \sum_{r=0}^{q} \left(\frac{\partial c_{ir}}{\partial \alpha}\right) L_r(t), \quad i = 1, 2,\ldots, m \tag{87}$$

In other words, if the temporal behavior of $\mathbf{x}$ is well described by the *q*th-degree polynomials, so is the temporal behavior of $\mathbf{w}$.

Since all polynomials of degree q are equivalent in approximating a given function,[106] we choose Lagrange interpolation polynomials to facilitate later numerical procedure:

$$L_r(t) = \prod_{\substack{l=0 \\ (l \neq r)}}^{q} \frac{t - t_l}{t_r - t_l} \tag{88}$$

Then, the coefficient $(\partial C_{ir}/\partial \alpha)$ in Equation 87 becomes simply $w_i(t_r)$. By letting $t_0 = c < t_1 < t_2 < \ldots < t_q \leq d$ and Equation 87 satisfy the sensitivity differential Equation 82 at t_r, $r = 1, 2, \ldots, q$, we obtain an algebraic equation:[73]

$$\mathbf{z} = \mathbf{A}^{-1}\,\mathbf{g} \tag{89}$$

where

$$\mathbf{z}^T = [w_1(u_1), w_1(u_2),\ldots, w_1(u_q), w_2(u_1), w_2(u_2),\ldots, w_2(u_q),\ldots, w_m(u_1), w_m(u_2),\ldots, w_m(u_q)] \tag{90}$$

$$\begin{aligned}\mathbf{g}^T = {} & [s_1(u_1), s_1(u_2),\ldots, s_1(u_q), s_2(u_1), s_2(u_2),\ldots, s_2(u_q),\ldots, s_m(u_1), s_m(u_2),\ldots, s_m(u_q)] \\ & - \frac{1}{d - c}[L_0'(u_1)\, w_1(c), L_0'(u_2)\, w_1(c),\ldots, L_0'(u_q)\, w_1(c), \\ & L_0'(u_1)\, w_2(c), L_0'(u_2)\, w_2(c),\ldots, L_0'(u_q)\, w_2(c),\ldots, \\ & L_0'(u_1)\, w_m(c), L_0'(u_2)\, w_m(c),\ldots, L_0'(u_q)\, w_m(c)]\end{aligned} \tag{91}$$

In Equations 90 and 91, the prime indicates differentiation with respect to t and

$$u_r = \frac{t_r - c}{d - c}, \quad r = 1, 2,\ldots, q \tag{92}$$

which transforms the interval [c,d] into [0,1].

The matrix **A** is a supermatrix composed of m^2 blocks $\mathbf{B}_{\mu\nu'\mu'\nu} = 1,2, \ldots, m$:

$$\mathbf{A} = \begin{pmatrix} \mathbf{B}_{11} & \mathbf{B}_{12} & \cdots & \mathbf{B}_{1m} \\ \mathbf{B}_{21} & \mathbf{B}_{22} & \cdots & \mathbf{B}_{2m} \\ \mathbf{B}_{m1} & \mathbf{B}_{m2} & \cdots & \mathbf{B}_{m\,m} \end{pmatrix} \tag{93}$$

Each block $\mathbf{B}_{\mu\nu}$ is a q × q matrix defined by

$$(\mathbf{B}_{\mu\nu})_{rl} = \delta_{\mu\nu}L_l'(u_r)/(d - c) - \delta_{rl}J_{\mu\nu}(u_r), \quad r,l = 1, 2,\ldots, q \tag{94}$$

Notice that **A** is a fairly sparse matrix with only $mq(m + q - 1)$ nonzero elements. The matrix **A** depends on the interval length d − c, on the *m*-vector **x**, and on the derivatives of Lagrange interpolation polynomials, but it is independent of the *mq*-vector **z**. No matter how many parameters the system has, and no matter that **w** in Equation 82 is a first-, second-, or higher order sensitivity coefficient, there is only one **A** in any given time interval to be inverted.[73,74] All higher order sensitivity coefficients can be generated consecutively through simple matrix-vector multiplications once first-order coefficients are known.

The implementation of PAM starts with solving the dynamic model (Equation 81) for species concentrations over the entire time domain of interest. (This is, in fact, the first step of any modeling study.) Then, based on the logic of mimicking the temporal behavior of the fastest varying component of the system by interpolation polynomials of degree q, the time domain is divided into subintervals. Note that the fastest varying component of the system may switch from one species to another as time evolves. To effect the division of a time domain, an automatic stepping algorithm[74] or the simple visual method of inspecting the plots of concentration-time profiles[73] may be adopted.

We present below an automatic division algorithm for reference:[74]

1. The algorithm requires as input the degree q of the interpolation polynomials, the q *collocation points* u_r, r = 1,2, . . . , q, in the (transformed) interval [0,1], and the species concentrations at a set of discrete time mesh points T_j, j = 0,1,2, . . . , where $T_0 = t_0$, generated by integrating Equation 81 with an ordinary differential equation solver.
2. Beginning with the first integration mesh $[t_0,T_1]$, which we call the trial interval, species concentrations at q interior collocation time points $t_r = t_0 + u_r(T_1 - t_0)$, r = 1,2, . . . , q, are interpolated from known concentration values at neighboring mesh points T_j.
3. The concentrations at the q + 1 points t_r, r = 0,1, . . . , q, are used to form *m* *q*th-degree polynomials, one for each species.
4. The *m* polynomials constructed in step 3 are then checked for their deviations by comparing values of the polynomials with known concentration values at the mesh points T_j encompassed by the trial interval. (For the first trial interval $[t_0,T_1]$, there is only one such point, T_1.) If all *m* deviations are smaller than a preset tolerance ϵ,

these *q*th-degree polynomials are considered successful in mimicking the temporal behavior of species concentrations. One then extends the trial interval by including one more integration mesh. The trial interval is now $[t_0,T_2]$ and the new collocation points are $t_r = t_0 + u_r (T_2 - t_0)$, $r = 1,2, \ldots, q$.

5. Steps similar to 2 to 4 are repeated. If the error test is again successful, one extends the trial interval further to $[t_0,T_3]$ by including still one more integration mesh. This procedure continues until the deviation of any *q*th-degree polynomial exceeds ϵ.
6. Let the error test fail for $[t_0,T_j]$. Then the largest successful trial interval is $[t_0,T_{j-1}]$. This constitutes a proper subinterval. The information is stored and the division continues with a new trial interval $[t_0 + u_q(T_{j-1} - t_0), T_j]$.

Note that in the above algorithm, the initial point of each subinterval is equated to the last collocation point of the previous subinterval for better computational efficiency. From our experience with PAM, the optimal polynomial degree is 2 to 3, while the proper positions of the collocation points are 0 of the second or third degree Legendre polynomial shifted to the interval [0,1]. The numerical stability of PAM has been established in Reference 73. A program listing of the PAM code is given in Reference 74.

As a final remark, we note that PAM is also well suited to the variable initial-time procedure of solving sensitivity differential equations. The use of other ordinary differential system solvers in the variable initial-time fashion may end up with using rather small integration step sizes throughout because of the accuracy requirement and the difficulty in step-changing mechanism.

2. *Inhomogeneous Systems*

We now extend the method of polynomial approximation to inhomogeneous systems[75] by considering the general reaction-diffusion Equation 95 as an illustrative example:

$$\partial \mathbf{x}/\partial t = \mathbf{D}(\partial^2 \mathbf{x}/\partial r^2) + \mathbf{f}(\mathbf{x}, \mathbf{k}) \tag{95}$$

In Equation 95, $\mathbf{x}$ denotes an *m* vector of species concentrations, $\mathbf{D}$ an $m \times m$ diagonal matrix of diffusion constants, $\mathbf{f}$ an *m* vector of chemical rate laws, and $\mathbf{k}$ a *p* vector of elementary rate constants (see Equation 53). The initial and boundary conditions of Equation 95 are usually given in the form:

$$\mathbf{x}(r, 0) = \mathbf{x}^{(0)}(r)$$

$$a_{i\rho}\left(\frac{\partial x_i}{\partial r}\right)(\rho, t) + b_{i\rho}\, x_i(\rho, t) = c_{i\rho}, \quad \rho = a, b, \quad i = 1, 2, \ldots, m \tag{96}$$

where $a \leq r \leq b$ is the spatial domain of interest and $a_{i\rho}$, $b_{i\rho}$ and $c_{i\rho}$ are three known functions of time.

The behavior of a first-order sensitivity coefficient $\mathbf{w} = \partial \mathbf{x}/\partial \alpha$, where α indicates any parameter appearing in Equations 95 and 96, is governed by the sensitivity differential equation:

$$\partial \mathbf{w}/\partial t = \mathbf{D}(\partial^2 \mathbf{w}/\partial r^2) + \mathbf{J}\mathbf{w} + \mathbf{s} \tag{97}$$

The nonhomogeneous term $\mathbf{s}$ of Equation 97 is $(\partial \mathbf{D}/\partial \alpha)(\partial^2 \mathbf{x}/\partial r^2) + (\partial \mathbf{f}/\partial \alpha)$ and $\mathbf{J}$ is the Jacobian matrix with elements $J_{ij} = (\partial f_i/\partial x_j)$. Equation 97 is obtained by differentiating Equation 95 with respect to α. The initial and boundary conditions of Equation 97 can be

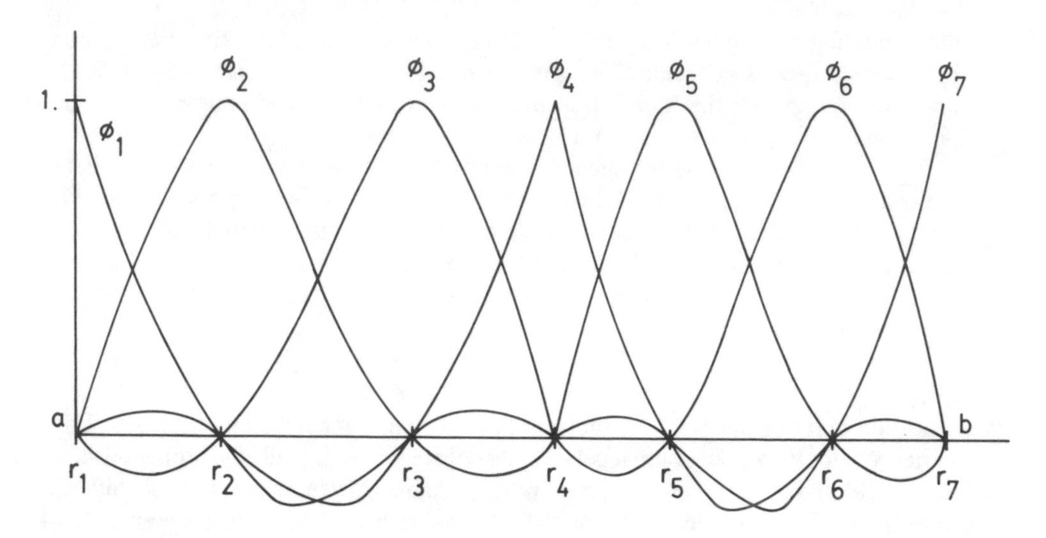

FIGURE 7. Seven piecewise cubic Lagrangian basis functions ϕ_l, $l = 1,2, \ldots ,7$, over two mesh elements $[a,r_4]$ and $[r_4,b]$.

obtained similarly by differentiating Equation 96 with respect to α. As with the case of homogeneous systems, Equation 97 is the generic form of sensitivity differential equations governing sensitivity coefficients of all orders.

To compute sensitivity coefficients, we first discretize the spatial domain [a,b] and transform Equation 97 into an ordinary differential system, which is then solved by PAM as developed in the previous section.

The spatial discretization is carried out by the finite element technique (Section III.B).[57-59] To fix ideas, we assume that, by employing cubic basis functions, two spatial mesh elements, $[a,r_4]$ and $[r_4,b]$, suffice (Figure 7). Then,

$$\mathbf{w}(r, t) \approx \mathbf{u}(r, t) = \sum_{l=1}^{7} \mathbf{w}(r_l, t)\, \phi_l(r)$$

$$\equiv \sum_{l=1}^{7} \mathbf{w}_l(t)\, \phi_l(r) \tag{98}$$

where r_l, $l = 1,2, \ldots ,7$ ($r_1 = a$, $r_7 = b$), are defined in Figure 7. The functions $\phi_l(r)$ are the cubic Lagrangian basis functions[57] which, in each mesh element, are simply the cubic Lagrange interpolation polynomials (see Equation 88 and Figures 3 and 7). Note that the derivative $(d\phi_4/dr)$ is discontinuous at the mesh boundary $r = r_4$ (Figure 7).

The expansion coefficients $\mathbf{w}_1$ and $\mathbf{w}_7$ of Equation 98 are known from the boundary conditions. The rest of the $\mathbf{w}_l$ (t), $l = 2,3, \ldots ,6$, can be determined by requiring that $\mathbf{u}(r,t)$ satisfy Equation 97 at $r = r_2, r_3, r_5, r_6$ [this is equivalent to choosing the weighting functions in Equation 58 as Dirac delta functions $\delta(r - r_l), l = 2,3,5,6$] and the flux continuity condition at the mesh boundary $r = r_4$:

$$(\partial \mathbf{u}/\partial r)_{r=r_4-\epsilon} = (\partial \mathbf{u}/\partial r)_{r=r_4+\epsilon} \tag{99}$$

where ϵ is an infinitesimal positive number. This constitutes a total of five equations relating five unknowns, and straightforward algebraic manipulation then leads to a set of four ($\mathbf{w}_2$, $\mathbf{w}_3$, $\mathbf{w}_5$, $\mathbf{w}_6$) coupled ordinary differential equations which can be solved by PAM. Note that if basis functions with continuous first-order derivatives throughout the spatial domain, like

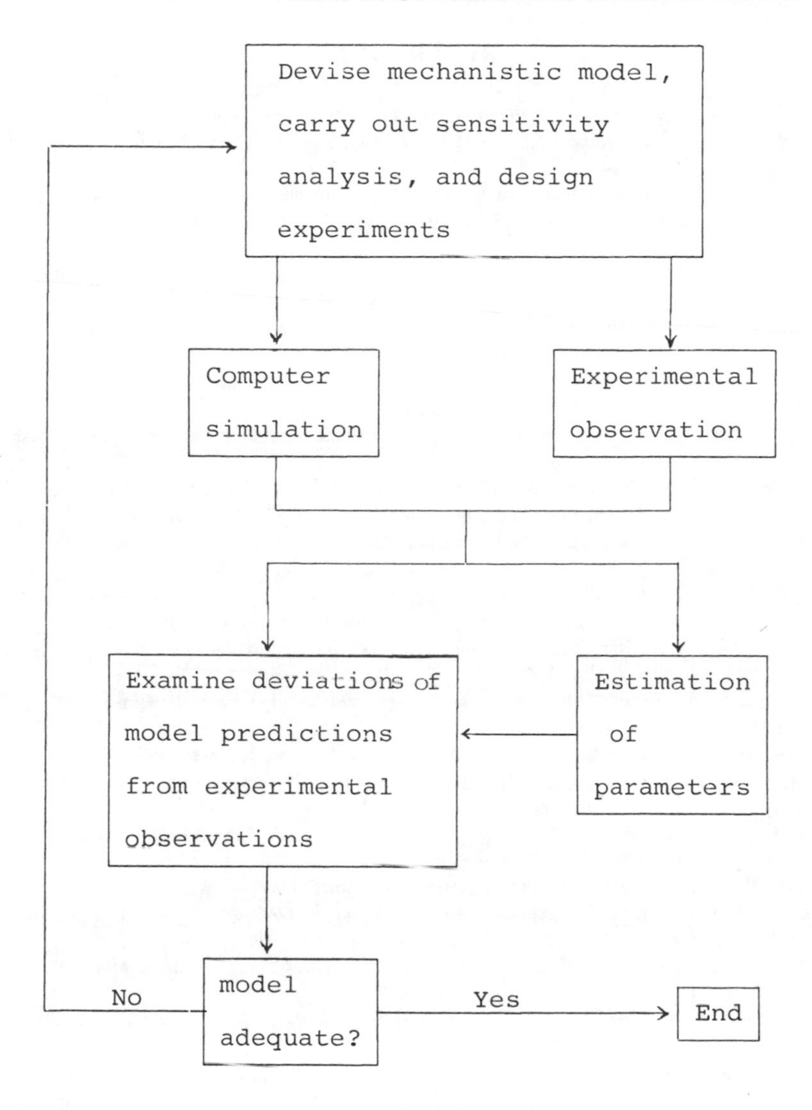

FIGURE 8. Flow diagram for the model development.

Hermite interpolation polynomials,[58] are employed, then the flux continuity condition (Equation 99) is automatically satisfied.

The application of this procedure has produced some encouraging results[75] on a simple three-species biological kinetic model involving enzymatically active membrane.

V. CONCLUDING REMARKS

The recent development in the computational tools has reached the impressive point where kinetic systems of great complexity can be satisfactorily modeled, and at the same time has stimulated active research on the subject of sensitivity analysis, whereby useful physical and chemical insights are extracted from the massive numerical data produced by complex modeling studies. It is now a general belief that systematic sensitivity analysis should be adopted as a routine tool when performing kinetic modeling.

In Figure 8, we conclude by summarizing the essential steps of a model-building process in a self-explanatory flow diagram, which also serves as a convenient reference for the interrelations among various topics discussed in the present chapter.

REFERENCES

1. **Bard, Y.,** *Nonlinear Parameter Estimation,* Academic Press, New York, 1974.
2. **Seinfeld, J. H. and Lapidus, L.,** *Process Modelling, Estimation, and Identification,* Prentice-Hall, Englewood Cliffs, N.J., 1974.
3. **Côme, G. M.,** The use of computers in the analysis and simulation of complex reactions, in *Comprehensive Chemical Kinetics,* Vol. 24, Bamford, C. H. and Tipper, C. F. H., Eds., Elsevier, New York, 1983, chap. 3.
4. **Frenklach, M.,** Modelling, in *Combustion Chemistry,* Gardiner, W. C., Jr., Ed., Springer-Verlag, New York, 1984, chap. 7.
5. **Brodlie, K. W.,** Unconstrained minimization, in *The State of the Art in Numerical Analysis,* Jacobs, D., Ed., Academic Press, New York, 1977, chap. 3-1.
6. **Dennis, J. E., Jr.,** Nonlinear least squares and equations, in *The State of the Art in Numerical Analysis,* Jacobs, D., Ed., Academic Press, New York, 1977, chap. 3-2.
7. **Marquardt, D. W.,** An algorithm for least squares estimation of nonlinear parameters, *J. Soc. Ind. Appl. Math.,* 11, 431, 1963.
8. **Papoulis, A.,** *Probability, Random Variables, and Stochastic Processes,* McGraw-Hill, New York, 1965.
9. **Shannon, C. E.,** A mathematical theory of communication, *Bell System Tech. J.,* 27, 379, 1948.
10. **Lindley, D. V.,** On a measure of the information provided by an experiment, *Ann. Math. Stat.,* 27, 986, 1956.
11. **Box, G. E. P., Hunter, W. G., and Hunter, J. S.,** *Statistics for Experimenters. An Introduction to Design, Data Analysis, and Model Building,* John Wiley & Sons, New York, 1978.
12. **Gardiner, W. C., Jr.,** Introduction to combustion modelling, in *Combustion Chemistry,* Gardiner, W. C., Jr., Ed., Springer-Verlag, New York, 1984, chap. 1.
13. **Engleman, V. S.,** Detailed approach to kinetic mechanisms in complex systems, *J. Phys. Chem.,* 81, 2320, 1977.
14. **Edelson, D.,** Computer simulation in chemical kinetics, *Science,* 214, 981, 1981.
15. **Ebert, K. H., Ederer, H. J., and Isbarn, G.,** Computer simulation of the kinetics of complicated gas phase reactions, *Angew. Chem. Int. Ed. Engl.,* 19, 333, 1980.
16. **Gear, C. W.,** *Numerical Initial Value Problems in Ordinary Differential Equations,* Prentice-Hall, Englewood, N.J., 1972, chap. 11.
17. **Gelinas, R. J.,** Stiff systems of kinetic equations — a practitioner's view, *J. Comp. Phys.,* 9, 222, 1972.
18. **Lambert, J. D.,** Stiffness, in *Computational Techniques for Ordinary Differential Equations,* Gladwell, I. and Sayers, D. K., Eds., Academic Press, New York, 1980, 19.
19. **Shampine, L. F.,** What is stiffness?, in *Proc. Int. Conf. Stiff Comp.,* Vol. 2, Aiken, R. C., Ed., Park City, Utah, 1982.
20. **Hall, G. and Watt, J. M., Eds.,** *Modern Numerical Methods for Ordinary Differential Equations,* Clarendon Press, Oxford, 1976.
21. **May, R. and Noye, J.,** The numerical solution of ordinary differential equations: initial value problems, in *Computational Techniques for Differential Equations,* Noye, J., Ed., North-Holland, Amersterdam, 1984.
22. **Gear, C. W.,** The automatic integration of ordinary differential equations, *Commun. ACM,* 14, 176, 1971.
23. **Gear, C. W.,** DIFSUB for solution of ordinary differential equations, *Commun. ACM,* 14, 185, 1971.
24. **Hindmarsh, A. C.,** GEAR: ordinary differential equation system solver, Rep. UCID-3001 (Rev. 3), Lawrence Livermore Laboratory, University of California, Livermore, 1974.
25. **Hindmarsh, A. C.,** A collection of software for ordinary differential equations, Rep. UCRL-82091, Lawrence Livermore Laboratory, University of California, Livermore, 1979.
26. **Hindmarsh, A. C.,** LSODE and LSODI, two new initial value ordinary differential equation solvers, *ACM SIGNUM Newsl.,* 15(4), 10, 1980.
27. **Bader, G. and Deuflhard, P.,** A semi-implicit mid-point rule for stiff systems of ordinary differential equations, Institute for Applied Mathematics Reprint No. 114, University of Heidelberg, West Germany, 1981.
28. **Kennealy, J. P. and Moore, W. M.,** A numerical method for chemical kinetics modelling based on the Taylor series expansion, *J. Phys. Chem.,* 81, 2413, 1977.
29. **Hesstvedt, E., Hov, Ö., and Isaksen, I. A.,** Quasi-steady-state approximations in air pollution modelling: comparison of two numerical schemes for oxidant prediction, *Int. J. Chem. Kinet.,* 10, 971, 1978.
30. **Aiken, R. C. and Lapidus, L.,** Pseudo steady state approximation for the numerical integration of stiff kinetic systems, *AIChE J.,* 21, 817, 1975.
31. **Warner, D. D.,** The numerical solution of the equations of chemical kinetics, *J. Phys. Chem.,* 81, 2329, 1977.
32. **McQuarrie, D. A.,** Stochastic approach to chemical kinetics, *J. Appl. Prob.,* 4, 413, 1967.

33. **McQuarrie, D. A.,** Stochastic theory of chemical rate processes, *Adv. Chem. Phys.*, 15, 149, 1969.
34. **Nicolis, G. and Prigogine, I.,** *Self-Organization in Nonequilibrium Systems,* Part 3, John Wiley & Sons, New York, 1977.
35. **Nicolis, G., Erneux, T., and Herschkowitz-Kaufman, M.,** Pattern formation in reacting and diffusing systems, *Adv. Chem. Phys.*, 38, 263, 1978.
36. **Prigogine, I.,** *From Being to Becoming: Time and Complexity in the Physical Science,* W. H. Freeman, San Francisco, 1980.
37. **Ortoleva, P. and Yip, S.,** Computer molecular dynamics studies of a chemical instability, *J. Chem. Phys.*, 65, 2045, 1976.
38. **Turner, J. S.,** Discrete simulation methods for chemical kinetics, *J. Phys. Chem.*, 81, 2379, 1977.
39. **Hanusse, P. and Blanche, A.,** A Monte Carlo method for large reaction-diffusion systems, *J. Chem. Phys.*, 74, 6148, 1981.
40. **Kramers, H. A.,** Brownian motion in a field of force and the diffusion model of chemical reactions, *Physica,* 7, 284, 1940.
41. **Gillespie, D. T.,** Exact stochastic simulation of coupled chemical reactions, *J. Phys. Chem.*, 81, 2340, 1977.
42. **Gillespie, D. T.,** A general method for numerically simulating the stochastic time evolution of coupled chemical reactions, *J. Comp. Phys.*, 22, 403, 1976.
43. **Gillespie, D. T.,** Concerning the validity of the stochastic approach to chemical kinetics, *J. Stat. Phys.*, 16, 311, 1977.
44. **Bunker, D. L., Garret, B., Kleindienst, T., and Long, G. S.,** III, Discrete simulation methods in combustion kinetics, *Combust. Flame,* 23, 373, 1974.
45. **Jeng, D.-C.,** Study on the Computer Simulation of Chemical Reactions, M.S. thesis, National Tsing Hua University, Hsinchu, Taiwan, 1985.
46. **Baras, F., Nicolis, G., Mansour, M. M., and Turner, J. W.,** Stochastic theory of adiabatic explosion, *J. Stat. Phys.*, 32, 1, 1983.
47. **Dixon-Lewis, G.,** Computer modelling of combustion reactions in flowing systems with transport, in *Combustion Chemistry,* Gardiner, W. C., Jr., Ed., Springer-Verlag, New York, 1984, chap. 2.
48. **Bird, R. B., Stewart, W. E., and Lightfoot, E. N.,** *Transport Phenomena,* John Wiley & Sons, New York, 1960.
49. **Crank, J.,** *The Mathematics of Diffusion,* 2nd ed., Oxford University Press, London, 1975, chaps. 1, 8, and 14.
50. **Britz, D.,** *Digital Simulation in Electrochemistry,* Springer-Verlag, Berlin, 1981.
51. **Banks, H. T.,** *Modeling and Control in the Biomedical Sciences,* Springer-Verlag, Berlin, 1975, chap. 2.
52. **Schryer, N. L.,** The state of the art in the numerical solution of time-varying partial differential equations, *J. Phys. Chem.*, 81, 2335, 1977.
53. **Noye, J.,** Finite difference techniques for partial differential equations, in *Computational Techniques for Differential Equations,* Noye, J., Ed., North-Holland, Amersterdam, 1984, 95.
54. **Mitchell, A. R. and Griffiths, D. F.,** *The Finite Difference Method in Partial Differential Equations,* John Wiley & Sons, Chichester, 1980, chap. 2.
55. **Liskovets, O. A.,** The method of lines, *J. Diff. Eqs.*, 1, 1308, 1965.
56. **Edelson, D. and Schryer, N. L.,** Modeling chemically reacting flow systems. I. A comparison of finite difference and finite element methods for one-dimensional reactive diffusion, *Comput. Chem.*, 2, 71, 1978.
57. **Zienkiewicz, O. C. and Morgan, K.,** *Finite Elements and Approximation,* John Wiley & Sons, New York, 1983, chap. 7.
58. **Chung, T. J.,** *Finite Element Methods in Fluid Dynamics,* McGraw-Hill, New York, 1978.
59. **Fletcher, C.,** The Galerkin method and Burger's equation, in *Computational Techniques for Differential Equations,* Noye, J., Ed., North-Holland, Amersterdam, 1984, 355.
60. **Gelinas, R., Doss, S. K., and Miller, K.,** The moving finite element method: applications to general partial differential equations with multiple large gradients, *J. Comp. Phys.*, 40, 202, 1981.
61. **Nikolakopoulou, G. A., Edelson, D., and Schryer, N. L.,** Modeling chemically reacting flow systems. II. An adaptive spatial mesh technique for problems with discontinuities and steep fronts, *Comput. Chem.*, 6, 93, 1982.
62. **Cukier, R. I., Fortuin, C. M., Schuler, K. E., Petschek, A. G., and Schaibly, J. H.,** Study of the sensitivity of coupled reaction systems to uncertainties in rate coefficients. I. Theory, *J. Chem. Phys.*, 59, 3873, 1973.
63. **Cukier, R. I., Levine, H. B., and Schuler, K. E.,** Nonlinear sensitivity analysis of multiparameter model systems, *J. Comp. Phys.*, 26, 1, 1978.
64. **Atherton, R. W., Schainker, R. B., and Ducot, E. R.,** On the statistical sensitivity analysis of models for chemical kinetics, *AIChE J.*, 21, 441, 1975.
65. **Dickinson, R. P. and Gelinas, R. J.,** Sensitivity analysis of ordinary differential equation systems — a direct method, *J. Comp. Phys.*, 21, 123, 1976.

66. **Hwang, J.-T., Dougherty, E. P., Rabitz, S., and Rabitz, H.,** The Green's function method of sensitivity analysis in chemical kinetics, *J. Chem. Phys.*, 69, 5180, 1978.
67. **Dougherty, E. P., Hwang, J.-T., and Rabitz, H.,** Further developments and applications of the Green's function method of sensitivity analysis in chemical kinetics, *J. Chem. Phys.*, 71, 1794, 1979.
68. **Hwang, J.-T.,** Nonlinear sensitivity analysis in chemical kinetic, *Proc. Natl. Sci. Council R.O.C.*, B6, 20, 1982.
69. **Hwang, J.-T.,** The scaled Green's function method of sensitivity analysis and its application to chemical reaction systems, *Proc. Natl. Sci. Council R.O.C.*, B6, 37, 1982.
70. **Hwang, J.-T. and Chang, Y. S.,** The scaled Green's function method of sensitivity analysis. II. Further developments and application, *Proc. Natl. Sci. Council R.O.C.*, B6, 308, 1982.
71. **Hwang, J.-T.,** On the proper usage of sensitivities of chemical kinetics models to the uncertainties in rate coefficients, *Proc. Natl. Sci. Council R.O.C.*, B6, 270, 1982.
72. **Hwang, J.-T.,** Nonlinear sensitivity analysis in chemical kinetics — the scaled Green's function method, in *Proc. Int. Conf. Stiff Comp.*, Vol. 2, Aiken, R. C., Ed., Park City, Utah, 1982.
73. **Hwang, J.-T.,** Sensitivity analysis in chemical kinetics by the method of polynomial approximations, *Int. J. Chem. Kinet.*, 15, 959, 1983.
74. **Hwang, J.-T.,** A computational algorithm for the polynomial approximation method of sensitivity analysis in chemical kinetics, *J. Chinese Chem. Soc.*, 32, 253, 1985.
75. **Chen, F.-T. and Hwang, J.-T.,** Sensitivity analysis of inhomogeneous chemical reaction systems by the method of polynomial approximation, presented at Int. Conf. Chem. Kinet., Gaithersburg, 1985.
76. **Kramer, M. A., Calo, J. M., and Rabitz, H.,** An improved computational method for sensitivity analysis: Green's function method with AIM, *Appl. Math. Model.*, 5, 432, 1981.
77. **Rabitz, H.,** Sensitivity analysis in chemical kinetics, *Ann. Rev. Phys. Chem.*, 34, 419, 1983.
78. **Miller, D. and Frenklach, M.,** Sensitivity analysis and parameter estimation in dynamic modeling of chemical kinetics, *Int. J. Chem. Kinet.*, 15, 677, 1983.
79. **Vajda, S., Valko, P., and Turányi, T.,** Principal component analysis of kinetic models, *Int. J. Chem. Kinet.*, 17, 55, 1985.
80. **Demiralp, M. and Rabitz, H.,** Chemical kinetic functional sensitivity analysis: elementary sensitivities, *J. Chem. Phys.*, 74, 3362, 1981.
81. **Yetter, R., Eslava, L. A., Dryer, F. L., and Rabitz, H.,** Elementary and derived sensitivity information in chemical kinetics, *J. Phys. Chem.*, 88, 1497, 1984.
82. **Dunker, A. M.,** Efficient calculation of sensitivity coefficients for complex atmospheric models, *Atmos. Environ.*, 15, 1155, 1981.
83. **Dunker, A. M.,** The decoupled direct method for calculating sensitivity coefficients in chemical kinetics, *J. Chem. Phys.*, 81, 2385, 1984.
84. **Koda, M., Dogru, A. H., and Seinfeld, J. H.,** Sensitivity analysis of partial differential equations with application to reaction and diffusion processes, *J. Comp. Phys.*, 30, 259, 1979.
85. **Koda, M. and Seinfeld, J. H.,** Sensitivity analysis of distributed parameter systems, *IEEE Trans. Autom. Control*, Ac-27, 951, 1984.
86. **Gardiner, W. C., Jr.,** The pC, pR, pP, pM and pS method of formulating the results of computer modeling studies of chemical reactions, *J. Phys. Chem.*, 81, 2367, 1977.
87. **Edelson, D. and Thomas, V. M.,** Sensitivity analysis of oscillating reactions. I. The period of the Oregonator, *J. Phys. Chem.*, 85, 1555, 1981.
88. **Larter, R.,** Sensitivity analysis of autonomous oscillator. Separation of secular terms and determination of structural stability, *J. Phys. Chem.*, 87, 3114, 1983.
89. **Hwang, J.-T.,** The Green's function method of sensitivity analysis in quantum dynamics, *J. Chem. Phys.*, 70, 4609, 1979.
90. **Eslava, L. A., Eno, L., and Rabitz, H.,** Further developments and applications of sensitivity analysis to collisional energy transfer, *J. Chem. Phys.*, 73, 4998, 1980.
91. **Bartine, D. E., Oblow, E. M., and Mynatt, F. R.,** Radiation-transport cross-section sensitivity analysis — a general approach illustrated for a thermonuclear source in air, *Nucl. Sci. Eng.*, 55, 147, 1974.
92. **Cacuci, D. G.,** Sensitivity theory for nonlinear systems. I. Nonlinear functional analysis approach, *J. Math. Phys.*, 22, 2794, 1981.
93. **Frank, P. M.,** *Introduction to System Sensitivity Theory,* Academic Press, New York, 1978.
94. **Tomović, R. and Vukobratović, M.,** *General Sensitivity Theory,* Elsevier, New York, 1972.
95. **House, W. J., Jr.,** Sensitivity Analysis of Making Capitol Investment Decisions, Rep. No. 3, National Association of Accountants, 1968.
96. **Yu, C. R. and Hwang, J.-T.,** Sensitivity analysis of a methane oxidation mechanism, presented at the 8th Int. Symp. Gas Kinet., Nottingham, 1984.
97. **Hwang, J.-T.,** Computational sensitivity study of oscillating chemical reaction systems — the overall temporal sensitivity behavior of the Oregonator, presented at the 8th Int. Symp. Gas Kinet., Nottingham, 1984.

98. **Lee, Y.-J., Hwang, J.-T., and Chao, K.-J.,** Study of a mechanism for the Fischer-Tropsch synthesis, presented at the 8th Int. Symp. Gas Kinet., Nottingham, 1984.
99. **Gelinas, R. J. and Vajk, J. P.,** Systematic sensitivity analysis of air quality simulation models, Rep. EPA-600/4-79-035, Environmental Protection Agency, Cincinnati, 1979.
100. **Gelinas, R. J. and Shewes-Cox, P. D.,** Tropospheric photochemical mechanisms, *J. Phys. Chem.*, 81, 2468, 1977.
101. **Kramer, M. A., Rabitz, H., Calo, J. M., and Kee, R. J.,** Sensitivity analysis in chemical kinetics: recent developments and computational comparisons, *Int. J. Chem. Kinet.*, 16, 559, 1984.
102. **Wann, G.-W. and Hwang, J.-T.,** unpublished data, 1985.
103. **Field, R. J. and Noyes, R. M.,** Mechanisms of chemical oscillators: conceptual bases, *Acc. Chem. Res.*, 10, 214, 1977.
104. **Chapman, S.,** A theory of atmospheric ozone, *Mem. R. Meteor. Soc.*, 3, 103, 1930.
105. **Edelson, D. and Rabitz, H.,** Numerical techniques for modeling and analysis of oscillating reactions, personal communication, 1984.
106. **Davis, P. J.,** *Interpolation and Approximation,* Dover Press, New York, 1975.

INDEX

A

B

C

D

M

N

O

P

Q

R

U

V

W

Z